Scattering of Light by Crystals

WILLIAM HAYES
University of Oxford

RODNEY LOUDON
University of Essex

Dover Publications, Inc.
Mineola, New York

Copyright

Copyright © 1978 by William Hayes and Rodney Loudon
All rights reserved.

Bibliographical Note

This Dover edition, first published in 2004, is an unabridged republication of the work first published by John Wiley and Sons, New York, in 1978.

Library of Congress Cataloging-in-Publication Data

Hayes, William.
    Scattering of light by crystals / William Hayes, Rodney Loudon.
       p. cm.
    Originally published: New York : Wiley, c1978.
    Includes bibliographical references and index.
    ISBN 0-486-43866-X (pbk.)
    1. Solids—Optical properties. 2. Light—Scattering. 3. Crystal optics.
I. Loudon, Rodney. II. Title.

QC176.8.O6H37 2004
530.4'12—dc22

                                                            2004058209

Manufactured in the United States of America
Dover Publications, Inc., 31 East 2nd Street, Mineola, N.Y. 11501

Preface

The experimental and theoretical study of light scattering has a lengthy history and some aspects of the phenomenon were well understood by the end of the nineteenth century. Its value as a tool for investigating the properties of matter, however, was fully realized only with the interpretation and observation of inelastic light scattering by Brillouin and Raman in the 1920s. More recently, the invention of the laser has extended and deepened the range of material properties accessible to measurement by light-scattering spectroscopy.

As with any field of science, the development and expansion of light-scattering work has necessarily led to a degree of fragmentation of the field. Thus, studies of gases, liquids, and solids are normally carried out by different groups of workers. There are further divisions associated with the size of the frequency shift of the light brought about by its inelastic scattering, the shifts being measured with increasing energy by intensity-fluctuation, Fabry-Perot, and diffraction-grating spectroscopy. A comprehensive account of the entire field would be extremely long.

This book is concerned with the scattering of light by solids, and its main emphasis is on the study of excitations in crystals by light-scattering spectroscopy. Chapter 1 surveys the scope of light-scattering experiments and presents the main theoretical tools available for calculation of scattering cross sections. The typical frequencies of excitations in crystals are such that they can normally be examined by either Brillouin scattering, using Fabry-Perot interferometric spectroscopy, or by Raman scattering, using grating spectrometers, and these techniques are described in Chapter 2.

The remainder of the book is devoted to a systematic account of the measurements and theories of light scattering by the various solid-state excitations. The major part of the published literature is concerned with light scattering by lattice vibrations and these receive the greatest emphasis here. Chapters 3 and 4 cover Raman scattering by nonpolar and polar optic vibrations, while Brillouin scattering by acoustic vibrations is discussed in Chapter 8. Vibrational effects are also of great importance in the light-scattering phenomena associated with structural phase changes; these are treated in Chapter 5. Raman scattering by magnetic and electronic excitations in crystals is described in Chapters 6 and 7.

The treatment is particularly intended for a graduate student or other research worker entering the field with the usual undergraduate background knowledge of quantum mechanics, electromagnetic theory, statistical mechanics, and solid-state physics, but without any prior knowledge of light-scattering theory. Some background in group theory would be useful but is not an essential prerequisite for reading the book. The book is also intended to serve as a reference text on the basic techniques and theory of light scattering by crystals. However, in view of its primary role as a textbook for students, there is no very serious attempt to assign due credit to those who made the first observations or first theories of the various phenomena. Rather, we have in each case chosen those experiments and theories that seem to illustrate and explain the features of interest in the simplest and most direct ways.

The book uses SI units throughout except for the retention of the inverse centimeter as a unit of frequency (1 cm^{-1} = 3×10^{10} Hz). This modest departure from the rules of the International System accords with almost universal usage in light-scattering research publications.

We are indebted to our colleagues M. G. Cottam, P. J. Dean, P. A. Fleury, R. T. Harley, J. F. Ryan, J. F. Scott, D. N. Stacey, and C. E. Webb for helpful comments on the manuscript. We also thank Mhairi Kimmitt, Mary Loudon, and Margaret Sherlock for their careful preparation of the typescript, Joan Hayes for assistance with the index, and Mike Kettle and David Manning for assistance with the figures.

WILLIAM HAYES
RODNEY LOUDON

Oxford, England
Colchester, England
April 1978

Contents

CHAPTER 1 BASIC FEATURES AND FORMAL THEORY OF LIGHT SCATTERING 1

   1.1 Historical Introduction, 2

   1.2 The Scattering Cross Section, 4

      1.2.1 Basic Definitions, 4

      1.2.2 Classical Theory of Elastic Scattering, 8

   1.3 Scope of Light-Scattering Experiments, 11

   1.4 Macroscopic Theory of Light Scattering, 16

      1.4.1 Susceptibility Derivatives, 17

      1.4.2 Radiation by the Stokes Polarization, 21

      1.4.3 The Cross Section, 25

      1.4.4 Fluctuation-Dissipation Theory, 27

      1.4.5 Relation between Stokes and Anti-Stokes Cross Sections, 31

   1.5 Microscopic Theory of Light Scattering, 33

      1.5.1 The Interaction Hamiltonians, 33

      1.5.2 Atomic-Scattering Cross Section, 39

      1.5.3 Scattering by Free Electrons, 41

   1.6 Symmetry Properties of Inelastic Cross Sections, 43

CHAPTER 2 EXPERIMENTAL ASPECTS 53

   2.1 Lasers, 54

      2.1.1 Conditions for Laser Operation, 54

      2.1.2 Optical Resonators and Mode Configurations, 58

      2.1.3 Gas Lasers, 64

- 2.1.4 *Dye Lasers,* 70
  - 2.1.5 *Solid-State Ionic Lasers,* 72
- 2.2 Spectrometers, 74
  - 2.2.1 *Diffraction-Grating Instruments and Sample Illumination,* 75
  - 2.2.2 *The Fabry-Perot Spectrometer,* 80
- 2.3 Detection of Scattered Light, 86
- 2.4 Measurement of Spectra, 88

CHAPTER 3 NONPOLAR VIBRATIONAL SCATTERING   95

- 3.1 First-Order Scattering, 96
  - 3.1.1 *Vibrational Symmetries,* 96
  - 3.1.2 *Lattice Dynamics,* 99
  - 3.1.3 *Internal and External Vibrations,* 103
  - 3.1.4 *The Scattering Cross Section,* 107
  - 3.1.5 *Randomly Oriented Scatterers,* 112
  - 3.1.6 *Experiments on Crystals,* 117
- 3.2 Second-Order Scattering, 121
  - 3.2.1 *Density of States and Selection Rules,* 121
  - 3.2.2 *Experiments and Calculations,* 124
- 3.3 Defect-Induced Scattering, 131
  - 3.3.1 *Light Scattering by Point Defects,* 133
  - 3.3.2 *Mixed Crystals and Amorphous Solids,* 137

CHAPTER 4 POLAR VIBRATIONAL SCATTERING   147

- 4.1 Macroscopic Theory, 148
  - 4.1.1 *Lattice Dynamics of Polar Modes,* 148
  - 4.1.2 *The Scattering Cross Section,* 153
  - 4.1.3 *Properties of the Susceptibility Derivatives,* 156
  - 4.1.4 *Polar-Mode Scattering in Cubic Crystals,* 162

4.1.5 Polar-Mode Scattering in Uniaxial Crystals, 169
4.1.6 Polar-Mode Scattering in Biaxial Crystals, 175
4.1.7 Scattering by Powdered Crystals, 177
4.2 Microscopic Theory, 179
    4.2.1 Electrons and Phonons in Crystals, 180
    4.2.2 The Scattering Cross Section, 183
    4.2.3 Resonance Scattering, 186
4.3 Light Scattering by Polaritons, 192

CHAPTER 5 STRUCTURAL PHASE CHANGES 201

5.1 Soft Modes, 206
5.2 Experimental Examples, 213
    5.2.1 Quartz, 213
    5.2.2 Perovskites, 214
    5.2.3 Hydrogen-Bonded Ferroelectrics and Order-Disorder Transitions, 219
    5.2.4 Cooperative Jahn-Teller Effects, 223
    5.2.5 Acoustic Anomalies, 228
    5.2.6 Light Scattering Near Zero Frequency, 232

CHAPTER 6 MAGNETIC SCATTERING 239

6.1 Scattering by Simple Paramagnets and Ferromagnets, 240
    6.1.1 Faraday Rotation, 240
    6.1.2 The Scattering Cross Section, 243
    6.1.3 Microscopic Theory, 251
6.2 First-Order Light Scattering by Antiferromagnets, 256
    6.2.1 Antiferromagnetic Magnons, 256
    6.2.2 Antiferromagnetic Cross Section, 262
6.3 Second-Order Light Scattering by Antiferromagnets, 266
    6.3.1 Scattering Mechanism, 266

6.3.2 Magnon Interaction Effects, 271

6.4 Magnetic Defect Scattering, 276

    6.4.1 Scattering by Point Defects, 277

    6.4.2 Scattering by Mixed Crystals, 279

CHAPTER 7 RAMAN SCATTERING BY ELECTRONS 287

7.1 Light Scattering by Rare-Earth Ions, 288

7.2 Light Scattering by Shallow Donors and Acceptors, 293

    7.2.1 Electronic Levels of Impurities in Semiconductors, 293

    7.2.2 Scattering by Donors, 296

    7.2.3 Scattering by Acceptors, 298

7.3 Light Scattering by Conduction Electrons, 301

    7.3.1 Some Properties of Free Carriers, 301

    7.3.2 General Aspects of Light Scattering by a Plasma, 305

    7.3.3 Plasmon Scattering, 310

    7.3.4 Single-Particle Scattering, 315

CHAPTER 8 RAYLEIGH AND BRILLOUIN SCATTERING 327

8.1 Kinematics of Brillouin Scattering and Determination of Elastic Constants, 332

    8.1.1 Centrosymmetric Crystals, 332

    8.1.2 Piezoelectric Crystals, 336

8.2 Brillouin Scattering Cross Section, 340

8.3 Some Experimental Examples, 345

    8.3.1 Rayleigh Scattering, 345

    8.3.2 Brillouin Scattering, 348

INDEX 355

CHAPTER ONE

Basic Features and Formal Theory of Light Scattering

1.1 Historical Introduction — 2

1.2 The Scattering Cross Section — 4
    1.2.1 Basic Definitions
    1.2.2 Classical Theory of Elastic Scattering

1.3 Scope of Light-Scattering Experiments — 11

1.4 Macroscopic Theory of Light Scattering — 16
    1.4.1 Susceptibility Derivatives
    1.4.2 Radiation by the Stokes Polarization
    1.4.3 The Cross Section
    1.4.4 Fluctuation-Dissipation Theory
    1.4.5 Relation between Stokes and Anti-Stokes Cross Sections

1.5 Microscopic Theory of Light Scattering — 33
    1.5.1 The Interaction Hamiltonians
    1.5.2 Atomic-Scattering Cross Section
    1.5.3 Scattering by Free Electrons

1.6 Symmetry Properties of Inelastic Cross Sections — 43

Much of the formal theory of light scattering is common to all varieties of measurement. It is the purpose of this first chapter to cover the common ground and to derive general results that can be applied in subsequent chapters to scattering by the various kinds of solid-state excitation. The main goals are a few basic formulas, summarized at the end of the chapter, for the kinematics and cross section of a light-scattering experiment.

1.1 HISTORICAL INTRODUCTION

We begin with a brief sketch of the historical development of light-scattering studies. Some of the earliest investigations were carried out by Tyndall (1868-1869). He found that white light, scattered at 90° to the incident light by very fine particles, was partly polarized and also slightly blue in color. He concluded that both the polarization and the blue color of light from the sky were caused by scattering of sunlight by dust particles in the atmosphere.

Lord Rayleigh (1899), following his earlier work (Strutt 1871a,b,c), treated the scattering of light by spherical particles of relative permittivity κ suspended in a medium of relative permittivity κ_0. If the particle separation is greater than the wavelength λ of the light so that the particles scatter independently of each other, and if in addition the particle radius is less than the wavelength of light, the intensity of the scattered light is (for a modern derivation, see Section 72 of Landau and Lifshitz 1960)

$$I_S = I\frac{9\pi^2 N v^2}{2\lambda^4 r^2}\left(\frac{\kappa - \kappa_0}{\kappa + 2\kappa_0}\right)^2 (1+\cos^2\phi), \tag{1.1}$$

where I is the intensity of the unpolarized incident light, N is the number of scattering particles of volume v, r is the distance to the point of observation, and ϕ is the angle through which the light is scattered. An important feature of this result, which we shall encounter many times in the course of the book, is the λ^{-4} dependence of the scattered intensity. This is known as Rayleigh's law, and it provides an explanation for the blueness of the sky. However, Rayleigh knew in 1899 that light is scattered by gas molecules in the air, and he suspected rightly at that time that particles of dust in the atmosphere are not essential for the blueness and polarization of light from the sky.

The treatment of molecular scattering encounters a fundamental problem of which Rayleigh was aware. In dense media such as liquids and solids, and even gases at atmospheric pressure, the molecular separation is small compared to the wavelength of light. There is now a coherence between the light beams scattered by different molecules, which no longer act as independent scatterers. Indeed, the light intensity scattered in any direction other than forward is zero for a perfectly homogeneous medium (see Sections 1.2 and 1.3). However, it was well known that apparently homogeneous fluids scatter light quite strongly. The scattering is especially pronounced when a fluid approaches its critical temperature, a phenomenon referred to as critical opalescence (Andrews 1869).

The problem of opalescence was explained by Smoluchowski (1908), who suggested that the density of an apparently homogeneous medium nevertheless varies from point to point because of thermal motions of the molecules. Light scattering is caused by density fluctuations, which become large at the critical point. Einstein (1910) showed that the wavevector of the scattering fluctuation conserves momentum between the incident and scattered photons.

These earlier workers were concerned with the intensity of the scattered light. Progress in understanding its frequency spectrum was first made by Brillouin (1914, 1922), who calculated the spectrum of light scattered by the density fluctuations associated with sound waves. He found that the spectrum of a fluid consists of a doublet split symmetrically around the frequency of the incident light. The splitting, which is very much smaller than the frequency of the incident light, is determined by the velocity of those sound waves whose wavelength is close to that of the light. The Brillouin doublets (called Mandelstam-Brillouin doublets in the Russian literature) were first observed by Gross (1930a, b, c, 1932) in liquid media. The liquids also showed a central unshifted component in the scattered light. This was explained by Landau and Placzek (1934) as the scattering by nonpropagating entropy fluctuations, although details of the calculation were not presented in the paper (see Chapter 8); it is generally referred to as Rayleigh scattering.

While progress was being made in the understanding of scattering by thermally induced density fluctuations, advances were also in progress on other fronts. Smekal (1923) studied the scattering of light by a system with two quantized energy levels and predicted the existence of sidebands in the scattered spectrum. This effect was subsequently observed in Raman's laboratory (Raman and Krishnan 1928a, b); it was found that the light scattered by liquids such as benzene contains sharp sidebands in pairs symmetrically disposed around the incident frequency with shifts identical to the frequencies of some of the infrared vibrational lines. At much the same time, Landsberg and Mandelstam (1928) discovered a similar phenomenon in quartz. This inelastic scattering of light by molecular and crystal vibrations is now known as the Raman effect. It is caused by modulation of the susceptibility (or, equivalently, polarizability) of the medium by the vibrations (and, as described in this book, the scattering by other excitations in solids, including plasmons, excitons, and magnons, occurs by the same mechanism). The vibrational Raman effect was well documented and well understood by 1934 (Placzek 1934).

After the intense activity of the late 1920s and early 1930s, the study of light scattering proceeded at a more sedate pace, receiving more attention in India, Russia, France, and Canada than in other countries (for reviews of this period, see Menzies 1953; Fabelinskii 1957; and Loudon 1964). This was due in part to the weakness of light scattering, a second-order effect, compared with infrared absorption as techniques for determining vibrational properties. This situation was dramatically altered by the invention of the helium-neon laser in 1961 and by the subsequent development of improved optical spectrometers and detection techniques. These advances produced a renaissance in light-scattering studies, which began in 1964 and still has considerable momentum.

1.2 THE SCATTERING CROSS SECTION

1.2.1 Basic Definitions

The main meeting point of light-scattering experiments and theory is the scattering cross section. We describe here a simple kind of light-scattering experiment and define the types of cross section that can in principle be measured.

Consider the idealized experiment shown in Figure 1.1. An incident parallel beam of light from a laser source passes through a volume V of some material and is scattered in all directions from all the illuminated part of the sample. A detector is set up to examine the light scattered at angle φ to the direction of the incident beam. The range of acceptance of the detector symbolized by the lens, is limited to a small solid angle $d\Omega$, and the detector field stop restricts the volume $\mathfrak{v}$ of the sample from which scattered light is received. The intensities of incident and scattered light are denoted $\mathcal{I}_I$ and $\mathcal{I}_S$, respectively.

The magnitude of the scattered intensity is determined by a variety of factors. One of these is the transmission of the light through the sample surfaces, which varies with φ in a complicated way, particularly for samples with edges or corners. The resulting dependence of $\mathcal{I}_S$ on φ obscures the basic physics of the scattering process. It is therefore customary to convert the directly measured intensities $\mathcal{I}_I$ and $\mathcal{I}_S$, solid angle $d\Omega$, and scattering angle φ into corresponding intensities I_I and I_S, solid angle $d\Omega$, and scattering angle ϕ *inside* the scattering sample. These

The Scattering Cross Section

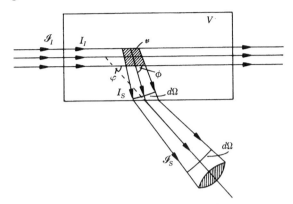

Figure 1.1 Idealized scattering experiment.

conversions can all be made from a knowledge of the sample shape and its refractive indices and the beam geometries (Lax and Nelson 1976). The variation of the scattering volume v with scattering angle can be similarly determined.

The experiment thus provides results for the variation of I_S with ϕ. An additional complication occurs for scattering by a crystalline medium, where the scattered intensity depends not only on the scattering angle but also on the orientations of the crystal symmetry axes relative to the light beams. The isotropy of a gas, liquid, or amorphous solid removes this latter dependence.

The spread of frequencies in the incident laser light is normally very small, and its intensity can be regarded as monochromatic with a single angular frequency ω_I. However, the scattered intensity is usually distributed across a range of frequencies, as shown in Figure 1.2. The peak in the center of the spectrum is the contribution of the incident photons that have been elastically scattered with no change in frequency. The remaining peaks correspond to inelastic scattering and their shifts from ω_I normally occur in two somewhat separate frequency ranges. The Brillouin component, resulting from scattering by sound waves, occurs close to the frequency of the incident light; typical shifts are approximately 1 cm^{-1} or smaller. The Raman component, resulting from scattering by internal vibrations of molecules or optic vibrations in crystals, lies at higher shifts, normally larger than 10 cm^{-1} and often of order 100–1000 cm^{-1}. The basic mechanisms for Brillouin and Raman scattering are essentially the same, but the experimental techniques are different; they are considered separately in Chapter 8 and Chapters 3 and 4, respectively.

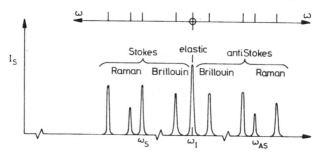

Figure 1.2 Schematic spectrum of scattered light.

The inelastic contributions are further subdivided; those scattered frequencies smaller than ω_I are denoted ω_S and are known as the Stokes component, while the scattered frequencies larger than ω_I are denoted ω_{AS} and they form the anti-Stokes component. Each scattered photon in the Stokes component is associated with a gain in energy $\hbar\omega$ by the sample, where

$$\omega = \omega_I - \omega_S. \tag{1.2}$$

Similarly, the sample loses energy $\hbar\omega$ for each scattered photon in the anti-Stokes component, where

$$\omega = \omega_{AS} - \omega_I. \tag{1.3}$$

The occurrence of scattered photons at particular frequencies ω_S and ω_{AS} depends upon the ability of the scattering sample to absorb or emit energy in quanta of magnitude $\hbar\omega$ determined by (1.2) or (1.3). The intensity peaks in the inelastic spectrum thus correspond to the various excited states of the sample. The most important application of inelastic light-scattering spectroscopy is the determination of excitation energies by measurement of frequency shifts from ω_I in the scattered light. Measurements of the frequency widths of the intensity peaks also provide information on the excited-state lifetimes.

Measurement of the Stokes part of the spectrum for a fixed scattering angle determines a function

$$\frac{d^2\sigma}{d\Omega d\omega_S} \equiv \text{spectral differential cross section}. \tag{1.4}$$

This is defined as the rate of removal of energy from the incident beam as

The Scattering Cross Section

a result of its scattering in volume v into a solid-angle element $d\Omega$ with a scattered frequency between ω_S and $\omega_S + d\omega_S$, divided by the product of $d\Omega \, d\omega_S$ with the incident-beam intensity. The various quantities are determined inside the scattering medium (see Figure 1.1). The spectral differential cross section has the dimensions of area divided by frequency.

An analogous definition can be made for the anti-Stokes part of the spectrum. However, it is shown in Section 1.4.5 that the Stokes and anti-Stokes cross sections for the great majority of scattering experiments are related to a very good approximation by

$$n(\omega)\frac{d^2\sigma}{d\Omega \, d\omega_S} = \{n(\omega)+1\}\frac{d^2\sigma}{d\Omega \, d\omega_{AS}}, \tag{1.5}$$

where the frequencies satisfy (1.2) and (1.3). The Bose-Einstein thermal factor is explicitly

$$n(\omega) = \frac{1}{\exp(\hbar\omega/k_B T)-1}, \tag{1.6}$$

where k_B is Boltzmann's constant and T is the sample temperature. Figure 1.3 shows the ratio $n(\omega)/\{n(\omega)+1\}$ as a function of $k_B T/\hbar\omega$. The anti-Stokes spectrum has a smaller strength than the Stokes spectrum, and it is usually more convenient to measure the latter. Most theoretical treatments consider only the Stokes component, relying on (1.5) to generate the corresponding results for the anti-Stokes component.

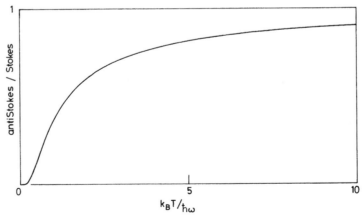

Figure 1.3 Ratio $n(\omega)/\{n(\omega)+1\}$ of the anti-Stokes to the Stokes cross section as a function of temperature and frequency shift.

The *differential cross section* is obtained by integration of the spectral differential cross section,

$$\frac{d\sigma}{d\Omega} = \int d\omega_S \frac{d^2\sigma}{d\Omega\, d\omega_S} \equiv \text{differential cross section.} \quad (1.7)$$

The integration can equally be taken over the frequency ω, since ω_S and ω differ by the constant frequency ω_I as in (1.2). The range of integration is usually restricted to include just a single intensity peak in the scattered spectrum. The differential cross section then determines the total scattering into solid angle $d\Omega$ associated with a particular excited state of the sample. It is often easier to calculate than the spectral differential cross section but it contains less information of potential value in interpreting the scattering process.

Finally, the *cross section* is obtained by integration of the differential cross section over all directions in space

$$\sigma = \int d\Omega \frac{d\sigma}{d\Omega} \equiv \text{cross section.} \quad (1.8)$$

The cross section determines the total scattering in all directions caused by a particular excited state, and its experimental evaluation requires measurements of the scattered intensity at a large number of scattering angles. Such series of measurements are rarely made because the sample excitation energies, and lifetimes are in most cases independent of the scattering angle used in their determination (for exceptions, see Sections 4.3, 7.3, and 8.1). The cross section, however, can be computed from a restricted set of measurements of the differential cross section if a theoretical expression for the angular variation of the latter is available.

The three varieties of cross section are introduced above in decreasing order of importance. All three are referred to simply as the cross section when the distinction is clear from the context. The cross sections for an extended medium are proportional to the scattering volume v and quoted results are usually expressed in terms of a unit volume of scatterer. Note that a cross section for 1 m^3 of scatterer expressed in square meters is numerically equal to 100 times the same cross section for 1 cm^3 of scatterer expressed in square centimeters.

1.2.2 Classical Theory of Elastic Scattering

Several typical features of light-scattering theory can be simply illustrated by a classical description of elastic scattering by an atom. Only the main results need be quoted, since a full account is given in Chapter 11 of

The Scattering Cross Section

Loudon (1973). The atom is represented by an electron of charge e and mass m in a harmonic oscillator potential. The natural frequency ω_O of the oscillator is chosen to match the main electronic absorption frequency of the atom. The electric field $\mathbf{E}_I$ of the incident light causes forced vibrations of the electron at frequency ω_I.

The oscillating charge radiates electromagnetic waves at frequency ω_I to produce scattered light whose intensity is found from the standard theory of electric-dipole radiation. The differential cross section for a single atom for scattering in which the electric vector $\mathbf{E}_S$ of the scattered light has the direction of a unit vector $\mathbf{\varepsilon}_S$ is

$$\frac{d\sigma}{d\Omega} = \frac{r_e^2 \omega_I^4}{\left(\omega_O^2 - \omega_I^2\right)^2 + \omega_I^2 \Gamma^2} (\mathbf{\varepsilon}_I \cdot \mathbf{\varepsilon}_S)^2, \qquad (1.9)$$

where Γ is the oscillator damping constant, $\mathbf{\varepsilon}_I$ is a unit vector parallel to $\mathbf{E}_I$, and the quantity

$$r_e = \frac{e^2}{4\pi\epsilon_O mc^2} \approx 2.8 \times 10^{-15} \text{ m} \qquad (1.10)$$

is called the classical electron radius. The assumed atomic model gives rise only to elastic scattering, because a purely harmonic oscillator in the steady state oscillates solely at the frequency ω_I of the driving field. To produce inelastic scattering, it is necessary to add anharmonic terms to the oscillator potential or couple it anharmonically to other oscillators.

Although based on a very simple model, the elastic differential cross section (1.9) reproduces many features of a more accurate calculation (for discussion, see Loudon 1973). A dependence on the polarization vectors of the light is a characteristic feature. In inelastic scattering, the contributions of different excited states of the sample often show different forms of variation with $\mathbf{\varepsilon}_I$ and $\mathbf{\varepsilon}_S$, and the experimental determination of this variation provides useful information about the excited states. Note that even for the simple form of polarization dependence in (1.9), and for unpolarized incident light, the scattered light is partially polarized for general scattering angles.

Averages of (1.9) over the two independent directions of $\mathbf{\varepsilon}_I$ and $\mathbf{\varepsilon}_S$, corresponding to unpolarized incident light and a detector that does not discriminate between the scattered polarizations, give

$$\frac{d\sigma}{d\Omega} = \frac{r_e^2 \omega_I^4}{\left(\omega_O^2 - \omega_I^2\right)^2 + \omega_I^2 \Gamma^2} \frac{1}{2}(1 + \cos^2\phi). \qquad (1.11)$$

The angular dependence of the atomic differential cross section is thus the same as that of Rayleigh's expression (1.1) for scattering by dielectric spheres. Integration over all directions of scattering, as in (1.8), leads to the cross section

$$\sigma = \frac{8\pi}{3} \frac{r_e^2 \omega_I^4}{(\omega_O^2 - \omega_I^2)^2 + \omega_I^2 \Gamma^2}. \tag{1.12}$$

The above expressions show that the intensity of scattering can be a strong function of the incident frequency ω_I, particularly in systems where the damping constant Γ is small. For ease of observation of scattering, it is clearly advantageous to have ω_I close to the atomic-transition frequency ω_O. This is known as resonance scattering and similar effects occur in all types of scattering, both elastic and inelastic, and for all kinds of scattering excitation.

The cross sections simplify when the incident frequency is much smaller than the main atomic absorption frequency. For example (1.9) becomes

$$\frac{d\sigma}{d\Omega} = r_e^2 \left(\frac{\omega_I}{\omega_O}\right)^4 (\varepsilon_I \cdot \varepsilon_S)^2 \qquad (\omega_I \ll \omega_O). \tag{1.13}$$

This fourth-power dependence on the incident frequency is the same as the λ^{-4} dependence of Rayleigh's law, mentioned in Section 1.1. It is a characteristic feature of cross sections at frequencies below the main absorption lines or bands of the scattering medium. It is shown in Section 1.4 that the corresponding frequency dependence for inelastic scattering is more accurately $\omega_I \omega_S^3$.

In the opposite limit of a high incident frequency, (1.9) becomes

$$\frac{d\sigma}{d\Omega} = r_e^2 (\varepsilon_I \cdot \varepsilon_S)^2 \approx 7.9 \times 10^{-30} (\varepsilon_I \cdot \varepsilon_S)^2 \, \text{m}^2 \qquad (\omega_I \gg \omega_O). \tag{1.14}$$

This is known as the Thomson cross section. It is valid for the scattering of light by free electrons, and it occurs for atoms in the limit where the incident frequency is much larger than all the characteristic frequencies of the atom. A more rigorous derivation for scattering by free electrons is given in Section 1.5.3. The Thomson cross section loses its validity for still higher incident frequencies when $\hbar \omega_I$ becomes comparable to the rest-mass energy mc^2 of the electron. A significant amount of energy is transferred to the electron in this case and the process becomes inelastic Compton scattering.

The scattering by N identical classical atoms is obtained upon multiplication of the cross sections given above by N. This apparently straightforward manipulation requires some justification, since as mentioned in Section 1.1 the scattering by a homogeneous assembly of closely spaced atoms vanishes by destructive interference except in the forward direction. However, it is not difficult to show (see for example section 93 of Landau and Lifshitz 1960 or Chapter 11 of Loudon 1973) that the spatial fluctuations in the density of a random distribution of atoms diminish the destructive interference and lead to a combined cross section that is N times the single-atom cross section. More generally, it is shown in Section 1.3 that scattering away from the forward direction by an extended medium is always associated with some spatially-varying excitation mode of the medium.

In summary, it is shown in the present section that the amount of light detected in a scattering experiment depends in general on the frequencies ω_I and ω_S and polarization vectors ε_I and ε_S of the incident and scattered light, on the orientation of the scattered beam relative to the incident beam (and relative to the symmetry axes for scattering by a crystalline sample), and on the sample temperature T. The aim of light-scattering theory is to provide expressions for cross sections that explicitly show the forms of dependence on these variables. The aim of experiments is to measure cross sections unencumbered by the effects of transmission through sample surfaces, detector responses that vary with frequency, and other distorting influences.

1.3 SCOPE OF LIGHT-SCATTERING EXPERIMENTS

Equations (1.2) and (1.3) express conservation of energy in the inelastic light-scattering process. These processes must also conserve momentum. Let $\mathbf{k}_I$, $\mathbf{k}_S$, and $\mathbf{k}_{AS}$ be the wavevectors inside the scattering medium of an incident photon, Stokes scattered photon, and anti-Stokes scattered photon, respectively. The photon momenta are obtained upon multiplication by $\hbar$. Conservation of momentum requires that the medium gains a momentum $\hbar\mathbf{q}$ in each Stokes-scattering event, where

$$\mathbf{q} = \mathbf{k}_I - \mathbf{k}_S, \qquad (1.15)$$

and loses a momentum $\hbar\mathbf{q}'$ in each anti-Stokes scattering event, where

$$\mathbf{q}' = \mathbf{k}_{AS} - \mathbf{k}_I. \qquad (1.16)$$

Figure 1.4 shows vector diagrams for the conservation of momentum in the two kinds of inelastic scattering.

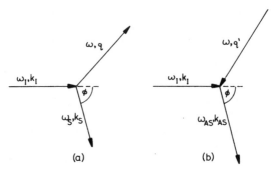

Figure 1.4 Vector diagrams for the conservation of momentum in (*a*) Stokes scattering and (*b*) anti-Stokes scattering.

It should be mentioned that the momentum-conservation conditions are strictly valid only for sufficiently large scattering samples, as discussed in the following section. It is also assumed that the photon wavevectors are real quantities, corresponding to the absence of any significant absorption of the light beams. These requirements are usually, but not always, satisfied (see Sections 4.1.7, 4.2.3, and 8.3.2 for examples of their occasional breakdown).

The influence of momentum conservation is especially important in the scattering of light by those excitations in crystals whose excitation frequency ω depends markedly on the excitation wavevector **q**, for example polaritons (Section 4.3), single-particle excitations in an electron plasma (Section 7.3.4), or acoustic phonons (Chapter 8). On the other hand, the effects of momentum conservation can normally be neglected in the scattering by liquids and gases (except for Brillouin scattering). Nevertheless, in the scattering of light by an extended medium of interacting atoms or molecules, the excitation involved is always strictly a travelling wave whose wavevector conserves momentum in accordance with (1.15) and (1.16). We consider here the resulting limitations on the scope of light-scattering experiments.

The wavevectors for the Stokes scattering, shown in Figure 1.4*a*, have the property

$$q^2 = k_I^2 + k_S^2 - 2k_I k_S \cos\phi. \tag{1.17}$$

The photon wavevectors are related to their frequencies by

$$k_I c = \eta_I \omega_I \tag{1.18}$$

$$k_S c = \eta_S \omega_S, \tag{1.19}$$

Scope of Light-Scattering Experiments

where η_I and η_S are the refractive indices of the medium for the incident and scattered light. Thus with the use of (1.2), (1.17) becomes

$$c^2 q^2 = \eta_I^2 \omega_I^2 + \eta_S^2 (\omega_I - \omega)^2 - 2\eta_I \eta_S \omega_I (\omega_I - \omega) \cos\phi. \tag{1.20}$$

This expression relates the frequency ω and the wavevector q of those excitations of the medium that scatter incident light of frequency ω_I through an angle ϕ.

In the special case where the refractive indices for the incident and scattered photons are equal, (1.20) can be written

$$\frac{c^2 q^2}{\eta^2 \omega_I^2} = \frac{\omega^2}{\omega_I^2} + 4\left(1 - \frac{\omega}{\omega_I}\right)\sin^2\frac{\phi}{2} \qquad (\eta_I = \eta_S = \eta). \tag{1.21}$$

This relation between ω and q is plotted in Figure 1.5 for various values of ϕ, where the refractive index is assumed to be constant for all values of the scattered frequency. It is then possible in principle to observe the Stokes scattering by all excitations of the medium whose frequency and wavevector lie within a triangular region of the ωq-plane, as ϕ varies continuously from 0 to 180°. The straight sides of the triangle become curved in a more realistic calculation that allows for the frequency dependence of the refractive index.

When the refractive indices are different for the incident and scattered photons, it is necessary to use the more general relation (1.20). This is the case for the example of scattering by a uniaxial or biaxial crystal when the incident and scattered photons have different polarization directions. The situation remains qualitatively similar to that shown in Figure 1.5 but the $\omega = 0$ baseline of the accessible region now covers the range

$$\left|1 - \frac{\eta_S}{\eta_I}\right| \leq \frac{cq}{\eta_I \omega_I} \leq 1 + \frac{\eta_S}{\eta_I}, \tag{1.22}$$

being diminished for $\eta_I > \eta_S$ and shifted for $\eta_I < \eta_S$. In the latter case, the left-hand side of the accessible triangle bends back where it strikes the $q = 0$ axis (see Figure 4.17b).

A similar analysis can be made for the wavevector in the anti-Stokes case illustrated in Figure 1.4b. For equal refractive indices at the incident and scattered frequencies, the wavevector is given by (1.21) but with the minus sign changed to a plus. The accessible region of the plane is now a parallel band for constant refractive index, but its baseline is the same as for Stokes scattering and the regions covered in the two cases are not very different for low-frequency excitations ($\omega \ll \omega_I$).

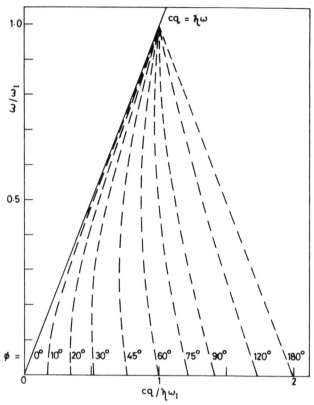

Figure 1.5 Accessible region of the frequency ω and wavevector q values of an excitation for Stokes scattering of incident light of frequency ω_I in a medium of refractive index η. The dashed lines show the experimental scans for fixed values of the scattering angle ϕ.

The information about a particular excitation of the medium that potentially is obtainable from a light-scattering experiment is assessed by superimposing Figure 1.5 on the ω versus q dispersion relation of the excitation concerned. The part of the dispersion relation investigated in an experiment with fixed incident frequency ω_I and scattering angle ϕ is determined by the intersection of the appropriate dashed line with the dispersion curve. The form of the experimental scan across ωq-space is particularly important for the crystal excitations whose frequencies vary rapidly with wavevector. On the other hand, the vibrational frequencies of weakly interacting molecules in liquids and gases, and of many lattice modes in crystals, are essentially independent of wavevector across the accessible region; the dispersion relation is then a horizontal line in Figure 1.5 and the consideration of momentum conservation becomes an unnecessary exercise.

Light-scattering experiments are most often performed with visible incident light, where a typical frequency is

$$\frac{\omega_I}{2\pi} \approx 5 \times 10^{14} \text{Hz}. \quad (1.23)$$

The excitations most conveniently studied by Raman scattering are those that cause a shift in the scattered frequency lying in the approximate range

$$3 \times 10^{11} < \frac{\omega}{2\pi} < 10^{14} \text{Hz}, \quad (1.24)$$

equivalent to an excitation frequency between 10 and 3000 cm^{-1}. The shifts in Brillouin scattering are smaller, approximately 1 cm^{-1} or less. Thus most experiments satisfy

$$\omega \ll \omega_I, \quad (1.25)$$

which is just the condition mentioned above for Stokes and anti-Stokes scattering to cover roughly the same range of wavevectors.

With a refractive index of about 1.5, the wavevector magnitude corresponding to the frequency (1.23) is

$$k_I \approx 1.5 \times 10^7 \text{m}^{-1}, \quad (1.26)$$

and the range of excitation wavevectors accessible to light-scattering experiments is approximately

$$0 < q < 3 \times 10^7 \text{m}^{-1}. \quad (1.27)$$

By comparison, the maximum wavevector q_M for crystal excitations that lie in the Brillouin zone is of order π/d where d is the lattice constant. This maximum is typically about 3×10^{10} m^{-1}, three orders of magnitude larger than the upper limit of light-scattering excitation wavevectors.

An approximation to (1.21) in the limit (1.25) of small excitation frequencies is

$$q = 2k_I \sin\frac{\phi}{2}, \quad (1.28)$$

and the anti-Stokes version of (1.21) gives the same wavevector. Figure 1.6

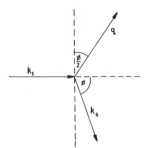

Figure 1.6 Wavevector orientations for Stokes scattering with small excitation frequency.

shows the resulting simple connection between the wavevector orientations in this limit. The approximate result breaks down for scattering close to the forward direction where $2\sin(\phi/2)$ becomes comparable with ω/ω_I, and the complete relation (1.21) must then be used. Except for near-forward scattering, (1.28) shows that in the limit (1.25)

$$cq \gg \omega, \tag{1.29}$$

and this inequality is also evident from Figure 1.5.

1.4 MACROSCOPIC THEORY OF LIGHT SCATTERING

There are basically two ways in which the classical harmonic-oscillator theory of elastic scattering in Section 1.2 can be extended to treat inelastic scattering. The two approaches are the microscopic method and the macroscopic method.

The microscopic method treats the atom quantum mechanically. The classical harmonic-oscillator variables are replaced by the atomic-energy levels and wavefunctions, and the interaction of the atom with light is expressed in terms of matrix elements of the electric-dipole operator. The resulting expression for the differential cross section, the Kramers-Heisenberg formula (see Chapter 11 of Loudon 1973 for a derivation; also Section 1.5.2), includes both elastic and inelastic contributions to the scattering.

More generally, the microscopic method can be used for scattering by all types of excitation in molecules, liquids, and solids. For example, the treatment of scattering by lattice vibrations or spin waves in crystals involves the quantum-mechanical electron-phonon interaction Hamiltonian or the matrix elements of the spin operator between atomic wavefunctions, respectively. Microscopic expressions for the cross section cannot as

Macroscopic Theory of Light Scattering

a rule be evaluated numerically for detailed comparison with experiment, but they often predict the variation of the cross section as the experimental parameters are changed, and they provide a deeper understanding of the fundamental nature of the light-scattering process.

The macroscopic method applies directly to the scattering by an extended medium, where the individual atomic dipole moments considered in the classical theory combine to form a macroscopic polarization vector. The scattered beam is radiated by the oscillatory macroscopic polarization, which is subject to the usual Maxwell equations. The macroscopic theory can be carried out using either classical variables or quantum-mechanical operators. The classical version is used in this book, although some use of quantum theory must be made in determining the temperature dependence of the cross section.

The macroscopic method can also be used for scattering by all kinds of excitation. For example, the theory of scattering by lattice vibrations or spin waves involves such macroscopic variables as the strain or the magnetization. The scattering cross sections are expressed in terms of other macroscopic properties of the medium, such as its electrooptic coefficients, elastic constants, elastooptic coefficients, and Faraday rotation.

The present section develops the macroscopic theory in a general form suitable for application to specific excitations in later chapters. The approach was originated by Van Hove (1954) in his theory of neutron scattering. A more elementary account can be found in Section 91 of Landau and Lifshitz (1960). The microscopic theory is discussed in Section 1.5.

1.4.1 Susceptibility Derivatives

Consider a scattering experiment in which the incident light can be regarded as strictly monochromatic with a well-defined frequency ω_I and wavevector $\mathbf{k}_I$. The jth Cartesian component of the incident macroscopic electric field at position $\mathbf{r}$ and time t is written

$$E_I^j(\mathbf{r},t) = E_I^j \exp(-i\omega_I t + i\mathbf{k}_I \cdot \mathbf{r}) + E_I^{j*} \exp(i\omega_I t - i\mathbf{k}_I \cdot \mathbf{r}), \quad (1.30)$$

where $\mathbf{E}_I$ is a complex amplitude vector. Symbols used as superscripts refer throughout the book to Cartesian components x, y, and z; all other kinds of label appear as subscripts.

The excitation of the medium responsible for the inelastic light scattering is characterized by a space- and time-dependent amplitude

$$X(\mathbf{r},t) = \sum_{\mathbf{q}} \{X(\mathbf{q},t)\exp(i\mathbf{q}\cdot\mathbf{r}) + X^*(\mathbf{q},t)\exp(-i\mathbf{q}\cdot\mathbf{r})\}, \quad (1.31)$$

where the complex Fourier amplitudes $X(\mathbf{q},t)$ include both the magnitude and phase of the excitation. Examples of $X(\mathbf{r},t)$ are the vibrational displacement in a molecular medium or the deviation of the magnetization from perfect alignment in a ferromagnetic crystal.

The discussion is restricted to scattering media in thermal equilibrium where these Fourier amplitudes are random quantities whose magnitudes and phases fluctuate on a time scale characteristic of the thermal excitation process. The phase angle can take any random value between 0 and 2π, and the occurrence of a specific value of the magnitude is governed by some statistical probability distribution. It is convenient to express the $X(\mathbf{q},t)$ in terms of their Fourier transforms with respect to the time,

$$X(\mathbf{q},t) = \int X(\mathbf{q},\omega)\exp(-i\omega t)\,d\omega. \qquad (1.32)$$

The statistical properties of the $X(\mathbf{q},t)$ are clearly shared by the $X(\mathbf{q},\omega)$.

Consider the quantity

$$\langle X^*(\mathbf{q},\omega)X(\mathbf{q},\omega')\rangle, \qquad (1.33)$$

where the brackets $\langle \cdots \rangle$ denote an average over the probability distribution. The two quantities in the brackets are independent random variables whose phases take all values between 0 and 2π. Their product therefore has a zero average except in the case $\omega' = \omega$, and we can write

$$\langle X^*(\mathbf{q},\omega)X(\mathbf{q},\omega')\rangle = \langle X^*(\mathbf{q})X(\mathbf{q})\rangle_\omega \delta(\omega-\omega'), \qquad (1.34)$$

where the Dirac delta-function has its usual properties. The quantity

$$\langle X^*(\mathbf{q})X(\mathbf{q})\rangle_\omega \qquad (1.35)$$

defined by (1.34) is the *power spectrum* of the fluctuations (for a fuller discussion of fluctuation theory, see Chapter 12 of Landau and Lifshitz, 1969). The power spectrum plays a central role in the macroscopic theory of light scattering; it is shown below that the cross section for scattering by an excitation is proportional to its power spectrum.

The polarization induced by the incident field (1.30) in the absence of any excitations of the scattering medium is

$$P^i(\mathbf{r},t) = \epsilon_0 \chi^{ij}(\omega_I) E_I^j(\mathbf{r},t), \qquad (1.36)$$

where $\chi^{ij}(\omega_I)$ is the first-order or linear susceptibility of the medium at frequency ω_I and the repeated superscript j is summed over x, y, and z.

The effect of the excitations is to modulate the wavefunctions and energy levels of the medium. The changes in these quantities are linear in $X(\mathbf{r},t)$ to the first order in perturbation theory, and their effect is represented macroscopically by an additional contribution to the susceptibility. Symbolically, (1.36) is replaced by an equation of the form

$$P = \epsilon_0(\chi E_I + \chi' X E_I), \qquad (1.37)$$

where χ' is a second-order susceptibility (discussed in detail below) that describes the modulation. A whole series of terms in increasing powers of X appears in the brackets if the perturbation theory is carried to higher orders.

The linear polarization from (1.36) or the first term of (1.37) oscillates at the same frequency as the incident field and contributes only to elastic scattering. The second-order polarization from the second term of (1.37) oscillates at frequencies different from ω_I because X is itself a time-dependent function. This part of the polarization radiates the inelastic contribution to the scattered light. In greater detail, the product of the incident field (1.30) with the excitation amplitude (1.31) produces four types of term, and the second-order polarization is conveniently separated into a sum of two contributions

$$P_S^i(\mathbf{r},t) = \sum_{\mathbf{K}_S} \left\{ P_S^i(\mathbf{K}_S,t) \exp(i\mathbf{K}_S \cdot \mathbf{r}) + P_S^{i*}(\mathbf{K}_S,t) \exp(-i\mathbf{K}_S \cdot \mathbf{r}) \right\} \quad (1.38)$$

and

$$P_{AS}^i(\mathbf{r},t) = \sum_{\mathbf{K}_{AS}} \left\{ P_{AS}^i(\mathbf{K}_{AS},t) \exp(i\mathbf{K}_{AS} \cdot \mathbf{r}) + P_{AS}^{i*}(\mathbf{K}_{AS},t) \exp(-i\mathbf{K}_{AS} \cdot \mathbf{r}) \right\},$$

$$(1.39)$$

where

$$\mathbf{K}_S = \mathbf{k}_I - \mathbf{q} \qquad (1.40)$$

$$\mathbf{K}_{AS} = \mathbf{k}_I + \mathbf{q}. \qquad (1.41)$$

The polarization (1.38) produces the Stokes component of the scattered light, whereas the polarization (1.39) generates the anti-Stokes component. The relation between $\mathbf{K}_S$ and the wavevector $\mathbf{k}_S$ introduced in Section 1.3 emerges later in (1.63).

The Fourier transform of the Stokes polarization with respect to time is defined by

$$P_S^i(\mathbf{K}_S,t) = \int P_S^i(\mathbf{K}_S,\omega_S) \exp(-i\omega_S t) d\omega_S. \qquad (1.42)$$

Then substitution into the second-order part of (1.37) of the expressions for E_I, X, and P_S from (1.30), (1.31), (1.32), (1.38) and (1.42) leads on comparison of terms with common time dependence to

$$P_S^i(\mathbf{K}_S, \omega_S) = \epsilon_0 \chi^{ij}(\omega_I, -\omega) X^*(\mathbf{q}, \omega) E_I^j, \qquad (1.43)$$

where the frequencies are related by (1.2). The second-order susceptibility is here written in an expanded notation to show its dependence on the Cartesian components of the field and polarization and on the frequencies of the incident light and the excitation. Its prime has been dropped, the dependence on two frequencies being sufficient to distinguish it from the linear susceptibility.

A similar treatment of the anti-Stokes polarization leads to

$$P_{AS}^i(\mathbf{K}_{AS}, \omega_{AS}) = \epsilon_0 \chi^{ij}(\omega_I, \omega) X(\mathbf{q}, \omega) E_I^j, \qquad (1.44)$$

where the frequencies are related by (1.3). The second-order susceptibilities for the Stokes and anti-Stokes polarizations are not the same but it can be shown (Loudon 1978) that for nonmagnetic materials

$$\chi^{ji}(\omega_S, \omega) = \chi^{ij}(\omega_I, -\omega), \qquad (1.45)$$

a relation that leads to physical consequences discussed in Section 1.4.5.

Both the Stokes and anti-Stokes polarizations are proportional to the Fourier components of the excitation amplitude. They therefore exhibit random fluctuations of a similar statistical nature to the $X(\mathbf{q}, \omega)$. The effect of the monochromatic incident light is to shift the random fluctuations of the excitation at frequency ω to higher frequencies $\omega_I \pm \omega$ where essentially the same fluctuations appear in the polarizations that radiate the Stokes and anti-Stokes light. The fluctuation spectrum is shifted without any distortion of its frequency distribution for strictly monochromatic incident light.

The excitation frequency is usually much smaller than the incident frequency as in (1.25), and it is often a good approximation to set ω equal to zero in theoretical expressions for the second-order susceptibility. The second-order contribution then expresses the change in susceptibility caused by a *static* excitation amplitude

$$X(\mathbf{r}) = X_\mathbf{q} \exp(i\mathbf{q} \cdot \mathbf{r}) + X_\mathbf{q}^* \exp(-i\mathbf{q} \cdot \mathbf{r}), \qquad (1.46)$$

similar to a Fourier component in (1.31) with the time dependence removed. The second-order susceptibilities can now be written

$$\chi^{ij}(\omega_I, 0) = \chi^{ji}(\omega_I, 0) = \frac{\partial \chi^{ij}(\omega_I)}{\partial X_q^*}. \tag{1.47}$$

These quantities are called *susceptibility derivatives*; the equivalent polarizability derivatives (defined for local electric fields) were first used by Born and Bradburn (1947) for light scattering by crystals (see also Cowley 1971).

1.4.2 Radiation by the Stokes Polarization

The characteristics of the scattered light are determined by the radiation fields generated by the Stokes and anti-Stokes polarizations. We consider the radiation of light by a single Fourier component of the Stokes polarization,

$$\mathbf{P}_S \exp(-i\omega_S t + i\mathbf{K}_S \cdot \mathbf{r}), \tag{1.48}$$

where $\mathbf{P}_S$ is shorthand for $\mathbf{P}_S(\mathbf{K}_S, \omega_S)$. Even with this restriction the calculation is in general quite arduous, particularly for scattering media that have anisotropic optical properties, and the results depend on the geometry of the region in which scattering takes place and on the external boundaries of the sample (Lax and Nelson 1976). Accordingly, it is assumed here that the medium has isotropic optical properties, the simplest possible shape of scattering region is used, and the effects of the sample boundaries are ignored.

The electric field generated by the Stokes polarization has frequency ω_S, and the spatial dependence of its amplitude $\mathbf{E}$ is determined from Maxwell's equations,

$$\nabla \times \nabla \times \mathbf{E} - \frac{\eta_S^2 \omega_S^2}{c^2} \mathbf{E} = \frac{\omega_S^2}{\epsilon_0 c^2} \mathbf{P}_S \exp(i\mathbf{K}_S \cdot \mathbf{r}). \tag{1.49}$$

The solution has the usual two contributions, the complementary function (homogeneous part) and the particular integral (inhomogeneous part),

$$\mathbf{E} = \mathbf{E}_h \exp(i\mathbf{k}_S \cdot \mathbf{r}) + \mathbf{E}_i \exp(i\mathbf{K}_S \cdot \mathbf{r}). \tag{1.50}$$

The homogeneous part corresponds physically to the free electromagnetic waves of frequency ω_S, which exist in the absence of any Stokes polarization. The solutions obtained by substituting zero on the right-hand side of (1.49) are transverse waves with

$$\mathbf{E}_h \cdot \mathbf{k}_S = 0 \tag{1.51}$$

and the magnitude of $\mathbf{k}_S$ is given by (1.19)

The inhomogeneous part corresponds physically to an electromagnetic wave driven by the Stokes polarization, with a wavevector $\mathbf{K}_S$ imposed by the polarization and not necessarily equal to the free wavevector $\mathbf{k}_S$. Substitution of the second term on the right of (1.50) into the complete wave equation (1.49), and use of the Maxwell equation

$$\nabla \cdot \mathbf{D} = i\mathbf{K}_S \cdot (\epsilon_0 \eta_S^2 \mathbf{E}_i + \mathbf{P}_S) = 0 \tag{1.52}$$

leads to the solution

$$\mathbf{E}_i = \frac{k_S^2 \mathbf{P}_S - (\mathbf{K}_S \cdot \mathbf{P}_S)\mathbf{K}_S}{\epsilon_0 \eta_S^2 (K_S^2 - k_S^2)}. \tag{1.53}$$

Note that neither $\mathbf{P}_S$ nor $\mathbf{E}_i$ is perpendicular to $\mathbf{K}_S$ in general.

The relation between the two parts of the general solution (1.50) depends on the geometrical details of the scattering experiment. We consider the two-dimensional arrangement shown in Figure 1.7, which approximates the scattering geometry of Figure 1.1. The scattering takes place in a limited region of the sample, represented by the shaded area, which is embedded in a much larger volume of the same medium. The scattering region is determined in an experiment by the part of the medium illuminated by the incident light. We assume here for simplicity that this is a slab of thickness L with faces perpendicular to the direction of observation of the scattered light, taken as the Z axis.

Scattered light is of course emitted in all directions from the illuminated region, but we ignore all contributions except that for which $\mathbf{K}_S$ is oriented in the positive Z direction. This component grows from zero intensity at the top of the illuminated region at $Z=0$ and its generation ceases at the bottom of the region at $Z=L$, where it passes into the unilluminated part of the medium. The scattering polarization $\mathbf{P}_S$ is there zero, and the field determined by (1.49) has the form

$$\mathbf{E}_S \exp(i\mathbf{k}_S \cdot \mathbf{r}), \tag{1.54}$$

where

$$\mathbf{E}_S \cdot \mathbf{k}_S = 0, \tag{1.55}$$

similar to (1.51). The various vector quantities associated with the scattered light are shown in Figure 1.7.

Macroscopic Theory of Light Scattering

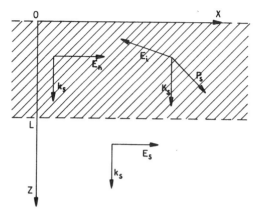

Figure 1.7 The shaded area represents a region of scattering medium illuminated by a beam of incident light propagated from left to right. The groups of arrows represent the two contributions to the scattered field inside the illuminated region and the single contribution outside. Only those contributions are considered whose wavevectors point vertically downward.

The symmetry of the slab arrangement restricts the wavevector $\mathbf{k}_S$ to be parallel to $\mathbf{K}_S$, and also causes all the vectors to lie in the plane defined by $\mathbf{K}_S$ and $\mathbf{P}_S$, taken as the ZX plane. The electric field vectors are subject to the usual boundary conditions at the surfaces of the slab. The Z component of the displacement is everywhere zero, and the tangential boundary conditions give respectively

$$E_h^X + E_i^X = 0 \tag{1.56}$$

$$E_h^X \exp(ik_S L) + E_i^X \exp(iK_S L) = E_S^X \exp(ik_S L). \tag{1.57}$$

Elimination of the homogeneous field and substitution of (1.53) for the inhomogeneous field gives

$$E_S = \frac{k_S^2 \boldsymbol{\varepsilon}_S \cdot \mathbf{P}_S}{\epsilon_0 \eta_S^2 (K_S^2 - k_S^2)} \left\{ \exp[i(K_S - k_S)L] - 1 \right\} \tag{1.58}$$

where $\boldsymbol{\varepsilon}_S$ is a unit vector parallel to $\mathbf{E}_S$ as before.

The cycle-averaged intensity of the scattered light is obtained from the field by forming

$$2\epsilon_0 c \eta_S E_S^* E_S. \tag{1.59}$$

However, the total scattered intensity $\bar{I}_S$ is obtained by integration of the scattered field over its frequency spectrum ω_S and summation of the intensity contributions over all the wavevectors $\mathbf{K}_S$ present in the Stokes polarization (1.38). The result must also be averaged over the fluctuations in the amplitude of the Stokes polarization, leading to

$$\bar{I}_S = 2\epsilon_0 c\eta_S \sum_{\mathbf{K}_S} \int d\omega_S \int d\omega'_S \frac{k_S^2 k_S'^2 \langle \boldsymbol{\varepsilon}_S \cdot \mathbf{P}_S^*(\mathbf{K}_S,\omega_S)\, \boldsymbol{\varepsilon}_S \cdot \mathbf{P}_S(\mathbf{K}_S,\omega'_S)\rangle}{\epsilon_0^2 \eta_S^4 (K_S^2 - k_S^2)(K_S^2 - k_S'^2)}$$
$$\times \{\exp[-i(K_S - k_S)L] - 1\}\{\exp[i(K_S - k'_S)L] - 1\}, \qquad (1.60)$$

where the full notation of the polarization Fourier amplitudes is restored, k'_S is related to ω'_S as in (1.19), and the angle brackets again denote an average over the fluctuation probability distribution. The average is equal to a product of the power spectrum of the polarization fluctuations with a delta function,

$$\langle \boldsymbol{\varepsilon}_S \cdot \mathbf{P}_S^*(\mathbf{K}_S,\omega_S)\boldsymbol{\varepsilon}_S \cdot \mathbf{P}_S(\mathbf{K}_S,\omega'_S)\rangle = \langle \boldsymbol{\varepsilon}_S \cdot \mathbf{P}_S^*(\mathbf{K}_S)\boldsymbol{\varepsilon}_S \cdot \mathbf{P}_S(\mathbf{K}_S)\rangle_{\omega_S}$$
$$\times \delta(\omega_S - \omega'_S), \qquad (1.61)$$

similar to (1.34). Thus (1.60) simplifies,

$$\bar{I}_S = \sum_{\mathbf{K}_S} \int d\omega_S \frac{2ck_S^4 \langle \boldsymbol{\varepsilon}_S \cdot \mathbf{P}_S^*(\mathbf{K}_S)\boldsymbol{\varepsilon}_S \cdot \mathbf{P}_S(\mathbf{K}_S)\rangle_{\omega_S}}{\epsilon_0 \eta_S^3 (K_S^2 - k_S^2)^2} |\exp[i(K_S - k_S)L] - 1|^2.$$
$$(1.62)$$

The well-known limit

$$\underset{L \to \infty}{\mathrm{Lt}} \frac{|\exp[i(K_S - k_S)L] - 1|^2}{(K_S - k_S)^2} = 2\pi L \delta(K_S - k_S) \qquad (1.63)$$

can be used to simplify the intensity still further in most cases. For a scattering region whose thickness L is not infinite, the delta function is replaced by a broader function whose value is small unless K_S differs from k_S by an amount less than $2\pi/L$. This amount can be significant for very small scattering samples, where there are observable departures from the delta function limit (see Sections 4.1.7 and 8.3.2). However, (1.63) is normally an adequate approximation, and it leads in conjunction with (1.40) to the momentum conservation condition (1.15).

Macroscopic Theory of Light Scattering

The wavevector summation in (1.62) is converted to an integration in the usual way

$$\sum_{\mathbf{K}_S} \to \frac{V}{(2\pi)^3} \iint dK_S \, d\Omega \, K_S^2, \quad (1.64)$$

where V is the volume of the scattering sample. The scattered intensity now becomes

$$\bar{I}_S = \int d\Omega \int d\omega_S \omega_S^4 V \eta_S \frac{\langle \boldsymbol{\varepsilon}_S \cdot \mathbf{P}_S^*(\mathbf{k}_S) \boldsymbol{\varepsilon}_S \cdot \mathbf{P}_S(\mathbf{k}_S) \rangle_{\omega_S} L}{8\pi^2 \epsilon_0 c^3}, \quad (1.65)$$

where (1.19) has been used.

1.4.3. The Cross Section

The cross-sectional area of the scattered beam in Figure 1.7 is $\mathfrak{v}/L$, where $\mathfrak{v}$ is again the volume of that part of the sample that contributes to the detected scattered light. The rate of energy flow in the scattered beam is thus $\mathfrak{v}\bar{I}_S/L$. Now each photon of energy $\hbar\omega_S$ in the scattered light is the result of a process in which the incident beam loses a larger quantum $\hbar\omega_I$. The rate of loss of energy by the incident beam therefore exceeds the rate of energy gain in the scattered beam by a factor ω_I/ω_S. Then in accordance with the definition that follows (1.4), the spectral differential cross section is

$$\frac{d^2\sigma}{d\Omega \, d\omega_S} = \frac{\omega_I \mathfrak{v}}{\omega_S L} \frac{d^2 \bar{I}_s}{d\Omega \, d\omega_S} \frac{1}{\bar{I}_I}, \quad (1.66)$$

where the cycle-averaged intensity of the incident beam is

$$\bar{I}_I = 2\epsilon_0 c \eta_I |\mathbf{E}_I|^2. \quad (1.67)$$

Insertion of (1.65) and (1.67) into (1.66) gives a general expression for the spectral differential cross section

$$\frac{d^2\sigma}{d\Omega \, d\omega_S} = \frac{\omega_I \omega_S^3 \mathfrak{v} V \eta_S \langle \boldsymbol{\varepsilon}_S \cdot \mathbf{P}_S^*(\mathbf{k}_S) \boldsymbol{\varepsilon}_S \cdot \mathbf{P}_S(\mathbf{k}_S) \rangle_{\omega_S}}{(4\pi\epsilon_0)^2 c^4 \eta_I |\mathbf{E}_I|^2}. \quad (1.68)$$

There are three somewhat trivial features of the cross-section expression worthy of brief mention. The factor $(4\pi\epsilon_0)^2$ is associated with the use of the

International System of Units; its removal converts the cross section to centimeter-gram-second units. Again, the refractive index ratio η_S/η_I is a common feature of cross-section expressions. It is not to be taken too seriously because it arises from the expressions (1.59) and (1.67) for the intensities of the light beams *inside* the scattering medium. However, as emphasized in Section 1.2.1, the beam intensities can only be directly measured *outside* the scattering medium. Inclusion of transmission coefficients at the sample surfaces replaces the simple ratio by a more complicated function of η_S and η_I, which depends on the geometrical arrangement of the experiment [see for example (4.100)]. Finally, there are two different volumes in the cross section. The total volume V of the sample, which enters in the conversion (1.64) from sum to integral, is always cancelled by a factor $1/V$, which appears in the power spectrum. It can be regarded as an arbitrary normalization volume. The cross section is thus proportional only to the scattering volume $\mathfrak{v}$.

The more important parts of the cross section (1.68) are the frequency factor $\omega_I \omega_S^3$, mentioned in the discussion of (1.13), and the power spectrum of the polarization fluctuations. These quantities determine the strength and shape of the scattered spectrum. The relation (1.43) between the polarization and excitation amplitudes enables the power spectrum of the polarization fluctuations to be written

$$\langle \varepsilon_S \cdot \mathbf{P}_S^*(\mathbf{k}_S) \varepsilon_S \cdot \mathbf{P}_S(\mathbf{k}_S) \rangle_{\omega_S} = |\epsilon_0 \varepsilon_S^i \chi^{ij}(\omega_I, -\omega) E_I^j|^2 \langle X(\mathbf{q}) X^*(\mathbf{q}) \rangle_\omega, \quad (1.69)$$

where the repeated superscript summation convention remains in force, and the absence of fluctuations from the incident beam is assumed. The cross section (1.68) thus becomes

$$\frac{d^2\sigma}{d\Omega d\omega_S} = \frac{\omega_I \omega_S^3 \mathfrak{v} V \eta_S |\epsilon_0 \varepsilon_S^i \varepsilon_I^j \chi^{ij}(\omega_I, -\omega)|^2}{(4\pi\epsilon_0)^2 c^4 \eta_I} \langle X(\mathbf{q}) X^*(\mathbf{q}) \rangle_\omega, \quad (1.70)$$

where ε_I is again a unit vector parallel to the incident field. There are two factors to be determined in the theory of scattering by a specific excitation. The second-order susceptibility or susceptibility derivative can often be related to an appropriate macroscopic property of the scattering medium. The power spectrum depends only on the fluctuation properties of the scattering excitation.

A simpler but less transparent way of writing the cross section is

$$\frac{d^2\sigma}{d\Omega d\omega_S} = \frac{\omega_I \omega_S^3 \mathfrak{v} V \eta_S}{16\pi^2 c^4 \eta_I} \langle |\varepsilon_S^i \varepsilon_I^j \delta \chi^{ij}|^2 \rangle_\omega \quad (1.71)$$

where

$$\delta\chi^{ij} = \chi^{ij}(\omega_I, -\omega) X^*(\mathbf{q}). \quad (1.72)$$

This form emphasizes the physical origin of the scattering in the fluctuations of the susceptibility around the constant value it would have in the absence of any excitations. Bearing in mind the relation

$$\kappa^{ij} = \delta^{ij} + \chi^{ij} \quad (1.73)$$

between relative permittivity and linear susceptibility, (1.71) leads straightforwardly to the Rayleigh formula (1.1) in the limit where κ differs only slightly from κ_0.

For the cross-section formula in an anisotropic scattering medium, including the effects of transmission through the sample surfaces, the reader is referred to the work of Lax and Nelson (1976)

1.4.4 Fluctuation-Dissipation Theory

We here consider the calculation of the power spectrum that appears in the cross section (1.70). The fluctuation properties of systems in thermal equilibrium form an entire field of study that can only be outlined here. A very clear account of the basic theory is to be found in Chapter 12 of Landau and Lifshitz (1969), whereas a more elementary review is given by MacDonald (1962).

The fluctuation-dissipation theorem provides a simple means of calculating the power spectra needed for light-scattering theory. Suppose that there exists a fictitious applied force $F(t)$ that couples to the amplitude $X(\mathbf{r})$ of the excitation whose power spectrum is required in such a way that the interaction Hamiltonian is

$$H = -X(\mathbf{r}) F(t). \quad (1.74)$$

The Hamiltonian must have the dimensions of energy and $F(t)$ thus does not in general have the dimensions of force; it is regarded as a generalized force.

The time dependence of the force is arbitrary, but it can be Fourier analyzed

$$F(t) = \int F(\omega) \exp(-i\omega t) d\omega. \quad (1.75)$$

The component $F(\omega)$ causes the average of the Fourier component $X(\mathbf{q}, \omega)$

of the excitation amplitude to differ from the zero value that it has in the absence of the applied force. The average is proportional to the force, and it is written

$$\overline{X}(\mathbf{q},\omega) = T(\mathbf{q},\omega)F(\omega), \quad (1.76)$$

where $T(\mathbf{q},\omega)$ is called a *linear response function* or admittance.

The fluctuation-dissipation theorem proved by Landau and Lifshitz (1969) is

$$\tfrac{1}{2}\langle X^*(\mathbf{q})X(\mathbf{q}) + X(\mathbf{q})X^*(\mathbf{q})\rangle_\omega = \frac{\hbar}{\pi}\{n(\omega) + \tfrac{1}{2}\}\operatorname{Im} T(\mathbf{q},\omega), \quad (1.77)$$

where the Bose-Einstein factor is defined in (1.6). The power spectrum of the amplitude fluctuations is therefore related to the imaginary, or dissipative, part of the linear-response function. The two contributions inside the angle brackets are of course identical, but they have been separated for a purpose that will become clear in the following discussion.

This form of the fluctuation-dissipation theorem is not quite correct for use in light-scattering calculations and it is necessary to make an excursion into quantum mechanics to understand the correction required. The quantum-mechanical form of (1.31) in the Heisenberg representation is

$$\hat{X}(\mathbf{r},t) = \sum_{\mathbf{q}} \{\hat{X}(\mathbf{q},t)\exp(i\mathbf{q}\cdot\mathbf{r}) + \hat{X}^\dagger(\mathbf{q},t)\exp(-i\mathbf{q}\cdot\mathbf{r})\}, \quad (1.78)$$

where the circumflexes indicate quantum-mechanical operators and the dagger denotes a Hermitian conjugate. The first term on the right includes the destruction operators in a second quantized formalism, whereas the second term includes the creation operators. The equations of the classical cross section are formally converted to quantum mechanics by notational changes like

$$X(\mathbf{q},t) \to \hat{X}(\mathbf{q},t) \qquad X^*(\mathbf{q},t) \to \hat{X}^\dagger(\mathbf{q},t)$$
$$X(\mathbf{q}) \to \hat{X}(\mathbf{q}) \qquad X^*(\mathbf{q}) \to \hat{X}^\dagger(\mathbf{q}). \quad (1.79)$$

Averages over the probability distribution, such as occur in the definitions of power spectra, are reinterpreted as quantum-mechanical averages.

The Stokes scattering involves a transition of the scattering medium in which a quantum $\hbar\omega$ of excitation is created. The quantum-mechanical transition rate, averaged over the states of the medium, is proportional to

$$\langle \hat{X}(\mathbf{q})\hat{X}^\dagger(\mathbf{q})\rangle_\omega. \quad (1.80)$$

This quantity, which is sometimes denoted by $S(\mathbf{q},\omega)$, replaces the classical power spectrum in the cross section (1.70), and a more explicit expression for it is given in (1.115). Similarly in the anti-Stokes scattering, the appropriate power spectrum is the quantity

$$\langle \hat{X}^\dagger(\mathbf{q})\hat{X}(\mathbf{q})\rangle_\omega, \qquad (1.81)$$

which occurs in the transition rate for destruction of a quantum $\hbar\omega$.

These quantum-mechanical power spectra satisfy fluctuation-dissipation theorems similar to (1.77),

$$\langle \hat{X}(\mathbf{q})\hat{X}^\dagger(\mathbf{q})\rangle_\omega = \frac{\hbar}{\pi}\{n(\omega)+1\}\operatorname{Im} T(\mathbf{q},\omega) \qquad (1.82)$$

$$\langle \hat{X}^\dagger(\mathbf{q})\hat{X}(\mathbf{q})\rangle_\omega = \frac{\hbar}{\pi}n(\omega)\operatorname{Im} T(\mathbf{q},\omega). \qquad (1.83)$$

Note that the sum of these equations gives the quantum-mechanical analogue of (1.77). The different ordering of operators in (1.82) and (1.83) is clearly very important, and the necessity of separating the fluctuation-dissipation theorem into two parts for light-scattering theory was pointed out by Butcher and Ogg (1965). Substitution of (1.82) for the power spectrum in (1.70) takes the determination of the cross section a step closer to the final explicit expression.

The final step is the determination of the response function, and it is possible to do this by a classical calculation based on equations (1.74) to (1.76). The quantum-mechanical considerations, which lead to the replacement of the classical fluctuation-dissipation theorem (1.77) by the appropriate form (1.82) or (1.83) do not require any change in the response function itself, but only affect the Bose-Einstein thermal factors. Having established the correct thermal factors, the remainder of the calculation can be carried out by classical methods. The corresponding quantum-mechanical calculation of response functions is briefly discussed in Section 1.5.1.

A simple illustration of the classical calculation of a linear response function is provided by a molecular vibration. It is shown in Section 3.1.2 that the motion can be represented by a harmonic oscillator equation

$$\ddot{W}_\sigma + \Gamma_\sigma \dot{W}_\sigma + \omega_\sigma^2 W_\sigma = 0, \qquad (1.84)$$

where W_σ measures the vibrational displacement, ω_σ is its natural frequency, Γ_σ is its damping constant, and σ labels the different modes of vibration. The interactions between the N identical molecules in the scattering medium are assumed sufficiently weak that the vibrational parameters are independent of $\mathbf{q}$.

The fictitious applied force is taken for simplicity to have a single frequency component, and the interaction Hamiltonian is

$$H = -NW_\sigma f \exp(-i\omega t). \tag{1.85}$$

The generalized force of frequency ω obtained by comparison with (1.74) and (1.75) is

$$F(\omega) = Nf. \tag{1.86}$$

The vibrational amplitude no longer has the zero steady-state value given by (1.84), but is obtained by insertion of the driving force on the right-hand side. The steady-state amplitude is given by

$$\left(\omega_\sigma^2 - \omega^2 - i\omega\Gamma_\sigma\right)\overline{W}_\sigma(\mathbf{q},\omega) = f. \tag{1.87}$$

The linear response function obtained by comparison of the last two equations with the general definition (1.76) is

$$T(\mathbf{q},\omega) = \frac{1}{N\left(\omega_\sigma^2 - \omega^2 - i\omega\Gamma_\sigma\right)}. \tag{1.88}$$

Thus

$$\operatorname{Im} T(\mathbf{q},\omega) = \frac{\omega\Gamma_\sigma}{N\left\{\left(\omega_\sigma^2 - \omega^2\right)^2 + \omega^2\Gamma_\sigma^2\right\}} \approx \frac{\pi g_\sigma(\omega)}{2N\omega_\sigma}, \tag{1.89}$$

where

$$g_\sigma(\omega) = \frac{\Gamma_\sigma/2\pi}{(\omega_\sigma - \omega)^2 + (\Gamma_\sigma/2)^2}. \tag{1.90}$$

The final form of the imaginary part of the response function is an approximation valid when Γ_σ is much smaller than ω_σ, as is usually the case for molecular vibrations. The resulting Lorentzian frequency dependence is here expressed for convenience in terms of a normalized lineshape function,

$$\int_{-\infty}^{\infty} g_\sigma(\omega)\,d\omega = 1, \tag{1.91}$$

which has a full width Γ_σ at half its peak magnitude.

Macroscopic Theory of Light Scattering

The power spectrum given by (1.82) for Stokes scattering by the molecular vibration is

$$\langle W_\sigma(\mathbf{q}) W_\sigma^*(\mathbf{q}) \rangle_\omega = \frac{\hbar}{2N\omega_\sigma} \{n(\omega_\sigma) + 1\} g_\sigma(\omega), \qquad (1.92)$$

where the narrow spread of the Lorentzian around its peak at ω_σ justifies the insertion of this frequency in the thermal factor. The power spectrum is independent of $\mathbf{q}$ for the weakly interacting molecules assumed here.

1.4.5 Relation between Stokes and Anti-Stokes Cross Sections

The molecular vibrational cross section for Stokes scattering obtained by substitution of (1.92) into (1.70) is

$$\frac{d^2\sigma}{d\Omega\, d\omega_S} = \frac{\hbar\omega_I \omega_S^3 \mathfrak{v} V\eta_S |\epsilon_0 \varepsilon_S^i \varepsilon_I^j \chi^{ij}(\omega_I, -\omega_\sigma)|^2 \{n(\omega_\sigma) + 1\}}{(4\pi\epsilon_0)^2 2c^4 \eta_I N \omega_\sigma} g_\sigma(\omega). \qquad (1.93)$$

The cross section is independent of the total volume of the scattering medium since the factor V/N is an intensive parameter determined by the molecular density. Because of the normalization property (1.91), the differential cross section defined in (1.7) is obtained by simple removal of the Lorentzian lineshape function from (1.93).

The anti-Stokes cross section is derived in a similar fashion. For the same incident beam subject to scattering by the same collection of noninteracting molecules, the cross section is given by the same expression (1.93) as for Stokes scattering except that the subscripts on ω, η, and ε must be changed from S to AS, the second-order susceptibility must be changed to $\chi^{ij}(\omega_I, \omega_\sigma)$, and the thermal factor is simply $n(\omega_\sigma)$ [compare (1.82) and (1.83)]. The Stokes and anti-Stokes cross sections considered here describe, respectively, the experiments represented in Figure 1.4a and b. There is no simple rigorous relation between these cross sections, because the second-order susceptibilities are in general different.

Consider, however, the pair of Stokes/anti-Stokes experiments represented in Figure 1.8. The Stokes experiment Figure 1.8a, is the same as that in Figure 1.4a, and its cross section is given by (1.93). The anti-Stokes experiment Figure 1.8b is represented by the same diagram as the Stokes experiment except that the directions of all the wavevectors are reversed. Thus the anti-Stokes experiment has incident light of frequency ω_S, wavevector $-\mathbf{k}_S$, and polarization ε_S producing scattered light of higher frequency ω_I, wavevector $-\mathbf{k}_I$, and polarization ε_I. The scattering angle is the same for both experiments; both satisfy the energy and momentum conservation laws (1.2) and (1.15).

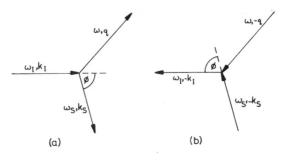

Figure 1.8 Wavevector directions for a time-reversed pair of (a) Stokes and (b) anti-Stokes experiments.

The cross section is obtained from (1.93) by interchange of the I and S subscripts, replacement of the second-order susceptibility by $\chi^{ij}(\omega_S, \omega_o)$, and removal of the unit contribution from the thermal factor. This anti-Stokes cross section is related to the Stokes cross section because the second-order susceptibilities that now occur are connected by (1.45). Indeed, the square modulus factors in the two cross sections are identical when account is taken of the interchange of the beam polarizations. The cross sections for the experiments of Figure 1.8 thus satisfy

$$\omega_I^2 \eta_I^2 n(\omega_o) \frac{d^2\sigma}{d\Omega\, d\omega_S} = \omega_S^2 \eta_S^2 \{n(\omega_o) + 1\} \frac{d^2\sigma}{d\Omega\, d\omega_I}. \quad (1.94)$$

Stokes   Anti-Stokes

This result can be proved more generally for scattering by all kinds of excitations (see Landau and Lifshitz 1960, Section 92; also 1965, Section 116). The scattering processes of Figure 1.8 are time reverses of each other, and the general proof uses the time-reversal properties of the quantum-mechanical states involved in the scattering. The scattering media to which the two cross sections refer are also strictly time reverses of each other. This consideration is important in the scattering by magnetic crystals where time reversal produces a reversal in the magnetization direction; thus (1.94) applies to experiments in which the magnetization vectors are oppositely directed for the Stokes and anti-Stokes cases (for discussion, see Loudon 1978).

Although the anti-Stokes cross section for the experiment of Figure 1.4b is not rigorously related to either of the cross sections in (1.94), it is of course not very different from the anti-Stokes cross section of the experiment in Figure 1.8b when the second-order susceptibilities vary only slightly for changes of order ω_o in the incident and scattered frequencies. This is just the condition for the validity of the susceptibility derivative approximation of (1.47). The approximate connection formula (1.5) is then obtained on removal of the frequencies and refractive indices from (1.94). The approximation breaks down seriously only in resonance scattering conditions when ω_I is close to a transition frequency of the scattering medium (see Section 4.2).

1.5 MICROSCOPIC THEORY OF LIGHT SCATTERING

1.5.1 The Interaction Hamiltonians

Microscopic calculations of the scattering cross section are based on the quantum-mechanical Hamiltonian of the coupled radiation field and scattering medium, which we write

$$\hat{H}_R + \hat{H} + \hat{H}_{ER}. \tag{1.95}$$

The contributions are, respectively, the radiation field Hamiltonian, including both incident and scattered fields, the total Hamiltonian of the scattering medium, and the coupling of the electrons of the scattering medium to the radiation. This is the most important coupling of the light to the medium for the optical frequencies normally used in experiments.

The Hamiltonian of the scattering medium is further divided into a sum $\hat{H}_O$ of the Hamiltonians of the various kinds of elementary excitation in the medium and the interactions $\hat{H}_I$ between these excitations

$$\hat{H} = \hat{H}_O + \hat{H}_I. \tag{1.96}$$

For example, in the theory of vibrational scattering $\hat{H}_I$ includes the electron-vibrational interaction.

Each scattering event in the quantum-mechanical theory corresponds to a transition between an initial state $|i\rangle$ and a final state $|f\rangle$ of the scattering medium, these being eigenstates of $\hat{H}_O$,

$$\hat{H}_O|i\rangle = \hbar\omega_i|i\rangle \tag{1.97}$$

$$\hat{H}_O|f\rangle = \hbar\omega_f|f\rangle. \tag{1.98}$$

The radiation field simultaneously undergoes a transition from an initial state with n_I incident photons and n_S scattered photons to a final state with $n_I - 1$ and $n_S + 1$ photons, where

$$\hat{H}_R|n_I, n_S\rangle = (n_I\hbar\omega_I + n_S\hbar\omega_S)|n_I, n_S\rangle \tag{1.99}$$

$$\hat{H}_R|n_I - 1, n_S + 1\rangle = \{(n_I - 1)\hbar\omega_I + (n_S + 1)\hbar\omega_S\}|n_I - 1, n_S + 1\rangle. \tag{1.100}$$

The energy gain by the scattering medium is $\hbar\omega$, where

$$\omega = \omega_f - \omega_i, \tag{1.101}$$

and (1.2) again expresses conservation of energy between the radiation and the medium.

Let $1/\tau$ be the rate of transitions between these initial and final states. The rate of removal of energy from the incident beam by the scattering process is $\hbar\omega_I/\tau$. If the scattering medium is assumed to be optically isotropic as before, the mean intensity of the incident beam is [see for example (12.50) of Loudon 1973]

$$\bar{I}_I = \frac{c^2 \hbar k_I}{\eta_I^2 \mathrm{v}} n_I, \tag{1.102}$$

where n_I is the number of incident photons in the scattering volume v. Then with the use of (1.18), the cross section defined as in Section 1.2 is

$$\sigma = \frac{\hbar\omega_I}{\tau \bar{I}_I} = \frac{\eta_I \mathrm{v}}{\tau n_I c}. \tag{1.103}$$

The determination of the cross section thus requires the calculation of an appropriate transition rate.

The standard expression for the transition rate $1/\tau$ obtained from time-dependent perturbation theory is given in (11.57) of Loudon (1973). The rate includes contributions of first, second, third, and higher orders in the matrix elements of the interaction parts of the Hamiltonian, namely $\hat{H}_{ER}$ and $\hat{H}_I$. It is of course necessary to retain all significant contributions to the transition rate for a given scattering process. The important contributions are different for scattering by different types of excitation, and examples treated in the book make various use of the first-, second-, and third-order terms in the matrix elements of the interaction Hamiltonians. The method can be used to find expressions for the differential and spectral differential cross sections from (1.103) by limiting the transition rate to final states in which the scattered photon has its wavevector direction within a solid angle $d\Omega$ and its frequency in a range $d\omega_S$.

The form of the electron-radiation interaction Hamiltonian is discussed in great detail in Chapters 6 and 8 of Loudon (1973), and the results are quoted here without proof. It is, however, necessary to make some small changes brought about by the immersion of the radiation in a scattering medium whose refractive index η may differ significantly from unity. The electromagnetic field energy in such a medium, whose magnetic permeability has the free-space value μ_0, is

$$U_R = \tfrac{1}{2} \int \left(\epsilon_0 \eta^2 E^2 + \mu_0 H^2 \right) dV. \tag{1.104}$$

Quantization of the radiation field based on this expression for the energy leads to vector-potential and electric-field operators

$$\hat{\mathbf{A}}(\mathbf{r}) = \sum_{\mathbf{k}} \left(\frac{\hbar}{2\epsilon_0 \eta^2 V \omega_{\mathbf{k}}} \right)^{1/2} \boldsymbol{\varepsilon}_{\mathbf{k}} \{ \hat{a}_{\mathbf{k}} \exp(i\mathbf{k} \cdot \mathbf{r}) + \hat{a}_{\mathbf{k}}^{\dagger} \exp(-i\mathbf{k} \cdot \mathbf{r}) \} \tag{1.105}$$

$$\hat{\mathbf{E}}(\mathbf{r}) = i \sum_{\mathbf{k}} \left(\frac{\hbar \omega_{\mathbf{k}}}{2\epsilon_0 \eta^2 V} \right)^{1/2} \boldsymbol{\varepsilon}_{\mathbf{k}} \{ \hat{a}_{\mathbf{k}} \exp(i\mathbf{k} \cdot \mathbf{r}) - \hat{a}_{\mathbf{k}}^{\dagger} \exp(-i\mathbf{k} \cdot \mathbf{r}) \} \tag{1.106}$$

in the Schrödinger representation. The creation and destruction operators for a photon of wavevector $\mathbf{k}$, frequency $\omega_{\mathbf{k}}$, and polarization $\boldsymbol{\varepsilon}_{\mathbf{k}}$ satisfy

$$\hat{a}_{\mathbf{k}}^{\dagger} | n_{\mathbf{k}} \rangle = (n_{\mathbf{k}} + 1)^{1/2} | n_{\mathbf{k}} + 1 \rangle \tag{1.107}$$

$$\hat{a}_{\mathbf{k}} | n_{\mathbf{k}} \rangle = n_{\mathbf{k}}^{1/2} | n_{\mathbf{k}} - 1 \rangle, \tag{1.108}$$

where $n_{\mathbf{k}}$ is the number of these photons. The volume V of the scattering medium is taken as the quantization volume.

Consider the interaction of the radiation with a collection of electrons, where the jth electron has momentum operator $\hat{\mathbf{p}}_j$ and position vector $\mathbf{r}_j$. The radiation field is included in the electronic Hamiltonian by making the replacement

$$\hat{\mathbf{p}}_j \to \hat{\mathbf{p}}_j + e\hat{\mathbf{A}}(\mathbf{r}_j), \tag{1.109}$$

and the resulting electron-radiation Hamiltonian has the form

$$\hat{H}_{ER} = \hat{H}'_{ER} + \hat{H}''_{ER}, \tag{1.110}$$

where

$$\hat{H}'_{ER} = \frac{e^2}{2m} \sum_j \hat{\mathbf{A}}(\mathbf{r}_j) \cdot \hat{\mathbf{A}}(\mathbf{r}_j) \tag{1.111}$$

$$\hat{H}''_{ER} = \frac{e}{m} \sum_j \hat{\mathbf{A}}(\mathbf{r}_j) \cdot \hat{\mathbf{p}}_j. \tag{1.112}$$

These two contributions are referred to as the A^2 and $\mathbf{A} \cdot \mathbf{p}$ parts.

The electron-radiation interaction can also be expressed in an alternative form involving the electric- and magnetic-field operators instead of the vector potential. The conversion is made by means of a gauge transformation. It is usual in this form of the interaction to expand the exponential factors that occur in the field operators. The lowest-order term, independent of **k**, is then the electric-dipole interaction

$$\hat{H}_{ED} = e \sum_j \hat{\mathbf{E}}(0) \cdot \mathbf{r}_j; \qquad (1.113)$$

the contributions of first-order in **k** are the magnetic-dipole and electric-quadrupole interactions.

This second form of the electron-radiation interaction is very convenient for electrons bound to atoms or molecules where the smallness of $a_0 k$ (a_0 = Bohr radius) for visible and infrared light ensures rapid convergence of the power series expansion of the exponentials about the atomic nucleus taken as coordinate origin. The electric-dipole part (1.113) is usually dominant, and its form has immediate physical appeal, since it represents the potential energy of the atomic or molecular dipole moment in the electric field of the radiation.

On the other hand, the first form of the electron-radiation interaction is generally more convenient for electrons in crystals. There are two reasons for this. Firstly, the Hamiltonian (1.110) is expressed in terms of exponentials $\exp(\pm i\mathbf{k} \cdot \mathbf{r}_j)$ and electron momenta $\hat{\mathbf{p}}_j$, whereas the Hamiltonian (1.113) involves the electron positions $\mathbf{r}_j$. The matrix elements of the electron momenta between the electronic wavefunctions in a crystal, the Bloch functions, have convenient properties, and they are much used in the electronic energy-band theory of crystals. By contrast, the corresponding matrix elements of the electron position vectors have awkward properties (Blount 1962).

Secondly, the exponential factors in (1.110) ensure that the conservation of momentum is built into calculations of the scattering cross section, as illustrated in Section 4.2.1. The expansion of the exponentials in the derivation of (1.113) removes the automatic provision of momentum conservation, and is in any case invalid for electrons whose Bloch wavefunctions are not confined to a dimension of the order of the Bohr radius but extend through the entire crystal. It is not feasible to retain the exponential factors intact in the transformation to the second form of interaction because the resulting Hamiltonian would then be very complicated.

The transition matrix element for any scattering process with initial and final photon states given by (1.99) and (1.100) must include the product of operators $\hat{a}_S^\dagger \hat{a}_I$, and in view of (1.107) and (1.108) the transition rate includes a factor $(n_S + 1)n_I$. The two contributions to the cross section that result from the parts n_S and 1 of the factor $n_S + 1$ describe, respectively, *stimulated* and *spontaneous* scattering. The mean number of scattered photons in a mode of the quantization volume is always much smaller than unity for the scattering of ordinary light beams; therefore, only the spontaneous contribution needs to be considered. Stimulated scattering is observed with the use of high-power laser light sources; it forms a branch of nonlinear optics and therefore lies beyond the scope of this book; its effects are removed from subsequent equations by setting $n_S = 0$.

Both parts of the first form (1.110) of the electron-radiation interaction contribute in general to the scattering matrix element. The matrix element of the first part (1.111) between the desired electronic and radiative initial and final states is

$$\langle f, n_I - 1, 1 | \hat{H}'_{ER} | i, n_I, 0 \rangle = \frac{\hbar e^2}{2m\epsilon_0 \eta_I \eta_S V(\omega_I \omega_S)^{1/2}} \boldsymbol{\varepsilon}_I \cdot \boldsymbol{\varepsilon}_S n_I^{1/2}$$

$$\times \left\langle f \Big| \sum_j \exp(i\mathbf{q} \cdot \mathbf{r}_j) \Big| i \right\rangle, \qquad (1.114)$$

where $\mathbf{q}$ is defined in (1.15), and (1.105), (1.107), and (1.108) have been used. For the scattering of light by a crystal, the smallness of $\mathbf{q}$ in comparison with Brillouin zone-boundary wavevectors results in a very small matrix element for excitations where the crystal electronic state is significantly changed, or more generally where other components of the crystal state are altered, as in scattering by lattice vibrations. The importance of this A^2 contribution is thus in scattering by electronic transitions with the same initial and final band states. These restrictions do not apply to the scattering of X-ray photons (not considered in this book) where the large transferred momenta $\hbar \mathbf{q}$ provide nonzero matrix elements (1.114) between quite different electronic states.

The second or $\mathbf{A} \cdot \mathbf{p}$ part of the electron-radiation interaction (1.110) is linear in the photon creation and destruction operators, and its contribution to the scattering is in second-order. The matrix elements include the electronic momenta and allow scattering in which the initial and final electronic states are different. The $\mathbf{A} \cdot \mathbf{p}$ part accordingly makes a significant contribution to the scattering for almost all kinds of excitations, and it often plays a more important role than the A^2 part. The ticks in Table 1.1 indicate where the contributions of the two parts of the electron-radiation interaction are important for inelastic scattering by the excitations considered in this book.

Table 1.1 Importance of the Two Parts of $\hat{H}_{ER}$ for Scattering by Various Excitations

Scattering Excitation	$\hat{H}'_{ER}\sim A^2$	$\hat{H}''_{ER}\sim \mathbf{A}\cdot\mathbf{p}$
Free electrons, including electrons in gaseous plasma	√	
Electronic transition of atom		√
Molecular or crystal lattice vibration		√
Polariton		√
Conduction electrons in crystal	√	√
Plasmon in crystal	√	√
Electron spin-flip or Landau-level transition in crystal		√
Magnon in ordered magnetic crystal		√

An alternative to the transition rate method of calculating cross sections outlined above is the quantum-mechanical version of the method of Section 1.4. The quantum-mechanical form of the power spectrum needed for the cross section (1.70) is explicitly given by

$$\langle \hat{X}(\mathbf{q})\hat{X}^\dagger(\mathbf{q})\rangle_\omega = \sum_{i,f} n_i \langle i|\hat{X}(\mathbf{q},0)|f\rangle\langle f|\hat{X}^\dagger(\mathbf{q},0)|i\rangle \delta(\omega-\omega_f+\omega_i), \quad (1.115)$$

where n_i is the thermal population of state $|i\rangle$ and the excitation amplitude operators are the same as used in (1.78)–(1.83). Thus the power spectrum can be calculated in principle from a knowledge of the wavefunctions and energy levels of the scattering medium. It is, however, more convenient to use a fully quantum-mechanical version of the fluctuation-dissipation theorem (1.82) of the form

$$\langle \hat{X}(\mathbf{q})\hat{X}^\dagger(\mathbf{q})\rangle_\omega = -\frac{\hbar}{\pi}\{n(\omega)+1\}\mathrm{Im}\,\mathcal{G}\{\hat{X}(\mathbf{q});\hat{X}^\dagger(\mathbf{q})\}_\omega, \quad (1.116)$$

where the quantity that replaces the negative of the classical response function is called a retarded thermal Green function. These have well-known properties (for example, see Zubarev 1960), and they are extensively used in light-scattering calculations. However, the classical response functions are used for the examples treated in this book.

The various ways of calculating cross sections are essentially different forms of the same basic theory, and results derived by the different methods are closely related. Thus, for example, in Chapters 3 and 4 essentially the same vibrational cross section is derived by the classical macroscopic method and by quantum-mechanical transition-rate theory; in Chapter 7 the two methods are combined in the treatment of electronic scattering.

1.5.2 Atomic Scattering Cross Section

The simplest example of a microscopic calculation of a cross section is provided by the scattering associated with an electronic transition of an isolated atom. The differential cross section in this case, derived in detail by Loudon (1973), is known as the Kramers-Heisenberg formula. There is no need to repeat the derivation here, but we consider briefly the effect of including the refractive index of the scattering medium and of using the two forms of the electron-radiation interaction Hamiltonian.

The cross section derived in the above reference uses the second form (1.113) of interaction and assumes the atom to be surrounded by free space. It is a simple matter to follow the same derivation keeping the additional refractive index factors, and the differential cross section obtained from the second-order contributions in $\hat{H}_{ED}$ to the transition rate is

$$\frac{d\sigma}{d\Omega} = \frac{e^4 \omega_I \omega_S^3 \eta_S}{(4\pi\epsilon_0)^2 \hbar^2 c^4 \eta_I} \left| \sum_l \left\{ \frac{\boldsymbol{\varepsilon}_S \cdot \mathbf{D}_{fl} \boldsymbol{\varepsilon}_I \cdot \mathbf{D}_{li}}{\omega_i + \omega_I - \omega_l} + \frac{\boldsymbol{\varepsilon}_I \cdot \mathbf{D}_{fl} \boldsymbol{\varepsilon}_S \cdot \mathbf{D}_{li}}{\omega_i - \omega_l - \omega_S} \right\} \right|^2, \quad (1.117)$$

where the summation runs over all electronic states $|l\rangle$ of the atom. The total electric dipole moment of the Z-electron atom is $e\mathbf{D}$, where

$$\mathbf{D} = \sum_{j=1}^{Z} \mathbf{r}_j, \quad (1.118)$$

and the electric-dipole matrix elements are written in the shorthand form

$$\mathbf{D}_{fl} = \langle f|\mathbf{D}|l\rangle. \quad (1.119)$$

The scattering medium is again assumed to be optically isotropic, and it is seen that the refractive indices enter the cross section in the same combination for the macroscopic and microscopic theories.

An alternative expression for the differential cross section is derived by using the first form (1.110) of the electron-radiation interaction. The contributions to the transition rate from the two parts of the interaction are represented diagrammatically in Figure 1.9. The initial states occur at the right-hand ends of the diagrams with the final states at the left-hand ends. These states are connected by the interactions indicated, with the convention that a line whose arrow is directed toward (away from) an interaction point corresponds to a photon or electronic state that is destroyed (created) in the interaction. Figure 1.9a represents the single contribution (1.114) of the A^2 part of the interaction, while Figures 1.9b and 1.9c represent two kinds of contribution from the $\mathbf{A} \cdot \mathbf{p}$ part.

40 **Basic Features and Formal Theory of Light Scattering**

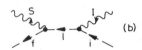

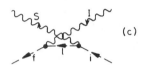

Figure 1.9 Diagrammatic representations of the three types of process that contribute to the scattering in (a) first order, (b) and (c) second order. The interaction in (a) is; $\hat{H}'_{ER}$; those in parts (b) and (c) are $\hat{H}''_{ER}$. Wavy lines represent photons, and the dashed lines represent electronic states.

Although the form of interaction is different, the structure of the cross-section calculation is very similar to that in Chapter 11 of Loudon (1973), and the result is

$$\frac{d\sigma}{d\Omega} = \frac{e^4 \omega_S \eta_S}{(4\pi\epsilon_0)^2 c^4 m^2 \omega_I \eta_I} \left| \boldsymbol{\varepsilon}_I \cdot \boldsymbol{\varepsilon}_S \left\langle f \left| \sum_j \exp(i\mathbf{q}\cdot\mathbf{r}_j) \right| i \right\rangle \right.$$

$$+ \frac{1}{\hbar m} \sum_l \sum_{j,j'} \left\{ \frac{\langle f|\exp(-i\mathbf{k}_S\cdot\mathbf{r}_j)\boldsymbol{\varepsilon}_S\cdot\hat{\mathbf{p}}_j|l\rangle \langle l|\exp(i\mathbf{k}_I\cdot\mathbf{r}_{j'})\boldsymbol{\varepsilon}_I\cdot\hat{\mathbf{p}}_{j'}|i\rangle}{\omega_i + \omega_I - \omega_l} \right.$$

$$\left. \left. + \frac{\langle f|\exp(i\mathbf{k}_I\cdot\mathbf{r}_j)\boldsymbol{\varepsilon}_I\cdot\hat{\mathbf{p}}_j|l\rangle \langle l|\exp(-i\mathbf{k}_S\cdot\mathbf{r}_{j'})\boldsymbol{\varepsilon}_S\cdot\hat{\mathbf{p}}_{j'}|i\rangle}{\omega_i - \omega_l - \omega_S} \right\} \right|^2, \quad (1.120)$$

where the terms are arranged in the same order as the diagrams of Figure 1.9. The electronic wavefunctions of an atom typically extend over a distance of the order of the Bohr radius a_0, and the exponentials in the matrix elements can be expanded in power series since $a_0 k_I$ and $a_0 k_S$ are small. Retention only of the zero-order terms gives

$$\frac{d\sigma}{d\Omega} = \frac{e^4 \omega_S \eta_S}{(4\pi\epsilon_0)^2 c^4 \hbar^2 m^4 \omega_I \eta_I} \left| \sum_l \left\{ \frac{\boldsymbol{\varepsilon}_S\cdot\mathbf{P}_{fl}\boldsymbol{\varepsilon}_I\cdot\mathbf{P}_{li}}{\omega_i + \omega_I - \omega_l} + \frac{\boldsymbol{\varepsilon}_I\cdot\mathbf{P}_{fl}\boldsymbol{\varepsilon}_S\cdot\mathbf{P}_{li}}{\omega_i - \omega_l - \omega_S} \right\} \right|^2, \quad (1.121)$$

Microscopic Theory of Light Scattering

where $\mathbf{P}_{fl}$ and $\mathbf{P}_{li}$ are matrix elements of the total electron momentum analogous to (1.119). The first term in the matrix element of (1.120) makes no contribution to zero order in $\mathbf{q}$, since the initial and final states of the atom are orthogonal. The expressions (1.117) and (1.121) are equivalent ways of writing the Kramers-Heisenberg formula, since the same electric-dipole approximation has been made in both.

A very rough numerical estimate of the atomic scattering cross section is easily derived for the special case where all the important intermediate states $|l\rangle$ have energies much larger than the incident and scattered photon energies, and the inequalities

$$\omega_f - \omega_i \ll \omega_l, \omega_S \ll \omega_l - \omega_i \qquad (1.122)$$

are satisfied. Then if ω_I and ω_S are neglected in the frequency denominators of (1.117), the D matrix elements are set equal to a_0, η_I is set equal to η_S, and the intermediate state excitation energies are all set equal to the Rydberg energy $\hbar\omega_R$, the differential cross section reduces to

$$\frac{d\sigma}{d\Omega} \approx \frac{e^4 \omega_I \omega_S^3 a_0^4}{(4\pi\epsilon_0)^2 \hbar^2 c^4 \omega_R^2} \approx r_e^2 \left(\frac{\omega_S^2}{2\omega_R^2}\right)^2 \qquad (1.123)$$

since

$$\omega_R = \frac{\hbar}{2ma_0^2}, \qquad (1.124)$$

and the classical electron radius is given by (1.10). Thus for a density ρ of atoms in a scattering volume $\mathfrak{v}$, the differential cross section for visible light is of the order

$$\frac{d\sigma}{d\Omega} \approx 10^{-32} \rho \mathfrak{v} \, \mathrm{m}^2. \qquad (1.125)$$

Light scattering by atoms in crystals is discussed in Section 7.1.

1.5.3 Scattering by Free Electrons

Another simple microscopic calculation is that of the Thomson cross section for scattering by free electrons. Consider an electron initially moving with a velocity that corresponds to a wavevector $\mathbf{k}$. The electron wavevector becomes $\mathbf{k} + \mathbf{q}$ after a light-scattering interaction with wavevector transfer $\mathbf{q}$. The shift in frequency between incident and scattered light is therefore

$$\omega = \omega_I - \omega_S = \frac{\hbar}{2m}\{(\mathbf{k}+\mathbf{q})^2 - k^2\} = \frac{\hbar}{m}\{\tfrac{1}{2}q^2 + \mathbf{k}\cdot\mathbf{q}\}. \qquad (1.126)$$

The scattering is clearly inelastic, but when the initial electron wavevector is much larger than the wavevector transfer, the relative frequency shift is of order

$$\frac{\omega}{\omega_I} \approx \frac{\hbar k q}{m\omega_I} \approx \text{electron velocity/velocity of light}. \qquad (1.127)$$

This is a very small fraction for electron velocities outside the relativistic region.

The calculation of the free-electron cross section parallels that for bound electrons. The contributions to the transition rate are again formally represented by the diagrams of Figure 1.9 when the electron-radiation interaction is taken in the form (1.110). The labels i and f now refer to the plane-wave states of an electron in free space, and the cross section given by (1.120) with appropriate insertion of free-electron energies and momenta is

$$\frac{d\sigma}{d\Omega} = \frac{e^4 \omega_S}{(4\pi\epsilon_0)^2 c^4 m^2 \omega_I} \bigg| \varepsilon_I \cdot \varepsilon_S$$

$$+ \frac{\hbar}{m}\left\{\frac{\varepsilon_S\cdot(\mathbf{k}+\mathbf{k}_I)\varepsilon_I\cdot\mathbf{k}}{\omega_I - (\hbar/m)(\tfrac{1}{2}k_I^2 + \mathbf{k}\cdot\mathbf{k}_I)} + \frac{\varepsilon_I\cdot(\mathbf{k}-\mathbf{k}_S)\varepsilon_S\cdot\mathbf{k}}{-(\hbar/m)(\tfrac{1}{2}k_S^2 - \mathbf{k}\cdot\mathbf{k}_S) - \omega_S}\right\}\bigg|^2. \qquad (1.128)$$

It is not difficult to show with the help of (1.15) and (1.126) that the denominators of the second and third terms are equal but opposite in sign for right-angle scattering ($\mathbf{k}_I \perp \mathbf{k}_S$). This equality holds approximately for any angle of scattering, and the contribution of these two terms taken together is in general smaller than that of the first term by a factor approximated in (1.127).

It follows that only the first, or A^2 term, in (1.128) need be retained, and since the difference between ω_S and ω_I is also of order (1.127), consistent neglect of terms of this order leads again to the Thomson cross section (1.14). Quantum mechanics thus reproduces the results of classical theory to zero order in the small quantity (1.127). The theory is generalized to scattering by electrons in crystals in Section 7.3.

1.6 SYMMETRY PROPERTIES OF INELASTIC CROSS SECTIONS

The symmetry properties of the scattering cross section are determined by the symmetry properties of the second-order susceptibility for the excitation concerned. The general symmetry property (1.45) applies to all kinds of excitation, and it leads to the time-reversal symmetry of the scattering expressed in (1.94). The spatial symmetry properties of the scattering medium lead to further connections between the cross sections measured in different experiments on the same sample. The Stokes cross sections for different polarizations of the incident and scattered light are often related by the spatial symmetry, and the cross section is sometimes required to vanish for certain polarizations that depend on the nature of the excitation.

The spatial symmetry of the scattering medium is formally specified by its symmetry group, the group of all spatial transformations that leave the medium invariant. Individual atoms and molecules have spatial symmetries characterized by a *point* group consisting of rotations and reflections that leave the atom or molecule invariant. The atomic arrangements in a regular crystal lattice are characterized by a *space* group that contains translations in addition to rotations and reflections. The effects of the translational invariance of a crystal are largely accounted for in the momentum conservation conditions (1.15) and (1.16), and the residual effects of the spatial symmetry derive from the crystal point group that remains on removal of translations from the space group (the space group itself must, however, be used in the treatment of second-order scattering, discussed in Sections 3.2 and 6.3). There are 32 different crystal point groups.

The effects of spatial symmetry are particularly important for scattering by crystal samples, and the anisotropy of the cross section is generally different for the different crystal symmetries. In a liquid or gas the random molecular orientations produce an averaging of the anisotropy of the single-molecule cross section, treated in detail in Section 3.1.5. We consider in the present section the restrictions imposed on the cross section by the point symmetry of a crystal or single molecule. The underlying theory of point symmetry groups on which the restrictions are based cannot adequately be summarized here, and a group-theory text (Heine 1960, Tinkham 1964, Cracknell 1968) should be consulted for further details.

The spatial properties of the excitations of the scatterer are described by irreducible representations of its symmetry group. Let Γ_X be the irreducible representation appropriate to the excitation considered in the derivation of the cross section in Section 1.4; we call Γ_X the excitation symmetry. In the microscopic theory of Section 1.5 with an initial state $|i\rangle$ of symmetry Γ_i and a final state $|f\rangle$ of symmetry Γ_f, the excitation symmetry

Table 1.2 Allowed Light-Scattering Symmetries and Second-Order Susceptibilities for the 32 Crystal Classes

Biaxial

Triclinic

1	C_1	$\begin{bmatrix} a & d & f \\ e & b & h \\ g & i & c \end{bmatrix}$	A	Γ_1
$\bar{1}$	C_i		A_g	Γ_1^+

Monoclinic

2	C_2	$\begin{bmatrix} a & d & \\ e & b & \\ & & c \end{bmatrix}$		A	Γ_1	
m	C_s			A'	Γ_1	
$2/m$	C_{2h}			A_g	Γ_1^+	
		$\begin{bmatrix} & & \\ & & \\ g & i & \end{bmatrix}$ $\begin{bmatrix} & & f \\ & & h \\ & & \end{bmatrix}$		B	Γ_2	
				A''	Γ_2	
				B_g	Γ_2^+	

Orthorhombic

222	D_2	$\begin{bmatrix} a & & \\ & b & \\ & & c \end{bmatrix}$	A	Γ_1	
$mm2$	C_{2v}		A_1	Γ_1	
mmm	D_{2h}		A_g	Γ_1^+	
		$\begin{bmatrix} & & \\ & & d \\ & e & \end{bmatrix}$	B_1	Γ_3	
			A_2	Γ_3	
			B_{1g}	Γ_3^+	
		$\begin{bmatrix} & & f \\ & & \\ g & & \end{bmatrix}$	B_2	Γ_2	
			B_1	Γ_2	
			B_{2g}	Γ_2^+	
		$\begin{bmatrix} & & \\ & & i \\ & h & \end{bmatrix}$	B_3	Γ_4	
			B_2	Γ_4	
			B_{3g}	Γ_4^+	

Uniaxial

Tetragonal

4	C_4	$\begin{bmatrix} a & c & \\ -c & a & \\ & & b \end{bmatrix}$	A	Γ_1	
$\bar{4}$	S_4				
$4/m$	C_{4h}	A_g	Γ_1^+		
		$\begin{bmatrix} d & e & \\ e & -d & \\ & & \end{bmatrix}$	B	Γ_2	
			B_g	Γ_2^+	
		$\begin{bmatrix} & & f \\ & & h \\ g & i & \end{bmatrix} \begin{bmatrix} & & -h \\ & & f \\ -i & g & \end{bmatrix}$	E	$\Gamma_3 + \Gamma_4$	
			E_g	$\Gamma_3^+ + \Gamma_4^+$	

422	D_4		$\begin{bmatrix} a & & \\ & a & \\ & & b \end{bmatrix}$	Γ_1	$\begin{bmatrix} & -c & \\ c & & \\ & & \end{bmatrix}$	A_2	Γ_2	$\begin{bmatrix} d & -d & \\ -d & & \\ & & \end{bmatrix}$	B_1	Γ_3	$\begin{bmatrix} & e & \\ e & & \\ & & \end{bmatrix}$ B_2 Γ_4 $\begin{bmatrix} & & f \\ & & g \\ g & f & \end{bmatrix}$ E Γ_5
$4mm$	C_{4v}	A_1									
$\bar{4}2m$	D_{2d}										
$4/mmm$	D_{4h}	A_{1g}		Γ_1^+		A_{2g}	Γ_2^+		B_{1g}	Γ_3^+	B_{2g} Γ_4^+ E_g Γ_5^+

Trigonal

3	C_3	A	$\begin{bmatrix} a & c \\ -c & a \\ & & b \end{bmatrix}$	Γ_1	$\begin{bmatrix} d & e & f \\ e & -d & h \\ g & i & g \end{bmatrix}$ E $\begin{matrix}\Gamma_2 + \Gamma_3 \\ \Gamma_2^+ + \Gamma_3^+\end{matrix}$		
$\bar{3}$	C_{3i}	A_g		Γ_1^+	$\begin{bmatrix} e & -d & -h \\ -d & -e & f \\ -i & g & \end{bmatrix}$ E_g		

32	D_3	A_1	$\begin{bmatrix} a & & \\ & a & \\ & & b \end{bmatrix}$	Γ_1	$\begin{bmatrix} & & \\ & & -c \\ & & \end{bmatrix}$	A_2	Γ_2	$\begin{bmatrix} d & e \\ d & f \\ -e & -f \end{bmatrix} \begin{bmatrix} & -e \\ -d & \\ & \end{bmatrix}$ E Γ_3	
$3m$	C_{3v}								
$\bar{3}m$	D_{3d}	A_{1g}		Γ_1^+		A_{2g}	Γ_2^+	E_g Γ_3^+	

Hexagonal

6	C_6	A	$\begin{bmatrix} a & c \\ -c & a \\ & & b \end{bmatrix}$	Γ_1	$\begin{bmatrix} & d & -f \\ f & & e \\ -g & e & \end{bmatrix}$ E_1 $\Gamma_5 + \Gamma_6$	$\begin{bmatrix} i & h & -i \\ h & -i & -h \\ & & \end{bmatrix}$ E_2 $\Gamma_2 + \Gamma_3$		
$\bar{6}$	C_{3h}	A'		Γ_1	E'' $\Gamma_5 + \Gamma_6$	E' $\Gamma_2 + \Gamma_3$		
$6/m$	C_{6h}	A_g		Γ_1^+	E_{1g} $\Gamma_5^+ + \Gamma_6^+$	E_{2g} $\Gamma_2^+ + \Gamma_3^+$		

Table 1.2 (continued)

622 6mm	D_6 C_{6v}	$\begin{bmatrix} a & & \\ & a & \\ & & b \end{bmatrix}$ A_1	$\begin{bmatrix} & -c & \\ c & & \\ & & \end{bmatrix}$ A_2	$\begin{bmatrix} & d & \\ d & & \\ & & \end{bmatrix} \begin{bmatrix} -e & \\ & -d \end{bmatrix}$ E_1		$\begin{bmatrix} f & \\ & -f \end{bmatrix} \begin{bmatrix} & f \\ f & \end{bmatrix}$ E_2	Γ_2 Γ_5	Γ_6
$\bar{6}m2$ 6/mmm	D_{3h} D_{6h}	A_1' A_{1g}	A_2' A_{2g}	E'' E_{1g}		E' E_{2g}	Γ_2 Γ_2^+ Γ_5^+	Γ_6 Γ_6^+

Isotropic

Cubic

23 $m3$	T T_h	$\begin{bmatrix} a & & \\ & a & \\ & & a \end{bmatrix}$ A A_g	$\begin{bmatrix} b & & \\ & b & -2b \\ & -3^{1/2}b & 3^{1/2}b \end{bmatrix}$ E E_g	$\begin{bmatrix} & d & c \\ d & & \\ c & & \end{bmatrix} \begin{bmatrix} & & c \\ & & d \\ c & d & \end{bmatrix}$ Γ_1 Γ_1^+	$\Gamma_2 + \Gamma_3$ $\Gamma_2^+ + \Gamma_3^+$	T T_g	Γ_4 Γ_4^+
432 $\bar{4}3m$ $m3m$	O T_d O_h	$\begin{bmatrix} a & & \\ & a & \\ & & a \end{bmatrix}$ A_1 A_{1g}	$\begin{bmatrix} b & & \\ & b & -2b \\ & -3^{1/2}b & 3^{1/2}b \end{bmatrix}$ E E_g	$\begin{bmatrix} & -c & \\ -c & & \\ & & \end{bmatrix} \begin{bmatrix} & & -c \\ & & \\ -c & & \end{bmatrix}$ Γ_1 Γ_1^+	Γ_3 Γ_3^+	T_1 T_{1g}	Γ_4 Γ_4^+	$\begin{bmatrix} & d & \\ d & & \\ & & \end{bmatrix} \begin{bmatrix} & & d \\ & & \\ d & & \end{bmatrix}$ T_2 T_{2g}	Γ_5 Γ_5^+

Symmetry Properties of Inelastic Cross Sections 47

is that of the operator $|f\rangle\langle i|$, which projects the initial state onto the final state. Thus

$$\Gamma_X = \Gamma_f \times \Gamma_i^*, \qquad (1.129)$$

where the asterisk denotes complex conjugation. The transformation properties of the incident and scattered light are described by the three-dimensional polar-vector representation Γ_{PV} of the point group considered, since the quantities $\mathbf{E}_I$, $\mathbf{E}_S$, and $\mathbf{P}_S$, which characterize the light, are all polar vectors.

The relation (1.43) between Stokes polarization, excitation amplitude, and incident field must be invariant under all the spatial transformations of the symmetry group of the scattering medium. This invariance condition (for a detailed discussion, see Nye 1957, pp. 20–21) is common to all equations that relate properties of a system with given spatial symmetry. It has two main consequences for light scattering.

The first consequence is the existence of selection rules. The invariance condition is satisfied only for those excitation symmetries where the two sides of (1.43) have the same transformation properties. In group-theoretical language, only those Γ_X are allowed for which the direct product $\Gamma_X \times \Gamma_{PV}$ includes the polar vector representation Γ_{PV} (or an irreducible part of it) in its decomposition. An equivalent statement is that those Γ_X that occur in the decomposition of $\Gamma_{PV}^* \times \Gamma_{PV}$ are the allowed excitation symmetries. The scattering of light by all excitations whose symmetries do not satisfy this condition is a forbidden process.

The second consequence of the invariance condition is the imposition of restrictions on the components of the second-order susceptibility for those excitation symmetries that are allowed in light scattering. For each allowed Γ_X, some of the Cartesian components ij are required to vanish while others are required to have related values. Nye (1957) gives details of similar determinations of the symmetry properties of a wide range of tensor quantities in various crystal symmetries.

Poulet and Mathieu (1976) give the fullest account of the calculation of selection rules and symmetry properties of the second-order susceptibilities for inelastic light scattering. These calculations are not repeated here, but the main results for the 32 crystal point groups are set out in Table 1.2, which is divided into three parts, for crystals with biaxial, uniaxial, and isotropic dielectric properties. The left-hand column shows a further subdivision into the seven crystal systems, and these are broken down into the 32 crystal classes in the second column. The irreducible representations that occur in the decomposition of $\Gamma_{PV}^* \times \Gamma_{PV}$ are listed against each crystal class. These excitation symmetries allowed by the selection rules are given in two different notations, the Γ (or Bethe) notation used for example by Koster et al. (1963) and the Mulliken notation A, B, E, and T (sometimes

F) used for example by Herzberg (1945) and Wilson et al. (1955). The latter notation is used in molecular and crystalline scattering, whereas the Γ notation is also used in the Raman spectroscopy of crystals. Both notations are used in later chapters in accordance with current usage in the various kinds of light-scattering research. Conversions between one and the other are easily made with the help of the table.

Consider for example the point group $4mm$ or C_{4v}. It is found with the help of the character tables of Koster et al. (1963) that

$$\Gamma_{PV} = \Gamma_1 + \Gamma_5 \quad \text{or} \quad A_1 + E \tag{1.130}$$

and

$$\Gamma_{PV}^* \times \Gamma_{PV} = 2\Gamma_1 + \Gamma_2 + \Gamma_3 + \Gamma_4 + 2\Gamma_5 \quad \text{or} \quad 2A_1 + A_2 + B_1 + B_2 + 2E. \tag{1.131}$$

All the irreducible representations of the group occur on the right-hand side of (1.131) and all excitation symmetries therefore are allowed to scatter light. Again, for the cubic group $m3m$ or O_h,

$$\Gamma_{PV} = \Gamma_4^-, \quad \text{or} \quad T_{1u} \tag{1.132}$$

and

$$\Gamma_{PV}^* \times \Gamma_{PV} = \Gamma_1^+ + \Gamma_3^+ + \Gamma_4^+ + \Gamma_5^+, \quad \text{or} \quad A_{1g} + E_g + T_{1g} + T_{2g}. \tag{1.133}$$

Only 4 of the 10 irreducible representations of the cubic group are allowed excitation symmetries for light scattering.

The decomposition $\Gamma_{PV}^* \times \Gamma_{PV}$ contains only even-parity representations for those point groups that contain the inversion operation, as in the cubic example above. By contrast, the selection rules for electric-dipole absorption restrict the participating excitations to those of odd parity (Heine 1960, Tinkham 1964, Cracknell 1968). Thus a rule of mutual exclusion applies for scatterers that are invariant under spatial inversion, in that light scattering and electric-dipole absorption have no allowed excitation symmetries in common.

The 3×3 matrix directly above each irreducible representation in Table 1.2 shows the symmetry-restricted form of the second-order susceptibility $\chi^{ij}(\omega_I, -\omega)$ for that excitation symmetry. The absence of an entry in the matrix position ij implies a zero component of the second-order susceptibility. The letters in the matrices indicate the nonzero components, and the occurrence of the same letter in different positions in the matrices for the same point group indicates equal components. The equality applies of course only for a *given* excitation in a *given* molecule or crystal.

The existence of degenerate excitations is a further consequence of the spatial symmetry and in general higher degeneracies are found in the more symmetric systems. Thus, all irreducible representations are nondegenerate for the biaxial point groups, there is at least one twofold degenerate representation for the uniaxial groups, and the isotropic or cubic point groups have both twofold and threefold degenerate irreducible representations. The corresponding entries in Table 1.2 have two or three matrices for a single irreducible representation.

The forms of the matrices in Table 1.2 depend on the orientations of the x, y, and z axes with respect to the rotation axes and reflection planes of the various point groups. The axes used here are the same as those defined by Koster et al. (1963). Other choices of axes are sometimes made, and in particular the twofold degenerate matrices for the groups 32, $\bar{3}m$, and $\bar{6}m2$ are often shown with different forms corresponding to an alternative choice of x and y axes. The twofold matrices for the groups 23 and $m3$ are sometimes shown with two apparently independent components, but the resulting cross section can always be expressed in terms of a single susceptibility component, equivalent to the cross section obtained from the simpler matrices given here.

The use of the matrices of Table 1.2 in predicting and elucidating the anisotropies of measured cross sections is described in Section 2.4.

SUMMARY

The *kinematics* of light-scattering processes are governed by energy conservation

$$\omega = \omega_I - \omega_S \qquad (1.2)$$

and momentum conservation

$$\mathbf{q} = \mathbf{k}_I - \mathbf{k}_S. \qquad (1.15)$$

These lead to the expression

$$\frac{c^2 q^2}{\eta^2 \omega_I^2} = \frac{\omega^2}{\omega_I^2} + 4\left(1 - \frac{\omega}{\omega_I}\right)\sin^2\frac{\phi}{2} \qquad (1.21)$$

for the scan of excitation ω and $\mathbf{q}$ values covered by an experiment with fixed scattering angle ϕ, reducing to

$$q = 2k_I \sin\frac{\phi}{2} \qquad (1.28)$$

for scattering angles larger than a few degrees.

Macroscopic Theory

The scattered light is produced by a polarization determined by a second-order susceptibility, an excitation amplitude, and the incident field

$$P_S^i(\mathbf{K}_S, \omega_S) = \epsilon_0 \chi^{ij}(\omega_I, -\omega) X^*(\mathbf{q}, \omega) E_I^j. \tag{1.43}$$

The spectral differential cross section is

$$\frac{d^2\sigma}{d\Omega\, d\omega_S} = \frac{\omega_I \omega_S^3 \mathfrak{v} V \eta_S |\epsilon_0 \epsilon_S^i \epsilon_I^j \chi^{ij}(\omega_I, -\omega)|^2}{(4\pi\epsilon_0)^2 c^4 \eta_I} \langle X(\mathbf{q}) X^*(\mathbf{q}) \rangle_\omega, \tag{1.70}$$

where the power spectrum is obtained from a fluctuation-dissipation theorem derived quantum mechanically

$$\langle \hat{X}(\mathbf{q}) \hat{X}^\dagger(\mathbf{q}) \rangle_\omega = \frac{\hbar}{\pi} \{ n(\omega) + 1 \} \operatorname{Im} T(\mathbf{q}, \omega). \tag{1.82}$$

Microscopic Theory

The cross section is related to a quantum-mechanical transition rate $1/\tau$ by

$$\sigma = \frac{\eta_I \mathfrak{v}}{\tau n_I c}. \tag{1.103}$$

The electron-radiation interaction to be used in calculating the transition rate has two components, an A^2 part

$$\hat{H}'_{ER} = \frac{e^2}{2m} \sum_j \hat{\mathbf{A}}(\mathbf{r}_j) \cdot \hat{\mathbf{A}}(\mathbf{r}_j) \tag{1.111}$$

and an $\mathbf{A} \cdot \mathbf{p}$ part

$$\hat{H}''_{ER} = \frac{e}{m} \sum_j \hat{\mathbf{A}}(\mathbf{r}_j) \cdot \hat{\mathbf{p}}_j. \tag{1.112}$$

Both parts make significant contributions in general.

REFERENCES

Andrews T. (1869), *Phil. Trans. Roy. Soc.* **159**, 575.
Blount E. I. (1962), *Solid State Phys.* **13**, 305.

References

Born M. and Bradburn M. (1947), *Proc. Roy. Soc.* **A188**, 161.

Brillouin L. (1914), *Compt. Rend.* **158**, 1331.

Brillouin L. (1922), *Ann. Phys. (Paris)* **17**, 88.

Butcher P. N. and Ogg N. R. (1965), *Proc. Phys. Soc.* **86**, 699.

Cowley R. A. (1971), in A. Anderson Ed., *The Raman Effect*, Vol. 1 (New York: Marcel Dekker), p. 95.

Cracknell A. P. (1968), *Applied Group Theory* (Oxford: Pergamon).

Einstein A. (1910), *Ann. Phys.* **33**, 1275.

Fabelinskii I. L. (1957), *Usp. Fiz. Nauk* **63**, 355.

Gross E. (1930a), *Nature* **126**, 201.

Gross E. (1930b), *Nature* **126**, 400.

Gross E. (1930c), *Nature* **126**, 603.

Gross E. (1932), *Nature* **129**, 722.

Heine V. (1960), *Group Theory in Quantum Mechanics* (Oxford: Pergamon).

Herzberg G. (1945), *Infrared and Raman Spectra of Polyatomic Molecules* (New York: Van Nostrand).

Koster G. F., Dimmock J. O., Wheeler R. G., and Statz H. (1963), *Properties of the Thirty-Two Point Groups* (Cambridge, Mass.: MIT Press).

Landau L. D. and Lifshitz E. M. (1960), *Electrodynamics of Continuous Media* (Oxford: Pergamon).

Landau L. D. and Lifshitz E. M. (1965), *Quantum Mechanics* (Oxford: Pergamon).

Landau L. D. and Lifshitz E. M. (1969), *Statistical Physics* (Oxford: Pergamon).

Landau L. D. and Placzek G. (1934), *Phys. Z. Sowjetunion* **5**, 172.

Landsberg G. and Mandelstam L. (1928), *Naturwiss.* **16**, 557.

Lax M. and Nelson D. F. (1976), in B. Bendow, J. L. Birman and V. M. Agranovich, Eds., *Theory of Light Scattering in Condensed Matter* (New York: Plenum Press), p. 371.

Loudon R. (1964), *Adv. Phys.* **13**, 423; erratum (1965), **14**, 621.

Loudon R. (1973), *Quantum Theory of Light* (Oxford: Clarendon Press).

Loudon R. (1978), *J. Raman Spect.* **7**, 10.

MacDonald D. K. C. (1962), *Noise and Fluctuations* (New York: Wiley).

Menzies A. C. (1953), *Rep. Prog. Phys.* **16**, 83.

Nye J. F. (1957), *Physical Properties of Crystals* (Oxford: Clarendon Press).

Placzek G. (1934), in E. Marx, Ed., *Handbuch der Radiologie*, Vol. 6, part 2 (Leipzig: Akademische Verlagsgesellschaft), p. 209.

Poulet H. and Mathieu J. P. (1976), *Vibration Spectra and Symmetry of Crystals* (Paris: Gordon and Breach).

Raman Sir C. V. and Krishnan K. S. (1928a), *Ind. J. Phys.* **2**, 387.

Raman Sir C. V. and Krishnan K. S. (1928b), *Nature* **121**, 501.

Rayleigh Lord (J. W. Strutt) (1899), *Phil. Mag.* **47**, 375.

Smekal A. (1923), *Naturwiss.* **11**, 873.

Smoluchowski M. (1908), *Ann. Phys.* **25**, 205.

Strutt J. W. (later Lord Rayleigh) (1871a), *Phil. Mag.* **41**, 107.

Strutt J. W. (later Lord Rayleigh) (1871b), *Phil. Mag.* **41**, 274.

Strutt J. W. (later Lord Rayleigh) (1871c), *Phil. Mag.* **41**, 447.
Tinkham M. (1964) *Group Theory and Quantum Mechanics* (New York: McGraw-Hill).
Tyndall J. (1868–1869), *Proc. Roy. Soc.* **17**, 223.
Van Hove L. (1954), *Phys. Rev.* **95**, 249.
Wilson E. B., Decius J. C., and Cross P. C. (1955), *Molecular Vibrations* (New York: McGraw-Hill).
Zubarev D. N. (1960), *Sov. Phys. Uspekhi* **3**, 320.

CHAPTER TWO

Experimental Aspects

2.1 Lasers _____ 54
    2.1.1 Conditions for Laser Operation
    2.1.2 Optical Resonators and Mode Configurations
    2.1.3 Gas Lasers
    2.1.4 Dye Lasers
    2.1.5 Solid-State Ionic Lasers

2.2 Spectrometers _____ 74
    2.2.1 Diffraction-Grating Instruments and Sample Illumination
    2.2.2 The Fabry-Perot Spectrometer

2.3 Detection of Scattered Light _____ 86

2.4 Measurement of Spectra _____ 88

Early experiments in Raman scattering were carried out with mercury arcs and prism spectrographs, using exposure times that sometimes required many hours. However, development of photoelectric recording techniques and of the laser changed the situation dramatically. The appearance of the laser in particular was responsible for the renaissance of light-scattering studies in the middle 1960s. The laser provides an intense, monochromatic, highly collimated, highly polarized beam of light that is ideal for scattering studies. The advent of continuously-operating tunable dye lasers has made the study of resonance Raman scattering more straightforward and also provides the possibility of discriminating against unwanted fluorescence in conventional Raman scattering.

There are now many textbooks on lasers available (e.g., Yariv 1976, Arecchi and Schulz–Dubois 1972, Willett 1974, Siegman 1974, Svelto 1976); we restrict our attention here to aspects of laser operation of interest to students of light scattering. Section 2.1 provides general background for the understanding of laser operation and deals with specific laser systems that are particularly useful in light-scattering studies.

In addition to considerably improved light sources, the dispersing elements available for studies of Raman and Brillouin scattering have also improved greatly in recent years. In the Raman case this improvement has come about by using coupled monochromators, thereby reducing sensitivity problems arising from the presence of unwanted background light. This discrimination against background light, sometimes referred to as increased contrast, can be achieved in Brillouin-scattering studies by passing the beam of scattered light a number of times through a single Fabry-Perot interferometer. These instrumental developments are described in Section 2.2.

Detection systems are discussed in Section 2.3. Attention is focused primarily on the use of the photomultiplier for photon counting. Finally, some aspects of spectral measurements are discussed in Section 2.4, particularly, the determination of the components of the Raman-scattering tensor.

2.1 LASERS

2.1.1 Conditions for Laser Operation

Explanation of laser action requires an explanation of coherent amplification of light. Spectral linewidths have a considerable influence on laser action, and at the outset we briefly consider some factors that affect

Figure 2.1 Schematic representation of an energy-level scheme for an atom.

linewidths (for a more extensive discussion see, e.g., Thorne, 1974).

Figure 2.1 is a schematic energy-level diagram representing atomic energy levels. If an atom is in state 2 at some point in time there is a finite probability per unit time that at some later time it will emit radiation to the lower level 1, radiating a photon of energy $h\nu = E_2 - E_1$ where $\nu = \omega/2\pi$. This process occurs even in the absence of an external radiation field and is described as spontaneous emission. It may be formally represented by the equation

$$-\frac{dN_2(t)}{dt} = A_{21} N_2(t), \qquad (2.1)$$

where A_{21} is the spontaneous transition rate, which can be calculated if the wavefunctions of the states of the system are known. It should be emphasized that spontaneous emission occurs only from higher to lower energy states so that A_{12} is zero. This emission has a finite width, related to the lifetime, which may be described by a line-shape function $g(\nu)$. If we consider the emission to be that of a damped harmonic oscillator, we obtain the normalized Lorentzian line-shape function $g(\nu)$ as defined in (1.90). This broadening, due to the finite lifetime of the atom in the emitting state, is described as homogeneous. It is characterized by the fact that all the emitting atoms are indistinguishable and hence all have the same $g(\nu)$. There are also other contributions to homogeneous broadening, for example, pressure broadening, arising from collisions between atoms in a gas.

There is another type of line broadening, referred to as inhomogeneous, which arises when the emitting atoms are distinguishable. For example, atoms in a particular type of lattice site in a crystal may be distinguishable because crystals are never perfect and imperfections change from one part of a crystal to another. Changes in the lattice environment will affect emitted frequencies and hence contribute to line broadening. Atoms in a gas are distinguishable because they move with different velocities and Doppler broadening of spectral lines is therefore inhomogeneous. The line shape due to the Doppler effect is Gaussian for a Maxwellian distribution of velocities and has the normalized form

$$g(\nu) = \frac{c}{\nu_0}\left(\frac{M}{2\pi k_B T}\right)^{1/2} \exp-\left[\frac{M}{2k_B T}\left(\frac{c}{\nu_0}\right)^2 (\nu-\nu_0)^2\right], \qquad (2.2)$$

where M is the mass of the emitting atom and ν_0 is the center frequency. The width Γ between half-intensity points is

$$\Gamma = 2\nu_0 \left[\frac{2k_B T}{Mc^2}\ln 2\right]^{1/2}. \qquad (2.3)$$

For the Ne atom emitting at 632.8 nm (see Section 2.1.3a), the value of Γ at room temperature is due largely to Doppler broadening and is about 1.5 GHz. This inhomogeneous width is of particular interest in connection with laser action (see Figure 2.5).

The problem of coherent amplification can be defined by considering the propagation of a plane electromagnetic wave of frequency ν through a collection of atoms with resonant frequency ν_0, such that $\nu \cong \nu_0$. The electric field at a point z inside the medium is

$$E^z(\nu) = E^0(\nu)\exp\frac{\gamma(\nu)z}{2}. \qquad (2.4)$$

The conditions under which $\gamma(\nu) > 0$ lead to coherent amplification and these conditions may be explored at various levels of sophistication (Barnes 1972). At a relatively high level one may treat the laser as a nonequilibrium problem in quantum statistical mechanics. However, a more straightforward approach uses Einstein's description of the interaction of radiation with matter (Einstein 1917). This description introduces the notion of stimulated emission, a concept that leads in a natural way to an understanding of laser action. It results in a value for $\gamma(\nu)$ (Yariv 1976)

$$\gamma(\nu) = \frac{1}{8\pi}\left[\left(N_2 - \frac{g_2}{g_1}N_1\right)A_{21}\lambda^2 g(\nu)\right], \qquad (2.5)$$

where g_2 and g_1 are the degeneracies of levels 2 and 1 and λ is the wavelength of the light. It follows that if $N_2 > (g_2/g_1)N_1$ the intensity of the radiation field will increase exponentially in its passage through the laser medium. This condition is essential for laser operation. Under conditions of thermal equilibrium ($N_2 < (g_2/g_1)N_1$), laser operation cannot occur.

The pathlength of the light beam in the medium and hence the amplification may be increased by enclosing the medium in a cavity with highly reflecting walls (see Section 2.1.2) so that the radiation traverses the medium many times. Although presently not incorporated in conventional lasers, the plane mirror–plane mirror Fabry-Perot etalon can be used as a resonant cavity for light waves. Oscillation will occur if the two mirrors enclose a medium with a population inversion and if the resultant amplification of a light beam matches the cavity losses. Under these conditions an increase in the radiation energy stored in the cavity will also occur, reducing the population inversion (a phenomenon referred to as gain saturation). Because of this the level of oscillation settles at a steady value, when the energy gain and the energy losses in the cavity are equal.

The plane mirror–plane mirror Fabry-Perot etalon is used for Brillouin scattering studies and some of its properties are discussed in Section 2.2.2. The sum E_t of the complex amplitudes of beams transmitted normally to the plates is (see, e.g., Cook 1971)

$$E_t = E_i t^2 \sum_{n=0}^{\infty} r^{2n} e^{-2i\chi l}, \qquad (2.6)$$

$$= \frac{E_i t^2}{1 - r^2 e^{-2i\chi l}}, \qquad (2.7)$$

where E_i is the incident amplitude, t and r are the fractions of the light amplitude transmitted and reflected at each surface, and l is the separation of the plates. The complex propagation constant χ may be written as (Yariv 1976)

$$\chi = a - i\frac{\alpha}{2} + c(\nu) + i\frac{\gamma(\nu)}{2}, \qquad (2.8)$$

where the first two factors describe dispersion and energy loss in the medium at frequencies far from the atomic line giving rise to laser action. The factor α takes account of general losses in the optical cavity, for example, losses by diffraction, losses in mirrors, and scattering by inhomogeneities in the laser medium. The second two terms take account of transmission of frequencies in the region of the laser line, $\gamma(\nu)$ being given by (2.5).

Oscillation occurs when the denominator of (2.7) vanishes, that is, when

$$r^2 \exp{-2i[a+c(\nu)]l} \times \exp[\gamma(\nu) - \alpha]l = 1. \qquad (2.9)$$

Equation (2.9) incorporates both an amplitude condition and a phase condition. The amplitude condition is

$$r^2 \exp[\gamma(\nu) - \alpha] l = 1 \tag{2.10}$$

or

$$\gamma(\nu) = \alpha - \frac{1}{l} \ln r^2. \tag{2.11}$$

It now follows, using (2.5), that the threshold population difference N_t for the onset of oscillation is

$$N_t = \left(N_2 - \frac{g_2}{g_1} N_1\right) = \frac{8\pi}{A_{21} \lambda^2 g(\nu)} \left(\alpha - \frac{1}{l} \ln r^2\right). \tag{2.12}$$

The condition (2.12) (orginally investigated by Schawlow and Townes 1958) now appears in laser textbooks in a variety of forms; it implies that achieving inversion becomes more difficult as the wavelength becomes shorter. For a line with a Lorentzian shape the value of the line-shape function $g(\nu_0)$ at the center of the line is $(0.637/\Gamma)$ where Γ is the full width at half-height. The corresponding value for a Doppler-broadened line that, being Gaussian, is sharper than a Lorentzian-broadened line, is $(0.939/\Gamma)$. For the helium-neon laser (Section 2.1.3a) with $\lambda_0 = 632.8$ nm, $A_{21} \simeq 10^7 \text{s}^{-1}$, $\Gamma \simeq 1.5$ GHz, $\alpha = 0$, $l = 0.1$ m, and $r = 0.995$, we find $N_t \simeq 10^{10}$.

The phase condition in (2.9) requires that

$$2[a + c(\nu)] l = 2n\pi \tag{2.13}$$

and a laser will oscillate if (2.13) and (2.10) are satisfied.

2.1.2 Optical Resonators and Mode Configurations

The purpose of a laser cavity is to direct the radiation produced by the active medium between the cavity plates back and forth many times through the medium so that the gain by stimulated emission exceeds the losses. Usually, laser cavities are used in such a way that standing-wave patterns occur in the cavity, and these exist only in well-defined modes. The dimensions of cavities that will operate in one mode only are of the order of one wavelength and are simple to make at microwave frequencies ($\lambda \sim 0.01$ m). However, at optical frequencies the dimensions of closed cavities are necessarily very much larger than the wavelength, resulting in multimode operation. For a bandwidth $d\nu$ the mode density $\rho(\nu)$ for electromagnetic radiation is given by

$$\rho(\nu) d(\nu) = \frac{8\pi \nu^2}{c^3} d\nu, \tag{2.14}$$

and this has the value $\sim 10^{16}/m^3$ for $\nu = 6 \times 10^{14}$ Hz and $d\nu = 3 \times 10^{10}$ Hz. Similar mode densities occur for atomic lines and amplification can occur for all possible modes. However, laser output in a single mode is quite often desirable, and we shall now see that the use of open resonators provides a solution to this problem. Because of the directional properties of light beams, a structure formed by two plane-parallel mirrors facing each other can act as a high-Q resonator, despite the fact that it is open (Q values of $\sim 10^8$ can be achieved with open optical resonators, compared with Q values of $\sim 10^4$, which are standard with microwave cavities). The open resonator discriminates strongly against modes not propagating along the axis; the high-Q modes take the form of nearly plane waves propagating back and forth between the reflectors and the number of such modes may be small. In fact, as we shall see, individual modes can be selected by fairly simple means.

Although a detailed discussion of the mode structure in optical resonators is outside the scope of this text (see Kaminow and Siegman 1973) some general background is desirable. At the outset, to establish a notation, we consider a closed rectangular metal cavity (Figure 2.2). We find that it can support two types of mode, referred to as transverse electric (TE) and transverse magnetic (TM). A field configuration is referred to as TE if the electric vector everywhere in the cavity is perpendicular to the long dimension, that is, $E^z = 0$ in Figure 2.2; correspondingly TM configurations are characterized by $H^z = 0$. A uniform plane wave is called transverse electromagnetic (TEM) if both the electric and magnetic vectors are perpendicular to the direction of propagation (TEM modes exist in free space and in coaxial transmission lines but not in a hollow transmission line or a hollow resonator).

Solving Maxwell's equations for the rectangular cavity one finds, for example, that the general TE mode polarized along the y axis (Figure 2.2)

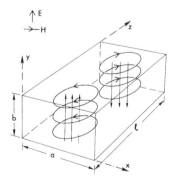

Figure 2.2 Schematic representation of the TE_{102} mode in a rectangular metal cavity.

has the spatial dependence

$$E^x = 0 \qquad E^y = \sin\frac{p\pi x}{a}\cos\frac{q\pi y}{b}\sin\frac{m\pi z}{l} \qquad E^z = 0, \qquad (2.15)$$

where p, q, and m are integers and the mode is specified as TE_{pqm}. This mode is shown in Figure 2.2 for $p=1$, $q=0$, and $m=2$. The field distribution contains no nodal lines of the electric field in the cross-sectional plane. In the general case there are $p-1$ nodal lines parallel to the y axis and q nodal lines parallel to the x axis. These solutions to the cavity problem are orthogonal if the walls are infinitely conducting and have the exact postulated geometrical shape. This implies that one of the modes may be excited without exciting others. However, orthogonality is only an approximation in real situations and precautions are generally necessary to prevent multimode operation.

The cavity theory discussed briefly above is not adequate for open structures. The open-structure resonators used with gas lasers consist of a pair of plane or curved mirrors (Figure 2.3) with an amplifying column of gas between, for example, a pair of plane circular mirrors about 0.02 m in diameter separated by a distance of about 1 m. With systems of this sort, diffraction loss may play an important role in determining the distribution of energy in the interferometer during oscillation. When operated as a passive device, that is, as a conventional Fabry-Perot interferometer, with uniform plane waves supplied by an outside source, the internal fields are also to a good approximation uniform plane waves. However, when power is supplied by the medium between the plates the loss of power from the edges of the wave by diffraction causes pronounced departures from uniformity.

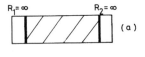

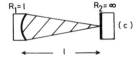

Figure 2.3 Some resonator configurations, (*a*) plane parallel, (*b*) confocal, and (*c*) hemispherical.

The modes of the plane mirror–plane mirror Fabry-Perot interferometer may be discussed in terms of self-reproducing field configurations over the mirror surfaces. A field configuration is said to be self-reproducing if after propagation from one reflector to another and back it returns to the same phase and amplitude pattern (but not necessarily the same amplitude). For each such field pattern, TEM in character, longitudinally propagating, and specified by subscripts p and q, there is a sequence of modes specified by the subscript m [see (2.13)] for which the round-trip phase shift is an integral multiple of 2π. Taking m to be the number of wavelengths in the distance $2l$, where l is the separation of the mirrors, and c to be the velocity of light in the medium in the cavity, then to a good approximation $\nu_m = mc/2l$ and the frequency separation of "longitudinal" modes is

$$\delta\nu = \nu_m - \nu_{m-1} = \frac{c}{2l}. \tag{2.16}$$

For a cavity of length 1 m the value of $\delta\nu$ is 150 MHz. Since m is a large number ($\gtrsim 10^6$), it is not normally specified and the optical cavity modes are described as TEM_{pq}.

There are various ways of calculating mode patterns for optical resonators, and we make some passing remarks here about the approach of Fox and Li (1961). They used the Huygens-Fresnel principle that enables one to calculate the distribution of amplitude and phase of the radiation at one point in the cavity, knowing the amplitude and phase at another point. Although complete analytical solutions are not generally available for the mode patterns of open resonators, the structure of the modes has nevertheless been fairly well established. As we have pointed out above, the solutions are classified in analogy with the modes of waveguides by their nodal lines, that is, the lines that divide the field into regions of opposite phase. The field distributions for some of these modes, found in cavities with plane mirrors, are shown in Figure 2.4. The TEM_{00} mode, usually dominant in laser operation, has no nodal lines. Only vertical polarization is shown in Figure 2.4, but in the absence of Brewster windows (see Figure 2.6) horizontal polarization is also possible.

Boyd and Gordon (1961) investigated the lower-order modes of a resonator made with confocal spherical mirrors (Figure 2.3) (see also Section 2.2.2). Here the field is concentrated more on the axis of the resonator and falls to a lower value at the edge than in the plane system so that diffraction losses are smaller. With this configuration the surface of the reflectors is in fact a phase front of the wave. It also transpires that the frequency differences of modes with different values of m are the same as in the plane case (2.16).

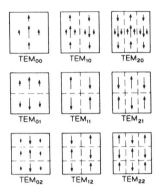

Figure 2.4 Electric field configurations for interferometers with plane mirrors.

We shall now make some comments on the relative merits of some resonator geometries (Figure 2.3). The plane-parallel resonator that we have used so far for illustrative purposes is largely of historical interest as far as conventional gas lasers are concerned. Even though the light fills the laser cavity (Figure 2.3a), making use of all the excited atoms, this geometry is not used because of the high alignment precision required for the mirrors and because of relatively high diffraction loss.

The confocal resonator (Figure 2.3b) gives the smallest possible mode dimension for a resonator of given length. Modes with larger p and q have a greater spatial extent and also greater power loss due to diffraction. To have operation in the TEM_{00} mode only, it is therefore desirable to have the aperture of the plasma tube between the mirrors (see Figure 2.6) just limiting on the TEM_{00} mode, thereby providing discrimination against TEM_{01} and higher modes. For radii of curvature $R_1 = R_2 = l = 1$ m and $\lambda = 632.8$ nm, a suitable plasma tube diameter is approximately 0.002 m. The atoms in this narrow-bore plasma tube are however efficiently used by the confocal system. Although it has low diffraction loss, the confocal configuration is close to being an unstable optical structure (Kogelnik and Li 1966). Nevertheless the unstable situation can be easily avoided by slight adjustment of the mirrors and a near-confocal configuration is readily usable in practice (for further discussion, see Section 2.2.2).

For most applications the combination of spherical and plane mirrors (Figure 2.3c) is the most desirable of the mirror configurations we have discussed. This system imitates the confocal arrangement but is a stable optical structure. The resultant mode has a large diameter at the spherical mirror and focuses to a diffraction-limited point at the plane mirror. Angular alignment of the mirrors is not critical and stable operation in the TEM_{00} mode is easily achieved by slight adjustments of mirror separation.

Lasers

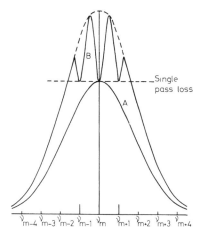

Figure 2.5 Laser line profiles for two different gain settings. For curve A oscillation sets in at ν_m. For curve B the modes $\nu_{m\pm 1}$ also come into oscillation. The dashed curve represents B in the absence of oscillation.

We conclude this section with some remarks on the mode structure of laser light. The active medium will oscillate over only a range of frequencies $\Delta\nu$ for which the unsaturated gain of the spectral line is greater than the loss. For a given transverse configuration, for example, TEM_{00}, we see that $N = \Delta\nu/\delta\nu$ modes can oscillate where $\delta\nu$ is given by (2.16). We have already pointed out that the inhomogeneous linewidth of the He-Ne laser is about 1.5 GHz (Section 2.1.1). For $l = 1$ m we find, using (2.16), that $N = 10$ and if the laser is operating well above threshold several of these modes may oscillate.

The situation is illustrated in Figure 2.5 for an inhomogeneously broadened line. The line-shape function $g(\nu)$ gives the distribution of frequencies for individual atoms and the gain $\gamma(\nu)$ is proportional to $g(\nu)$ up to the threshold for oscillation [curve A in Figure 2.5; see also (2.5)]. We assume, for simplicity, that the frequency ν_m of one of the TEM_{00} modes coincides with the center of the atomic line. When the threshold for oscillation is reached (peak of curve A), the gain at ν_m remains clamped at the threshold value. Further pumping (curve B) leads to oscillation at $\nu_{m\pm 1}$. The gain at each oscillating frequency is clamped, and hence as more modes oscillate the gain profile acquires depressions at the oscillating frequencies. This phenomenon, referred to as hole burning, is a characteristic of inhomogeneously broadened lines. The width of the holes is of the order of the homogeneous linewidth.

In the gas lasers most commonly used for light scattering (Section 2.1.3) the aperture of the plasma tube is in fact sufficiently narrow to discourage oscillation in any but the TEM_{00} mode. In addition, it is possible to pick out a single ν_m from the number oscillating within the gain profile of a single line by inserting a Fabry-Perot etalon into the cavity whose free spectral range (for a definition, see Section 2.3.2) is less than the separation $\delta\nu$ (2.16). This filter passes only one ν_m, and lasers with such inserts are commercially available. This single moding is highly desirable for Brillouin scattering studies (Chapter 8).

2.1.3 Gas Lasers

In this section we describe some actual gas lasers used in light-scattering studies. For a more detailed discussion the reader should consult Willett (1974). The principal component of a gas laser is a discharge tube, sometimes referred to as a plasma tube (Figure 2.6), which provides optical gain at the laser frequency. In addition, one requires a power supply for exciting the discharge and a mechanical structure for supporting the plasma tube and resonator mirrors. The resonator provides feedback to the amplifying plasma tube, producing oscillation. The resonator mirrors are external to the plasma tube to avoid damage by the discharge. The discharge is contained by end windows on the discharge tube and these are generally aligned at Brewsters angle θ_B (Figure 2.6). This angle is given by $\tan\theta_B = \eta$ where η is the refractive index of the window material at the wavelength concerned. Fused silica is commonly used for windows in the ultraviolet, visible, and near-infrared regions; for this material $\theta_B = 55.6°$ in the visible region. At Brewster's angle the electric-field component in the plane of incidence is transmitted with no reflection and the perpendicular component is partly reflected. This results in an effectively low-Q system for the perpendicular component, so that the laser output is completely plane polarized in the plane of incidence.

With a laser operating at different wavelengths, an internal cavity prism can be used for wavelength selection (Figure 2.6). A Brewster-angle prism

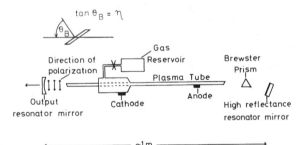

Figure 2.6 Schematic representation of a gas laser. The insert shows the Brewster angle θ_B.

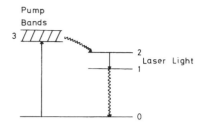

Figure 2.7 Schematic representation of a four-level laser system. Level 3 is normally broad for a solid and sharp for a gas.

has no more loss than a Brewster-angle window.

Figure 2.7 shows a four-level scheme characteristic of many laser systems. In a gas laser the pump levels are normally populated by collision with electrons in an electric discharge. Atoms raised into the upper state of the pump level may be excited to higher levels or they may decay directly to the ground state. They may also decay to level 2 (Figure 2.7). The laser transition occurs between levels 2 and 1, and this is followed by decay to the ground state. For laser action to start the populations of levels 1 and 2 must be inverted so that (2.12) applies.

The Helium-Neon Laser

The first gas laser made to operate involved transitions between levels of the neutral Ne atom in a gas discharge (dc or rf). The active medium consists of a gas mixture composed of about 1 torr of He and 0.1 torr of Ne. This medium was the first to operate as a continuous-wave (cw) system and the mechanisms leading to laser action are now reasonably well understood. The discharge raises the He atoms into highly excited states, many of which decay into the metastable levels 2^3S and 2^1S (Figure 2.8). These have energies close to the levels of the $2p^54s$ and $2p^55s$ configurations of Ne and resonant transfer of energy occurs with high efficiency to these states during collision (cross section $= 4 \times 10^{-21}$ m^2). The relatively small energy differences (200–400 cm^{-1} for the various levels) are taken up by the kinetic energy of the atoms involved in the collision. This resonance transfer is the main excitation mechanism for the $2p^55s$ and $2p^54s$ configurations of Ne (partly because of the high He to Ne ratio).

The characteristic emission of the He-Ne laser is a red line at 632.8 nm.*
The upper level, which has a lifetime of $\sim 10^{-7}$ s is one of the $2p^55s$ states ($j=2$) and the terminal level is one of the $2p^53p$ states ($j=4$) (Figure 2.8) (for notation and for actual energy levels of atoms and ions, see Moore

---

*Figures 2.8–2.14 contain a wide range of energies, and spectroscopists use different units to cover this range. Note that 10,000 Å = 1000 nm = 1 μm and that 1 μm is the equivalent of 10,000 cm^{-1}; 1 eV = 8060 cm^{-1}.

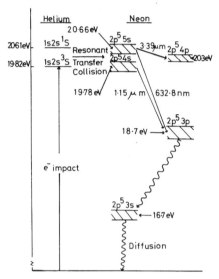

Figure 2.8 Schematic representation of the energy levels of He and Ne atoms relevant to action of the He-Ne laser.

1949, 1952, 1958). The terminal level radiates to the metastable $2p^53s$ states with a lifetime of $\sim 10^{-8}$ s. This shorter lifetime of the $2p^53p$ level allows population inversion to occur.

Oscillation also occurs in the infrared, at 1.15 and 3.39 μm (Figure 2.8). The latter line, which originates in the same level as the 632.8 nm emission, has a high gain, and oscillation would normally occur at 3.39 μm rather than 632.8 nm. This happens because the threshold condition (2.12) is reached first for 3.39 μm, and this prevents further increase of population in $2p^55s$. However, the 3.39-μm oscillation can be quenched by using optical components that absorb at 3.39 μm, but not at 632.8 nm, thereby raising the threshold for 3.39 μm above that of 632.8 nm. Standard commercial lasers provide up to about 20 mW of power in the 632.8 nm line, although powers of up to 100 mW are possible.

The single isotope ^{20}Ne may be used to eliminate isotopic broadening; the ^{20}Ne$-^{22}$Ne line splitting is about 820 MHz.

The Argon-Ion Laser

The argon-ion (Ar$^+$) laser produces intense output in the blue (488.0 nm) and green (514.5 nm) and is now the most commonly used laser in

Table 2.1 Some Laser Lines in Singly Ionized Argon*

Wavelength (nm)	Transition	Relative Intensity
454.5	$4p\,^2P^0_{3/2} \leftrightarrow 4s\,^2P_{3/2}$	Weak
457.9	$4p\,^2S^0_{1/2} \leftrightarrow 4s\,^2P_{1/2}$	Moderate
465.8	$4p\,^2P^0_{1/2} \leftrightarrow 4s\,^2P_{3/2}$	Moderate
472.7	$4p\,^2D^0_{3/2} \leftrightarrow 4s\,^2P_{3/2}$	Weak
476.5	$4p\,^2P^0_{3/2} \leftrightarrow 4s\,^2P_{1/2}$	Moderate
488.0	$4p\,^2D^0_{5/2} \leftrightarrow 4s\,^2P_{3/2}$	Strong
496.5	$4p\,^2D^0_{3/2} \leftrightarrow 4s\,^2P_{1/2}$	Moderate
501.7	$4p\,^2F^0_{5/2} \leftrightarrow 3d\,^2D_{3/2}$	Weak
514.5	$4p\,^4D^0_{5/2} \leftrightarrow 4s\,^2P_{3/2}$	Strong
528.7	$4p\,^4D^0_{3/2} \leftrightarrow 4s\,^2P_{1/2}$	Weak

*Taken from Bridges and Chester (1971)

light-scattering studies. Some lines in which oscillation has been observed are given in Table 2.1 with transition assignments and relative intensities. With one exception they arise from $4s$ to $4p$ transitions and their upper levels are approximately 35 eV above the ground state of the neutral atom (Figure 2.9). Additional lines appear in the ultraviolet and the yellow when high current densities are used. In these circumstances Ar^{++} may also be produced giving rise to oscillation at 514.3 nm. The energy-level structure shown in Figure 2.9 is typical of laser transitions occurring in all singly ionized noble gases.

Although the excitation process for Ar^+ is not selective (Figure 2.9) a steady-state inversion is possible because of the very fast radiative decay of the $3p^4 4s,\,^2P_{1/2,3/2}$ levels (Table 2.1 and Figure 2.9) to the Ar^+ ground state. This decay is fast because (a) it is a resonance transition, (b) the frequency is high, and (c) the population of the ground state of Ar^+ is not high enough to give rise to appreciable resonance trapping at normal current densities (for discussions of resonance trapping, see Mitchell and Zemansky 1934, Holstein 1947). The lifetime of the ^2P levels is about 3×10^{-10} s, whereas that of $^4D^0$ (Table 2.1) is approximately 7.5×10^{-9} s, giving a lifetime ratio of ~ 20.

The cw argon-ion laser operates with a gas pressure of approximately 0.3 torr; the discharge voltage is about 240 V, and discharge currents of up to about 30 A are commonly used, providing continuous operation in the range 1–5 W. Considerably higher output powers may be achieved with higher discharge currents. The power dissipation in the capillary tube may be about 7 kW, depending on operating conditions, and this is transferred directly to the walls by ion bombardment. The plasma tube is therefore

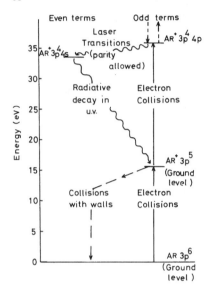

Figure 2.9 Configurations involved in laser action by the Ar^+ ion.

made from graphite or beryllia to withstand the ion bombardment and high temperature and is cooled with water. An axial magnetic field of about 0.1 T along the plasma tube improves the performance of the laser although the reasons for this are not fully understood. The plasma tube is about 3 mm in diameter and $\gtrsim 0.6$ m long and is, in effect, a rugged version of that used with the He-Ne laser.

Krypton-ion lasers provide useful dc outputs in a range of wavelengths extending from the ultraviolet to the near infrared. The output is as high as 2 W in the ultraviolet, 2.5 W in the violet, and 4.6 W in the red wavelengths. This laser is especially useful for driving tunable infrared lasers (see Section 2.1.4).

The Carbon Dioxide-Nitrogen Laser

The lasers we have described earlier emit between different electronic levels, giving light output in the visible region and at somewhat higher energies. The CO_2-N_2 laser emits in the near infrared, at about 10.6 μm, as a result of transitions between vibrational levels of the ground electronic state ($^1\Sigma$) of CO_2. In light-scattering studies it has found particular use in the investigation of electronic Raman scattering by semiconductors (see Chapter 7).

In its ground state CO_2 is a linear symmetric molecule with three normal modes of vibration. These are the symmetric stretching mode (ν_1), the

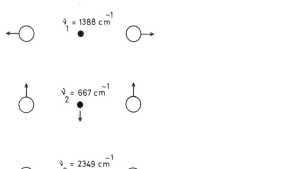

Figure 2.10 Normal modes of vibration of the CO_2 molecule.

doubly-degenerate bending mode (ν_2), and the asymmetric stretching mode (ν_3) (Figure 2.10). The double degeneracy of ν_2 arises from the possibility of vibrating in two orthogonal planes that intersect on the molecular axis. To the extent that we neglect anharmonicity we can regard ν_1, ν_2, and ν_3 as independent oscillators, and we can then write the ground state as $|000\rangle$ and excited vibrational states as $|n_1 n_2 n_3\rangle$ (appreciable mixing of ν_1 and ν_2 does in fact occur because $\nu_1 \cong 2\nu_2$; this effect is referred to as Fermi resonance).

The $^{12}C^{16}O_2$ molecule oscillates in approximately 100 transitions between 9.1 and 10.7 μm. The lasing states are complex involving, among others, the rotational-vibrational bands $|001\rangle \rightarrow |100\rangle$ and $|001\rangle \rightarrow |020\rangle$ (Figure 2.11). The laser is operated at a pressure of about 10 torr, and the main processes determining population inversion are not radiative lifetimes but rather near-resonant collisions between CO_2 molecules and other molecules added to the discharge. The population of the $|001\rangle$ state of CO_2 is generated largely by an efficient near-resonant energy transfer between vibrationally excited N_2 molecules and CO_2 in its ground state (Figure 2.11). The excited nitrogen molecules N_2^* are readily produced in the discharge through the reaction

$$N_2 + e^- \rightarrow N_2^- \rightarrow N_2^* + e^-,$$

which has an unusually large cross section ($\sigma \cong 4 \times 10^{-20}$ m^2).

In normal use the gas in a CO_2 laser consists of a mixture of CO_2 (~ 2 torr) + N_2 (~ 10 torr) + He (~ 5 torr). The addition of helium can increase the laser output at 10.6 μm by up to a factor of five. Helium has a thermal conductivity that is an order of magnitude greater than that of either CO_2 or N_2 so that the presence of He increases the rate of removal of heat from the discharge. Heating of the gas in a carbon dioxide cw laser

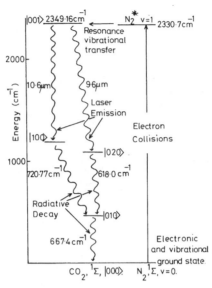

Figure 2.11 Energy-level diagram for the CO_2-N_2 laser.

reduces population inversion and limits the output of a plasma tube 1 m long to about 200 W. However, if the hot gas is removed by forced flow through the discharge region higher power outputs can be obtained. In this way cw laser output at 10.6 µm of about 20 kW is attainable with commercial lasers and cw outputs greater than 100 kW have been achieved. Although depopulation of the lower laser levels by radiative decay is allowed (Figure 2.11), collisions are much more effective.

2.1.4 Dye Lasers

Many organic compounds with conjugated double bonds absorb in the visible region of the spectrum and are described as dyes. They may also fluoresce in the visible region (Figure 2.12) and are suitable laser media (for a review see Schäfer 1973). A schematic energy-level diagram for a standard dye is shown in Figure 2.13. The ground electronic state is a spin singlet (S_0), and the first excited electronic state is a spin triplet (T_0). These are associated with excited ladders of singlet and triplet states. The color of a dye is determined by the transition $S_0 \rightarrow S_1$.

Vibrational and rotational motion of the dye molecule occurs in each electronic state, the vibrational energies being ~ 1500 cm^{-1}. The potential minima of S_0 and S_1 are displaced resulting in broad absorption bands.

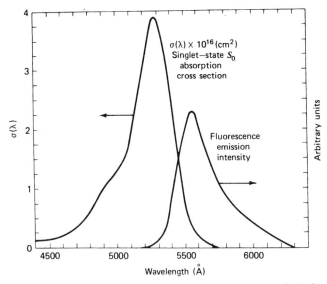

Figure 2.12 Singlet-state absorption and fluorescence spectra of rhodamine 6G obtained from measurements with a 10^{-4} molar ethanol solution of the dye (after Snavely 1969).

After absorption into S_1 the molecule relaxes rapidly to the lowest level of S_1 and then decays with a large Stokes shift to S_0 (Figure 2.13) with emission of a broad band (Figure 2.12). The molecule may also decay via the spin-forbidden transitions $S_1 \rightarrow T_0$ and $T_0 \rightarrow S_0$. The lifetime of T_0 is relatively long against decay to S_0 ($\sim 10^{-3}$ to 10^{-7} s) so that T_0 may act as a sink for excited molecules. It follows that if the allowed transition $T_0 \rightarrow T_1$ (Figure 2.12) overlaps laser emission, the dye is less effective as a

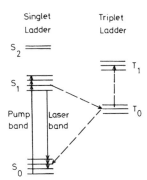

Figure 2.13 Schematic representation of the electronic energy levels of a dye laser showing vibrational structure.

laser medium. The dye is also less effective as a laser medium in the region of overlap of absorption and emission bands so that large Stokes shifts are desirable.

The broad emission band of dyes means that dye lasers incorporating a single dye can be tuned to operate over wavelength ranges of about 100 nm. There are now commercially available cw dye lasers using an argon-ion laser as a light pump (the dye laser will, of course, operate only at wavelengths longer than those of the pumping light). A feature of particular interest is the use of a stream of dye open to the atmosphere, thus avoiding damage to containing windows by pump light; oxidation of the dyes does, however, occur. Prisms or diffraction gratings may be used as tuning elements or, more recently, birefringent single-crystal quartz filters. An input of 4 W from an argon-ion laser will result in an output of $\sim$ 0.8 W from rhodamine 6 G covering the spectral range 550–650 nm. The spectral region for continuous operation covered at the time of writing extends from approximately 420 to 950 nm. Wavelength resolution of better than 0.025 nm can be achieved. Single-mode operation of dye lasers is possible giving a linewidth of $\sim$5 MHz in favorable cases.

The spectroscopist should be aware that gas lasers emit relatively weak plasma lines in addition to the oscillating line, and that dye lasers emit a background fluorescence.

2.1.5 Solid-State Ionic Lasers

Laser action was first demonstrated in the solid state, using ruby in a pulsed mode. A large number of solid-state lasers emitting in the visible and near infrared were later discovered in the 1960s based on transitions in rare-earth and actinide ions. At the present time the most highly developed solid-state lasers are ruby, neodymium in glass, and also neodymium in yttrium aluminium garnet. The ruby laser normally operates in a pulse mode, emitting at 694.3 nm (R_1 line); it has not been used extensively in light scattering in recent years. The Nd laser emits strongly at 1.06 μm and has found wider application; we shall discuss it further below. Although semiconductor lasers are now fairly highly developed (Arecchi and Schulz-Dubois), they usually have a low power output. They have not been used extensively in light scattering, and we shall not concern ourselves further with them.

Solid-state ionic lasers differ in their operational characteristics from gas lasers. The ion density in solid-state systems is generally higher than in gas systems and quite often provides a good compromise between requirements of compactness and cooling capacity. The lifetime for population inversion in solid-state systems tends to be longer than in gas systems. This means that solid-state systems are well suited to the production of high-power pulses using such techniques as Q switching.

Solid-state ionic lasers are optically pumped, and the performance of such lasers is determined to a considerable degree by the efficiency of the pumping lamp. Mercury- or xenon-filled lamps of cylindrical shape are commonly employed. A polished elliptic cylinder focuses the light from the lamp, placed along one focal axis, on the laser rod placed on the other focal axis. The threshold pump power for a four-level laser depends on the population of level 1 (Figure 2.7; the pump level in solids is normally a broad band) and in some cases lower temperatures lead to a lower threshold. High thermal conductivity of the solid is generally advantageous; as well as affecting populations it reduces optical distortions of the laser medium resulting from thermal gradients.

Lasers rods vary in size but are typically approximately 0.004 m in diameter and approximately 0.05 m long. The ends of the rod are polished flat and parallel to about 1 min of arc. Mirror coatings may be deposited on the end faces of the rod, but more generally separate resonator mirrors are used.

The Nd^{3+} ion dissolved in yttrium aluminium garnet (YAG) or in glasses is an important solid-state laser system. Pump bands each approximately 30.0 nm in width occur at 525.0, 585.0, 750.0, 810.0, and 870.0 nm (Snitzer and Young 1968). Laser action has been observed for a number of transitions in the near infrared. The most important of these transitions ($^4F_{3/2} \to {}^4I_{11/2}$; Figure 2.14) occurs at 1.064 μm at room temperature. The lower level is about 2000 cm^{-1} above the ground state ($^4I_{9/2}$) so that we are dealing with a four-level system. The lifetime of the $^4F_{3/2}$ level depends on the host somewhat, being of the order of 10^{-4} s.

Neodymium lasers now produce pulses of higher power than any other laser source. Of more interest from the light-scattering viewpoint is the fact

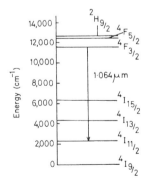

Figure 2.14 The main laser transition and energy-level scheme for Nd^{3+} in YAG (after Geusic et al. 1964).

that they are readily operated in the cw mode, giving a power of about 10 W for a lamp input of about 1 kW. In this way the Nd laser is different from the ruby laser, which has a much higher threshold for laser oscillation, and is much more difficult to operate as a cw system. It is of course possible to carry out Raman spectroscopy using pulsed lasers but studies using such sources are not extensive.

Finally we mention the ABC:YAG or alphabet-YAG laser. This is the name given to yttrium aluminium garnet doped with various concentrations of Er, Tm, Ho, and Yb (Doehler and Solin 1972). At liquid-nitrogen temperature it lases at 2.098 μm and, with the Nd:YAG laser, is useful for Raman-scattering studies of semiconductors (Chapter 7).

2.2 SPECTROMETERS

In light-scattering experiments the difference in frequency between the scattered and unscattered light extends from a few hertz to approximately 10^{14} Hz ($\sim$3000 cm^{-1}) and a range of dispersing elements is required to resolve these differences. Although different types of dispersing system depend on different physical principles, for example, differential refraction (prisms), diffraction (gratings), or interference (interferometers), they have some characteristics in common. The resolving power ρ is a property of a dispersing system of major importance. It is determined by the apparatus function W (the representation the system gives of an incident beam of monochromatic light). If $\delta\nu$ is the width of W then, for a given frequency ν, the resolving power may be taken to be $\rho = \nu/\delta\nu$. The dispersion of a grating spectrometer with a 1-m focal length is adequate, with slit widths of a few microns, to accommodate diffraction gratings with $\rho \simeq 10^5$ in first order. However, this magnitude of resolution is not easy to achieve for Raman scattering in solids with excitations of energy less than $\sim$10 cm^{-1}, because of the presence of unwanted scattered light from static imperfections in crystals and from optical components. Ways of improving this situation are discussed in Section 2.2.1.

For the study of Rayleigh and Brillouin scattering in solids, resolving powers as high as 10^8 are desirable, and values approaching this can be conveniently achieved with the Fabry-Perot interferometer. The practical limit of resolution of this instrument is in fact about 10 MHz (Section 2.2.2). Higher resolution still (up to 10^{14}) required, for example, for the study of diffusive motion in liquids, may be achieved using the techniques of photon-correlation spectroscopy. These techniques have not, however, been extensively applied to solids and will not be discussed here (for detailed information see Cummins and Pike 1974). In many situations detector sensitivity (Section 2.3) is a critical factor in determining ultimate resolution.

2.2.1 Diffraction-Grating Instruments and Sample Illumination

In light-scattering experiments we are usually interested in collecting the maximum amount of wanted inelastically scattered light. Diffraction gratings are generally blazed for a given order of interference, and this may result in up to about 70% of the incident light going into a given order for a given wavelength region. It should be emphasized that the reflection factor of a grating depends on the polarization of the incident light relative to the grating rulings and this effect must be kept in mind when obtaining polarization data (manufacturers supply efficiency curves, measured as a function of wavelength, for light polarized parallel and perpendicular to the grating rulings). If polarization is not of interest, effects of polarization on transmitted intensity can be avoided by using a polarization scrambler. In a simple form this consists of a quartz wedge inserted in the scattered beam in front of the entrance slit of the spectrometer so that different parts of the image travel through different thicknesses of quartz.

Diffraction gratings depart from perfection in other ways. Random ruling errors and surface defects contribute to unwanted scattering. In addition, periodic errors in the ruling engine result in the production of spurious lines in the spectrum, referred to as ghosts. If one cycle of the periodic error approximately covers the grating width we find Rowland ghosts, symmetric about real lines and close to them. The intensities of such ghosts can make measurements of Raman lines close to the Rayleigh line (< 20 cm^{-1}) quite difficult. However, holographic gratings now available are free of such ghosts (e.g., see Landon and Mitteldorf 1972). Ghost lines may also occur at relatively large energies from the parent line if there are a number of cycles of ruling errors across the grating. These are referred to as Lyman ghosts and, once recognized, can in fact be useful as frequency markers.

The contrast* of a single-grating spectrometer (a measure of its ability to detect weak signals in the presence of background) increases with increasing energy separation from the position of the Rayleigh line. Raman scattering by optic phonons is readily detected by such instruments because the scattered intensity is generally fairly high and optic phonons normally occur at fairly high energies. However, scattering by excitons and spin waves is not usually so strong and the contrast of a single-grating instrument does not generally suffice. A greatly increased contrast can be obtained, at the cost of some reduction in overall transmission, by using two single-grating instruments in series (van Cittert, 1926). Good instruments of this type are available commercially and are now standard equipment for Raman-scattering studies.

---

*The contrast of a monochromator at $\omega_0 + \Delta\omega$ may be defined to be the ratio of throughput at $\omega_0 + \Delta\omega$ to that at ω_0 when the instrument is illuminated with monochromatic light at ω_0.

An instrument in common use (Spex 1401) consists of two Czerny-Turner $\frac{3}{4}$ m grating spectrometers in series (Figure 2.15), with an aperture $f/6.8$. This particular mounting is the most common variant of the Ebert mounting and is frequently used for the construction of high-performance spectrometers using plane reflection gratings (for discussions of grating mounts, see Harrison et al. 1948, Sawyer 1963, James and Sternberg 1969). There are two mirrors associated with each grating (Figure 2.15): one to collimate and one to focus the light. Even though the mirrors are used off-axis, the aberrations generated by the first mirror are compensated to some extent by the second mirror, and, if the off-axis angle is kept small, it is possible to realize most of the theoretical resolving power of the grating. Compensation for coma decreases with increasing off-axis angle but effects of coma on throughput may be reduced by the use of curved slits.

The two gratings in the double instrument are driven by a common screw. The dispersion of the double instrument is twice that of the single instrument (Christensen and Potter 1963; these authors also discuss the resolution function for the double instrument) and the light transmission is only about a factor of two smaller than that of the single-grating instrument. The contrast at 100 cm^{-1} from the Rayleigh line approaches 10^{11} and the resolution obtainable with 20 μm slits which are straight and a few millimeters in height can be better than 1 cm^{-1}. Tuning can be done with either a sine drive giving an output linear in λ or a cosecant drive giving an output linear in $1/\lambda$ (for a discussion of grating drives see, e.g., James and Sternberg 1969, p. 191).

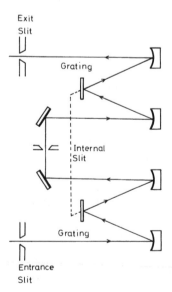

Figure 2.15 Schematic representation of a double monochromator using two ganged gratings mounted to give double dispersion.

The requirements for greater contrast to facilitate the study of low-energy excitations (< 10 cm^{-1}) have motivated the use of triple-grating spectrometers despite their complexity and high cost. Before the advent of the laser, when mercury line sources were used for light scattering, unwanted light at the Rayleigh peak could be filtered out by placing a mercury-vapor cell between the sample and the spectrometer. The 514.5-nm argon laser line may also be filtered using an iodine vapor cell, leading to increased contrast (Devlin et al. 1971). This is possible because, by chance, a sharp absorption line of molecular iodine falls within the gain profile of the argon line. It follows that a single-mode argon-ion laser can be tuned to the iodine absorption line and, with a cell 0.1 m long containing iodine at 80°C, an attenuation of the Rayleigh line by a factor of about 10^3 can be achieved with only a small reduction in intensity of a Raman spectrum at frequencies greater than 300 MHz from the Rayleigh line. With this filtering lines $\lesssim 10$ cm^{-1} from the Rayleigh line can be resolved with a double monochromator. However, if the laser frequency is not locked to the iodine absorption line, it gradually drifts with time so that locking is essential for continuous use.

Caution is necessary when using the iodine cell in experiments requiring spectral-shape and intensity analysis because the iodine molecule has weak absorption lines near the filtering line that can distort the appearance of the Raman spectrum (Fleury and Boon 1973). In addition, fluorescence from iodine in the cell can produce spurious lines in the region under investigation.

It is also possible to reduce the intensity of elastic scattering at crystal surfaces by using the technique of index matching. The crystal is immersed in a mixture of transparent organic liquids with an identical refractive index (benzene derivatives have refractive indices varying from 1.3 to 1.6 and fine variation of the refractive index can be achieved by mixing). However, this method is limited to temperatures greater than about 100 K, because the liquids freeze and become optically unclear.

Light-scattering studies in transparent crystals are generally carried out with the incident and scattered beams at right angles to each other so that the direct beam does not enter the spectrometer (Figure 2.16). If there are absorption bands present in the material under investigation that attenuate the laser light (e.g., salts of transition metal ions), the efficiency of the Raman-scattering process may fall (but see Section 4.2.3 for a discussion of resonant enhancement of Raman scattering), and the temperature of the scattering medium will rise. When the absorption is very intense, as in the case of materials whose bandgap occurs at wavelengths longer than the laser light, backscattering techniques are used. A schematic arrangement of backscattering geometry is shown in Figure 2.17. The mirrors M_1 and M_2 direct the beam on to a highly polished surface of the sample S. The cylindrical lens L_1 forms a line image of the laser light on the sample

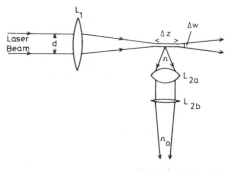

Figure 2.16 Illumination of a sample and collection of scattered light using 90° scattering geometry. The direction of the laser beam is parallel to the slit of the spectrometer.

surface. The collecting lens L_2 focuses the backscattered Raman light in the usual way on the entrance slit of the monochromator. Although the mirror M_2 stops some Raman light entering the monochromator, it also blocks direct reflections from the crystal surface and cryostat windows. The crystal surface will be damaged if too much laser power is used; the optimum amount will depend on the particular material under investigation and is determined by trial and error.

Sometimes, for example, in the study of polaritons (Section 4.3) it is necessary to study light scattered at a few degrees to the incident beam, a situation referred to as forward scattering. In this case the path of the scattered beam must be long enough so that the direct beam does not strike the entrance slit.

In concluding this section we point out that sensitivity is affected by the

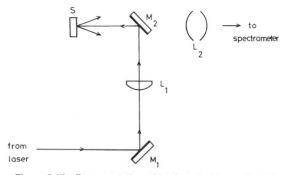

Figure 2.17 Representation of backscattering geometry.

method of sample illumination. Arrangements that involve multiple passing of the laser beam through the sample, although useful for gases, are not generally used in solids. An alternative arrangement requires focusing the laser beam into the sample (Figure 2.16), giving a small scattering source of high brightness. In a medium of good optical quality the dimensions of the focal region are determined by diffraction. In a very approximate sense (see Born and Wolf 1964) this region may be regarded as a bright cylindrical source whose length Δz and width Δw are given by

$$\Delta z \sim \lambda \left(\frac{f}{d}\right)^2 \qquad (2.17a)$$

$$\Delta w \sim \lambda \frac{f}{d}, \qquad (2.17b)$$

where f is the focal length of lens L_1, d is the diameter of the incident laser beam (the actual distribution of light in the laser beam is not considered here), and λ is the wavelength of the laser light (Figure 2.16). The collecting lens system $L_{2a,b}$ focuses this effective source on the entrance slit of the spectrometer. It is in principle desirable to have the value of the magnification M of $L_{2a,b}$ given by

$$\Omega = M^2 \Omega_0, \qquad (2.18)$$

where Ω is the solid angle of the light collected by $L_{2a,b}$ and Ω_0 is the acceptance solid angle of the spectrometer. At first sight it appears from (2.18) that it would be an advantage to make M very large. However, each of the dimensions (2.17) is related to the corresponding dimension of the image on the spectrometer slit by the factor M, and so increasing M eventually makes the image greater than the slit size. In addition, a large value of M affects the precision of polarization studies (Section 2.4), and in fact a small value of M will be necessary if the excitation frequency varies with wavevector [see, e.g., (1.21) and Figure 1.5].

In practice the optimum value of M is determined by trial and error. If, for example, we take $f(L_1) = 0.06$ m, $d = 3$ mm, and $\lambda = 500$ nm, we find $\Delta z \sim 0.2$ mm and $\Delta w \cong 10.0$ μm. However, because of optical inhomogeneities in solids Δw is likely to be greater than this estimate, and the width of the image on the slit will be overmagnified before the height. For many experiments with a Spex 1401 spectrometer in which polarization properties and small wavevector transfer are not a concern, a value of $M \cong 7$ will in fact be suitable. To achieve this the lens L_{2a} (Figure 2.16) should be a good-quality camera lens ($\sim f/1$).

2.2.2 The Fabry-Perot Spectrometer

Spectrometers incorporating the plane mirror–plane mirror Fabry-Perot interferometer are used for high-resolution optical studies from the ultraviolet to the near infrared (for a detailed discussion, see Jacquinot 1960, Cook 1971, Thorne 1974). The interferometer is a multiple-beam instrument consisting of two fused silica plates, from approximately 0.02 to 0.15 m in diameter, held accurately parallel at a fixed distance l. Using a broad source of monochromatic incident light, we obtain a system of concentric rings in the focal plane of the collecting lens L_2 (Figure 2.18). The lens L_1 increases the illumination since it images the source in the plane of the fringes. The transmitted intensity I_t for normal incidence is given by the square of (2.6), and, if we neglect effects of the medium between the plates, we obtain the Airy distribution:

$$I_t = I_i \left(\frac{T}{1-R} \right)^2 \frac{1}{1 + \left[4R/(1-R)^2 \right] \sin^2 \delta/2}, \quad (2.19)$$

where $T = t^2$ and $R = r^2$ are the fractions of the incident intensity I_i transmitted and reflected at each surface and $\delta = 4\pi l/\lambda$ is the phase change associated with the path $2l$. Interference maxima occur for orders of interference n given by $n\lambda = n/\sigma = 2l$ where λ is the wavelength of the light and σ is the corresponding wavenumber. The sharpness of the fringes and hence the resolving power of the instrument depend critically on R and, in particular, on the parameter

$$\beta = \frac{4R}{(1-R)^2} \quad (2.20)$$

[see (2.19)]. For a fixed l we may define the free spectral range $\Delta\sigma$ as the energy separation (in cm^{-1}) between neighboring orders of interference and for normal incidence this is simply $1/2l$, where l is measured in centimeters. If we take $\delta\sigma$ to be the full width at half height of an

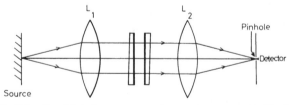

Figure 2.18 Fabry-Perot interferometer used as a spectrometer with photoelectric detection.

interference maximum, then it is straightforward to show that

$$F_R = \frac{\Delta\sigma}{\delta\sigma} = \frac{\pi\beta^{1/2}}{2}, \qquad (2.21)$$

where F_R is called the reflectivity finesse (it is a measure of the fineness of the fringes). For $R=0.70$ we find $F_R=8.3$, whereas if R is increased to 0.95 F_R becomes 61.2. Almost all textbooks dealing with the Fabry-Perot interferometer derive an expression for the resolving power

$$\rho = \frac{\sigma}{\delta\sigma} = nF, \qquad (2.22)$$

where F is the effective finesse.* The order of interference n is a maximum at the center of the ring system and for $l=0.10$ m and $\lambda=500$ nm has the value 4×10^5; operating with an effective finesse $F=25$ we have $\rho=10^7$.

In light-scattering studies using a Fabry-Perot interferometer, photoelectric detection of the scattered light is usually used. A circular pinhole at the center of the screen (Figure 2.18) plays the same part as the exit slit of a grating spectrometer. The photocurrent I is measured as a function of plate separation l or as a function of the refractive index η of the medium between the plates. This type of operation is sometimes referred to as the center-spot scanning technique. It can be used only when few wavelengths are present in the scattered light, a situation usual in Brillouin-scattering studies with a laser. One form of scanning involves pumping out the gas between the plates and measuring I_t as a function of pressure. For $\lambda=500$ nm ($\eta-1=2.7\times 10^{-4}$ in air at 15°C and a pressure of 1 atm) and $l=0.01$ m, the order of interference changes by 11.1 fringes as the pressure is reduced from 1 atm to zero. However, mechanical scanning of the plate separation is now commonly used, although this may give rise to a reduction in finesse; in this type of operation λ varies linearly with l for a given order of interference.

The light-gathering power (or etendue) of an optical instrument is defined to be

$$L = \frac{SA}{f^2}, \qquad (2.23)$$

where S is the area of the slit, A is the area of the limiting aperture, and f is

---

*The effective finesse is usually limited by defects in the reflecting surfaces. A roughness of $\lambda/100$ will spread the local path difference by $\lambda/50$ and hence the phase by $1/50$ of an order. The defect finesse F_D is therefore <50. Making $F_R > F_D$ may lead to a considerable reduction of effective finesse; if both are made equal the effective finesse is about 0.6 of either and the working finesse tends to be ~ 30.

the focal length of the collimating lens. The flux reaching the detector is proportional to L and L is conserved as the light passes through the entire optical system (L can equally well be defined for entrance and exit optics). A is usually determined by the size of a diffraction grating or interferometer plate. Write

$$L = \Omega A, \qquad (2.24)$$

where Ω is the admission solid angle and remember (see Thorne 1974, p. 203) that

$$\rho \Omega = \text{constant}, \qquad (2.25)$$

where ρ is the resolving power (the magnitude of the constant depends on the particular instrument; for a Fabry-Perot interferometer with perfectly reflecting surfaces it has the value 2π); thus it is possible to show (Thorne 1974) that for a given ρ and similar values of A the value of L for a Fabry-Perot interferometer is about two orders of magnitude greater than for a grating spectrometer. However, in Brillouin-scattering experiments we require ρ to be about three orders of magnitude greater than in Raman-scattering experiments, and this means that Ω and hence L are about an order of magnitude smaller. Hence there is a greater premium on sensitivity in Brillouin-scattering experiments than in Raman-scattering experiments. In the Fabry-Perot spectrometer (Figure 2.18) the value of Ω and hence ρ [see (2.25)] are controlled by the size of the pinhole and generally, for Brillouin studies, the radius of the pinhole will have a value of about 10 μm.

Since solids give rise to more spurious scattering of light than liquids the contrast of the Fabry-Perot interferometer is particularly important in studies of the solid state. In this case we may define contrast C to be I_{max}/I_{min} and using (2.19) we obtain

$$C = \frac{I_{max}}{I_{min}} = 1 + \beta \cong \beta. \qquad (2.26)$$

From (2.21) we get for a finesse F

$$C = \frac{4F^2}{\pi^2}, \qquad (2.27)$$

and with $F \cong 40$ this leads to a contrast of about 10^3.

A greater effective finesse and hence a higher contrast may be achieved by using two Fabry-Perot interferometers in series. If, for example, a

Spectrometers

second shorter interferometer is placed in series with a longer one, having a length that is an integral fraction of the long one, regions of low transmittance of the shorter instrument coincide with peak transmissions of the longer one. If the ratio of the lengths is exactly $n:1$ only every nth peak of the longer instrument is transmitted with high intensity by the combination, leading to an increase in effective finesse by a factor of $\sim n$. However, maintaining the correct length ratio of the two instruments is troublesome and the alternative scheme of using two or more passes through the same interferometer (Figure 2.19) is more attractive.

In the multipass scheme all passes are to a good approximation identical. Since the instrumental functions for the separate passes are multiplied together, the contrast C for n passes is related to the contrast C_1 for one pass by

$$C_1 = C^{1/n}. \qquad (2.28)$$

From (2.27) the corresponding single-pass finesse is

$$F_1 = \frac{\pi}{2} C^{1/2n}. \qquad (2.29)$$

From our definition of finesse (2.21) and the fact that the transmitted intensity for n passes is I^n, where I is given by (2.19), we find that the overall finesse F for n passes is related to F_1 and C by

$$F = \frac{F_1}{(2^{1/n} - 1)^{1/2}} = \frac{(\pi/2) C^{n/2}}{(2^{1/n} - 1)^{1/2}}. \qquad (2.30)$$

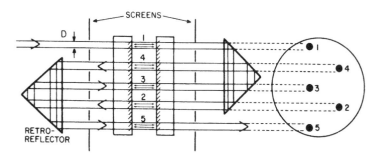

FIVE – PASS INTERFEROMETER

Figure 2.19 Five-pass interferometer (after Sandercock 1971).

To achieve optimum performance of a multipass system requires a careful balance of various design factors, including plate loss, reflectivity, and flatness and also overall transmission. If, for example, we consider the factor $[T/(1-R)]^2$ in (2.19) and write $T = 1 - R - A$, where A is the absorption in the reflecting film, we see that this factor becomes $[1-(A/1-R)]^2$. Hence the peak transmission remains large only if A is small compared with $(1-R)$ and attempts to increase contrast by increasing R may be counterproductive.

A high-performance multipass system was constructed by Sandercock (1971) using mirrors flat to $\lambda/100$ and with $R = 0.86$. For five passes (Figure 2.19) the overall finesse was about 40 and the contrast of the order of 10^9. The maximum transmission, at about 50%, was quite tolerable. The increase in contrast obtainable with only two passes of this instrument was remarkable (Figure 2.20).

The requirement that each beam uses a different area of the plates provides some restrictions on light-gathering power. However, this can be

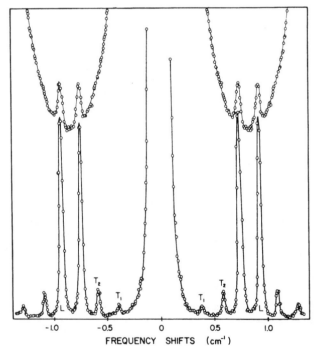

Figure 2.20 Increased contrast in the spectrum of light scattered from SbSI on going from single pass (upper trace) to double pass (lower trace) (after Sandercock 1971).

compensated for by using repetitive piezoelectric scanning and multichannel storage. Maintenance of mirror parallelism is the main stabilization problem. This can be done by using a servo system that continuously adjusts the alignment of the mirrors to maintain maximum peak signal. This system is easier to construct and to control than the tandem system, although it does not have the advantage of near elimination of neighboring orders of interference. It is now coming increasingly into use, especially for solid-state studies.

In concluding this section we shall briefly discuss the confocal (or spherical) Fabry-Perot (SFP) interferometer. We have already pointed out that the plane Fabry-Perot (PFP) etalon has a low tolerance for mirror misalignment (Section 2.1.2), but this can be obviated by using a SFP, two identical spherical mirrors separated by a distance r equal to the radius of curvature (Figure 2.21). This confocal configuration is much more tolerant of misalignment and even though it is close to being an unstable optical structure it is readily usable (Section 2.1.2).

The two rays emerging at B (Figure 2.21) have a path difference of $\Delta = 4r$ if d/r is small enough to neglect spherical aberration; also ∇ is independent of the angle of incidence. The free spectral range for the SFP is $c/4r$ and if a large free-spectral range is required, for example, 7.5 GHz ($r = 0.01$ m), the restriction on d/r results in a smaller etendue L than for a PFP. In addition, fabrication of spherical mirrors with this relatively small r is much more difficult than for plane mirrors of the same roughness finesse. However, d can be increased in proportion to r and hence to the resolving power ρ, and we have the unusual result that the etendue L is proportional to ρ rather than to $1/\rho$ [see (2.24) and (2.25)]. The usefulness of the SFP is therefore greatest for large values of r. In practice, for a free-spectral range of 0.75 GHz ($r = 0.1$ m), the L value of the SFP will be about 60 times greater than that of a PFP with 1-in. diameter plates (Connes 1958).

The SFP can be used only with photoelectric recording and requires a form of scanning that does not appreciably alter the geometric conditions (e.g., pressure variation). Hercher (1968) has given a detailed account of the theory, design, and use of this instrument.

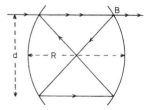

Figure 2.21 Spherical Fabry-Perot interferometer.

2.3 DETECTION OF SCATTERED LIGHT

It appears from Chapter 1 that first-order Raman and Brillouin scattering are two-photon processes and therefore weak. Effects of stray light, which determine our ability to detect small signals, can be reduced in the Raman case by using coupled monochromators (Figure 2.15) and equivalently in the Brillouin case by using multipassing (Figure 2.19). Photoelectron recording techniques have now almost entirely superseded photographic recording because of the very high sensitivity of the photomultiplier in the visible and near-infrared regions.

In a light-scattering experiment the laser is a source of noise. In this case the limit of noise is determined by spontaneous emission, a limit not detectable in normal use. Also in modern gas lasers noise arising from instability of the discharge is not normally a problem. A potentially more troublesome source of noise arises, in a multimode laser, from mode interference. This, of course, does not occur with a laser operating in a single mode, but even here there is the possibility of mode jumping.

It is perhaps useful to consider a typical Raman experiment with about 100 mW of laser power in a single line ($\sim 3 \times 10^{17}$ quanta s^{-1}) incident on the sample. With excitations that scatter strongly, generally phonons, approximately 10^9 quanta s^{-1}, enter the collecting lens ($L_{2a,b}$ in Figure 2.16). If the spectrometer transmits about 10% of the collected light and the resolving power is fixed so that the bandpass contains approximately 1% of the transmitted light, we expect to have $\sim 10^6$ quanta s^{-1} incident on the cathode of the photomultiplier. Assuming that the quantum efficiency Q of the cathode is 20%* and that the gain is $\sim 10^6$, we get a signal current of $\sim 2 \times 10^{-8}$A. With a gain of $\sim 10^6$ it is in fact possible to achieve dark currents of $\sim 10^{-13}$A so that strong Raman scatterers can be readily investigated and dc detection methods are suitable. However, for weak Raman scatterers, for example, spin waves (Chapter 6), giving anode currents of the order of 10^{-11}–10^{-12} A or for Brillouin scattering experiments where the solid angle available for light scattering is small (Section 2.2.2), photon counting techniques are more appropriate and in fact are commonly used (detection techniques are discussed at length by Hathaway 1971).

The Q value of a photomultiplier varies with wavelength, depending on the photocathode material. The most common photocathodes used in Raman spectroscopy are bialkalies (e.g., CsNa$_2$KSb), which have a response referred to as S-20. The S-20 response is the most efficient available for the visible region; the Q value peaks at $\sim 25\%$ in the region of 400 nm,

---

*The quantum efficiency is the probability that a single photon incident on the cathode will eject an electron.

but the response is still high at 780 nm. Ag-0-C_5 photocathodes are used for wavelengths in the range 780-1000 nm; this response is referred to as S-1.

The ITT-FW130 photomultiplier, which has an S-20 response, is commonly used for light-scattering studies. The photomultiplier should have the smallest possible cathode consistent with matching to the exit optics of the monochromator since this reduces dark current. The standard FW-130 tube used in photon counting has a cathode diameter of 2.5 mm. Cooling the S-20 photocathode to $-20°C$ produces a noticeable reduction of dark current compared with room temperature operation and the FW-130 tube is normally operated at $-20°C$. Similar cooling with S-1 photocathodes is even more effective. Photocathodes coated with GaAs have a good response from the ultraviolet to about 900 nm and are coming increasingly into use in light-scattering studies.

The photon counting technique is particularly suited for very low light levels, since the output of the photomultiplier consists of resolved pulses*; the signal pulses as distinct from noise pulses are correlated with individual incident photons. The pulses from the photomultiplier anode are fed into a pulse-shaping preamplifier. The noise pulses from electrons originating on the dynode chain will be smaller than the signal pulses, since they do not travel the full length of the dynode chain. The output will also contain pulses larger than the signal pulses arising from ionization of residual gas in the photomultiplier tube. If therefore the output of the preamplifier is fed into a single-channel pulse-height analyzer that discriminates against pulses below a height h_1 and above a height h_2, some improvement in signal-to-noise results. With a cooled photocathode it is possible to achieve a dark count of ~ 1 s^{-1}.

For normal operation the output of the discriminator is fed into a linear count-rate meter with a variable time constant. This converts the pulses into a dc output that is fed to a chart recorder. However, for very weak signals a multiple-scanning system may be used, storing the information in a multichannel analyzer.

It should perhaps be said that the photomultiplier itself provides the best high-gain, wide-band amplification, and the tube should be operated close to the highest specified supply voltage to minimize external amplification. Photomultiplier gains can be as high as 10^6-10^8 and when used with an amplifier with a gain of 10-10^2 give the necessary pulse height to exceed the threshold level of discriminators. The amplifier should have a rise time approaching that of the photomultiplier (~ 1 ns).

---

*For a discussion of the relative merits of synchronous detection and photon counting see, for example, Nakamura and Schwarz (1968).

When very weak Raman spectra are of interest, it may be desirable to count photon pulses with a much longer time constant than is practicable with ratemeters ($\lesssim 50$ s). This can be done by using a minicomputer both to control the experiment and as a multichannel accumulator of photon pulse data, making an arbitrarily long time for signal averaging possible (Ushioda et al. 1974).

The sensitivity of photomultiplier tubes falls off in the infrared, and they are not used at wavelengths longer than ~ 1 μm. Cooled lead sulfide detectors are suitable for the region 1–2.5 μm (Smith et al. 1957) and are available commercially. They can be used with Nd:YAG or ABC:YAG lasers (Section 2.1.3).

2.4 MEASUREMENT OF SPECTRA

Since the properties of a Raman spectrum may depend on the frequency of the laser light (Section 4.2.3), it is customary to quote this frequency. The properties of the scattered radiation of principal interest are the frequency, polarization, and intensity. The measurement of frequency is relatively straightforward and is not discussed here. Much of the detailed work in the study of light scattering by crystals is concerned with the polarization of the scattered light. The focusing of the laser beam and the finite aperture of the collecting optics affect the precision of these studies; normally collecting optics with a cone of $\lesssim 5°$ is required, considerably less than the maximum achievable (Section 2.2.1). Single crystals of good optical quality are desirable for polarization studies, and these are usually prepared with polished faces simply related to the crystal principal axes. Analysis of the scattered light is straightforward for crystals of orthorhombic or higher symmetry where the principal axes of the crystal coincide with the principal axes of the optical indicatrix (Nye 1957). Interpretation of the measurements is simplest if the directions of the light beams and polarization coincide with principal axes. However, measurements made with respect to an arbitrary set of axes in a crystal can be transformed to the principal axes (Nye 1957). It should be emphasized that when a beam of polarized light travels in an arbitrary direction through a birefringent crystal it is split into two perpendicularly polarized components with different velocities and the polarization properties of the light gradually change on passing through the crystal (Porto et al. 1966). This also applies to scattered light and occurs, of course, with backscattering.

We have already pointed out that the scattering cross section for phonons is related to the second-order susceptibility components χ^{ij} by,

Measurement of Spectra

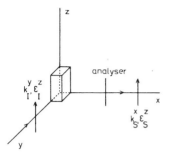

Figure 2.22 Right-angled scattering geometry $y(zz)x$.

for example, (1.70) which is proportional to

$$|\varepsilon_S^i \varepsilon_I^j \chi^{ij}|^2 \to |\boldsymbol{\varepsilon}_S \cdot \boldsymbol{\chi} \cdot \boldsymbol{\varepsilon}_I|^2. \tag{2.31}$$

Table 1.2 of susceptibility components combined with (2.31) provides expressions for the dependence of cross sections on the polarization vectors $\boldsymbol{\varepsilon}_I$ and $\boldsymbol{\varepsilon}_S$ of the incident and scattered light. A given component of the susceptibility may be found experimentally by arranging the geometry such that the incident light is polarized in the i direction and only the scattered light polarized in the j direction is observed. For example, the 90° scattering geometry shown in Figure 2.22 will measure only $|\chi^{zz}|^2$ (note that the sign of χ^{ij} is not determined).*

To illustrate the use of Table 1.2 we consider first the orthorhombic groups where the four types of excitation symmetry give rise to dependence of the cross sections on the polarization orientations given by

$$A(\Gamma_1): \quad |\varepsilon_S^x \varepsilon_I^x a + \varepsilon_S^y \varepsilon_I^y b + \varepsilon_S^z \varepsilon_I^z c|^2, \tag{2.32}$$

$$B_1(\Gamma_3): \quad |\varepsilon_S^x \varepsilon_I^y d + \varepsilon_S^y \varepsilon_I^x e|^2, \tag{2.33}$$

$$B_2(\Gamma_2): \quad |\varepsilon_S^x \varepsilon_I^z f + \varepsilon_S^z \varepsilon_I^x g|^2, \tag{2.34}$$

$$B_3(\Gamma_4): \quad |\varepsilon_S^y \varepsilon_I^z h + \varepsilon_S^z \varepsilon_I^y i|^2. \tag{2.35}$$

Excitations of the four different symmetries can be distinguished by the

---

*A notation commonly used to describe polarization data takes the form $y(zx)x$ where the first letter gives the direction of incidence of the light, the letters in the bracket give the direction of polarization of the incident and scattered light and the last letter gives the direction of the scattered light.

different dependence of their cross sections on the Cartesian components of the incident and scattered polarizations. A good example of a symmetry analysis of phonons is shown later in Figure 3.5 for MnF_2.

An example of a degenerate excitation is provided by the $E(\Gamma_5)$ symmetry of the tetragonal groups 422, 4mm, or $\overline{4}2m$ (Table 1.2). The total cross section is obtained by summing the contributions of the degenerate members of the excitation. The polarization dependence of the cross section is therefore given by

$$E(\Gamma_5): \left|\varepsilon_S^x \varepsilon_I^z f + \varepsilon_S^z \varepsilon_I^x g\right|^2 + \left|\varepsilon_S^y \varepsilon_I^z f + \varepsilon_S^z \varepsilon_I^y g\right|^2. \tag{2.36}$$

The forms of the second-order susceptibility shown by the matrices in Table 1.2 are those enforced by the spatial symmetry alone, without any consideration of the nature of the excitation. It is shown in later chapters that further restrictions apply approximately for some kinds of physical excitation. For example, the susceptibility is required to be symmetric,

$$\chi^{ji} = \chi^{ij}, \tag{2.37}$$

for vibrational excitations (see Section 3.1.4), but antisymmetric,

$$\chi^{ji} = -\chi^{ij}, \tag{2.38}$$

for magnons in ordered magnetic crystals (see Section 6.1.2). These requirements introduce additional selection rules; for example, a vibration of $T_1(\Gamma_4)$ or $T_{1g}(\Gamma_4^+)$ symmetry in one of the cubic groups 432, $\overline{4}3m$, or $m3m$ is not allowed to participate in inelastic light scattering. They also provide an additional tool for elucidating the natures of the excitations responsible for observed light scattering, and it is sometimes important to distinguish experimentally the symmetric and antisymmetric scattering.

As an example, consider the antisymmetric $A_2(\Gamma_2)$ excitation and the symmetric $B_2(\Gamma_4)$ excitation in the tetragonal groups 422, 4mm, and $\overline{4}2m$. The cross sections are proportional to (see Table 1.2)

$$A_2(\Gamma_2): \left|(\varepsilon_S^x \varepsilon_I^y - \varepsilon_S^y \varepsilon_I^x)c\right|^2, \tag{2.39}$$

$$B_2(\Gamma_4): \left|(\varepsilon_S^x \varepsilon_I^y + \varepsilon_S^y \varepsilon_I^x)e\right|^2. \tag{2.40}$$

Both types of excitation can be observed experimentally if one polarization has a nonzero x component and the other has a nonzero y component.

Figure 2.23 shows an experimental arrangement for distinguishing the two types of cross section. The incident wavevector $\mathbf{k}_I$ bisects the x and y

Measurement of Spectra

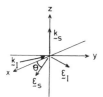

Figure 2.23 Arrangement for distinguishing between symmetric and antisymmetric scattering.

axes, while the scattered wavevector $\mathbf{k}_S$ is parallel to the z axis. The incident polarization lies in the xy plane,

$$\varepsilon_I = (2^{-1/2}, 2^{-1/2}, 0). \tag{2.41}$$

Let the scattered polarization make an angle θ with the x axis,

$$\varepsilon_S = (\cos\theta, \sin\theta, 0); \tag{2.42}$$

then from (2.39) and (2.40) the cross sections are proportional to

$$A_2(\Gamma_2): \tfrac{1}{2}(1 - \sin 2\theta)|c|^2, \tag{2.43}$$

$$B_2(\Gamma_4): \tfrac{1}{2}(1 + \sin 2\theta)|e|^2. \tag{2.44}$$

The excitations can thus be distinguished by recording the dependence of the scattered intensity on the orientation of a polarizer placed in the scattered light beam. Further discussion of the angular dependence of scattering cross sections of single crystals is given in Sections 3.1.6 and 4.1.4.

It will be shown (Section 3.1.5) that in general the polarization of light scattered by vibrations of polycrystalline material is also different from that of the incident beam. This is explained by orientational averaging, a process that loses much of the detailed information provided by crystal symmetry. In fact only totally symmetric [that is $A_1(\Gamma_1)$] vibrations can be unambiguously identified in such materials.

Measurements of intensity in all forms of spectroscopy tend to be imprecise. However, in Raman scattering the measurement of temperature dependence of intensity is not very demanding experimentally and is useful, for example, in distinguishing between one- and two-phonon (Section 3.2) excitations. The temperature dependence of one- and two-phonon spectra is given in Table 2.2.

Measurement of absolute Raman cross sections is difficult partly because of the large difference in intensity between the incident and

Table 2.2 Temperature Dependence of Phonon Excitations

Excitation		Temperature Dependence
1 phonon	S	$1+n$
	AS	n
2 phonon summation	S	$(1+n_1)(1+n_2)$
	AS	$n_1 n_2$
2 phonon difference	S	$(1+n_1)n_2$
	AS	$n_1(1+n_2)$

$n = (\exp(\hbar\omega/k_B T) - 1)^{-1}$ $S =$ Stokes $AS =$ antiStokes

scattered radiation (Section 2.3). Nevertheless the absolute scattering cross sections for a few Raman lines are known with reasonable accuracy. The Raman line of benzene at 992 cm^{-1} has been studied by Skinner and Nilsen (1968) who report the spectral differential cross section

$$\frac{d^2\sigma}{d\Omega \, d(1/\lambda)} = 1.05(\pm 0.08) \times 10^{-33} \text{ m}^2/\text{steradian}/\text{wavenumber}, \quad (2.45)$$

at the peak of the Raman line for one molecule and one plane of polarization.

The differential cross section $d\sigma/d\Omega = 3.8 \times 10^{-33}$ m^2/steradian for a single molecule, and a single plane of polarization has been obtained by integrating the spectral differential cross section over the peak (Nilsen 1969). The benzene line may be used as a secondary standard for measuring the cross sections of other materials.

Considerable care is required even when making comparison with a secondary standard (Fabelinskii 1968, Kato and Stoicheff 1975). It may be necessary, for example, to make corrections for variation of detector sensitivity with frequency.

REFERENCES

Arecchi F. and Schulz-Dubois E. O., Eds. (1972), *Laser Handbook* Vol. 2 (Amsterdam: North-Holland.)

Barnes F. S., Ed. (1972), *Laser Theory* (New York: IEEE Press), this book is a collection of original papers.

Born M. and Wolf E. (1964), *Principles of Optics*, 2nd ed. (New York: MacMillan) p. 435.

Boyd G. D. and Gordon J. P. (1961), *Bell System Tech. J.* **40**, 489.

Bridges W. B. and Chester A. N. (1971), R. J. Pressley, Ed., *Handbook of Lasers* (Cleveland: Chemical Rubber Co.).

Christensen R. L. and Potter R. J. (1963), *Appl. Opt.* **2**, 1049.

References

Connes P. (1958), *J. Phys. Radium,* **19**, 262.
Cook A. H. (1971), *Interference of Electromagnetic Waves* (Oxford: Clarendon Press).
Cummins H. Z. and Pike E. R., Eds. (1974), *Photon Correlation and Light Beating Spectroscopy* (New York: Plenum Press).
Devlin G. E., Davis J. L., Chen L., and Geschwind S. (1971), *Appl. Phys. Lett.* **19**, 138.
Doehler J. and Solin S. A. (1972), *Rev. Sci. Instrumen.* **43**, 1189.
Einstein A. (1917), *Phys. Z.* **18**, 121; A translation of this paper into English is contained in Barnes 1972.
Fabelinskii I. L. (1968), *Molecular Scattering of Light* (New York: Plenum Press), Ch. 3.
Fleury P. A. and Boon J. P. (1973), *Adv. Chem. Phys.* **24**, 1.
Fox A. G. and Li T. (1961), *Bell System Tech. J.* **40**, 453.
Geusic J. E., Marcos H. M. and Van Uitert L. G. (1964), *Appl. Phys. Lett.* **4**, 182.
Harrison G. R., Lord R. C., and Loofbourow J. R. (1948), *Practical Spectroscopy* (Englewood Cliffs, N. J.: Prentice Hall).
Hathaway C. E. (1971), in H. Anderson, Ed. *The Raman Effect*, Vol. 1 (New York: Marcel Dekker) p. 183.
Hercher M. (1968), *Appl. Opt.* **7**, 951.
Holstein T. (1947), *Phys. Rev.* **72**, 1212.
Jacquinot P. (1960), *Rep. Prog. Phys.* **23**, 267.
James J. F. and Sternberg R. S. (1969), *The Design of Optical Spectrometers* (London: Chapman and Hall).
Kaminow I. P. and Siegman A. E., Eds. (1973), *Laser Devices and Applications* (New York: IEEE Press).
Kato Y. and Stoicheff B. P. (1975), *Phys. Rev.* **11**, 3984.
Kogelnik H. and Li T. (1966) *Appl. Opt.* **5**, 1550.
Landon D. O. and Mitteldorf A. J. (1972) *The Spex Speaker* **XVII**, 1.
Mitchell A. C. G. and Zemansky M. W. (1934), *Resonance Radiation and Excited Atoms* (New York: Cambridge University Press).
Moore, C. E. (1949) Vol. 1, (1952) Vol. 2, (1958) Vol. 3, Atomic Energy Levels, Circular 467, National Bureau of Standards.
Nakamura J. K. and Schwarz S. E. (1968), *Appl. Opt.* **7**, 1073.
Nilsen W. G. (1969), *Phys. Rev.* **182**, 838.
Nye J. F. (1957), *Physical Properties of Crystals*, (Oxford: Clarendon Press).
Porto S. P. S., Giordmaine J. A., and Damen T. C. (1966), *Phys. Rev.* **147**, 608.
Sandercock J. R. (1971) in M. Balkanski, Ed., *Light Scattering in Solids* (Paris: Flammarion Press), p. 9.
Sawyer R. A. (1963), *Experimental Spectroscopy* (New York: Dover).
Schäfer F. P. (1973) *Dye Lasers* (New York: Springer-Verlag).
Schawlow A. L. and Townes C. H. (1958), *Phys. Rev.* **112**, 1940.
Siegman A. (1974), *Masers and Lasers* (New York: McGraw-Hill).
Skinner J. G. and Nilsen W. G. (1968), *J. Opt. Soc. Am.* **58**, 113.
Smith R. A., Jones F. E., and Chasmar R. P. (1957) *The Detection and Measurement of Infra-Red Radiation* (Oxford: Clarendon Press).
Snavely B. B. (1969), *Proc. IEEE* **57**, 1374.

Snitzer E. and Young C. G. (1968), in A. K. Levine, Ed., *Lasers*, Vol. 2 (New York: Marcel Dekker), p. 191.
Svelto O. (1976), D. C. Hanna, translator, *Principles of Lasers* (New York: Plenum Press).
Thorne A. A. (1974), *Spectrophysics* (London: Chapman and Hall).
Ushioda S., Valdez J. B., Ward W. H., and Evans A. R. (1974), *Rev. Sci. Instrumen.* **45**, 479.
van Cittert P. H. (1926), *Rev. Opt.* **5**, 393.
Willet C. S. (1974), *Introduction to Gas Lasers* (Oxford: Pergamon Press).
Yariv A. (1976), *Introduction to Optical Electronics*, 2nd ed. (New York: Holt, Rinehart and Winston).

CHAPTER THREE

Nonpolar Vibrational Scattering

3.1 First-Order Scattering _____ 96
    3.1.1 **Vibrational Symmetries**
    3.1.2 **Lattice Dynamics**
    3.1.3 **Internal and External Vibrations**
    3.1.4 **The Scattering Cross Section**
    3.1.5 **Randomly Oriented Scatterers**
    3.1.6 **Experiments on Crystals**

3.2 Second-Order Scattering _____ 121
    3.2.1 **Density of States and Selection Rules**
    3.2.2 **Experiments and Calculations**

3.3 Defect-Induced Scattering _____ 131
    3.3.1 **Light Scattering by Point Defects**
    3.3.2 **Mixed Crystals and Amorphous Solids**

Most light-scattering experiments are concerned with the vibrational properties of molecules and crystals. Vibrational frequencies normally lie in the range (1.24) where light scattering is a convenient tool, and the vibrational scattering cross section is usually larger than those of other excitations in the same frequency range. The spread (1.27) of accessible wavevectors limits the direct study of crystal-lattice vibrations to only a small part of the Brillouin zone. However, the small-wavevector region is the part of the zone most easily amenable to theoretical treatment, and the spectral resolution available in light scattering exceeds that of neutron scattering.

The lattice vibrations of polar crystals fall into two distinct categories corresponding to whether or not there is an associated electric-dipole moment. Vibrations that carry an electric-dipole moment are called *polar* modes; the presence of the electric dipole complicates the small-wavevector properties of polar vibrations and these are given separate consideration in Chapter 4. This chapter is concerned with the light-scattering properties of *nonpolar* vibrations, that is, the vibrations of homopolar crystals and those vibrations of polar crystals that do not carry an electric-dipole moment.

The main concern of the chapter is first-order scattering where single vibrational quanta (phonons) are created or destroyed in the scattering process. However, there is also some consideration of second-order scattering, where two phonons are involved. The second-order cross sections are usually small, and the theoretical interpretation is difficult, but the spectra contain information about the vibrations at wavevectors throughout the Brillouin zone. The chapter concludes with an account of the influence of crystal defects on light scattering.

3.1 FIRST-ORDER SCATTERING

3.1.1 Vibrational Symmetries

The first problem in the theory of first-order light scattering by a molecule or crystal is the determination of the excitation symmetries Γ_X introduced in Section 1.6. The Γ_X for molecules are irreducible representations of the molecular point group. In the case of a crystal, advantage can be taken of the smallness of the vibrational wavevector (1.27) compared to the Brillouin zone-boundary wavevector q_M; the wavelengths of the vibrations involved in first-order light scattering are at least 1000 crystal-lattice constants. For many purposes, the properties of these small-wavevector, long-wavelength vibrations are identical to the properties of the corresponding zero-wavevector, infinite-wavelength vibrations. The vibrational symmetries can then be classified by irreducible representations of the crystal point group, the group obtained by removal of all translations from the crystal space group (the point group is strictly a factor group of the crystal space group with respect to its invariant subgroup of pure translations).

The method of determination of vibrational symmetries of polyatomic molecules is described in detail by Tinkham (1964); Poulet and Mathieu (1976) give a full account for the case of a crystal lattice. The methods in the two cases are very similar, and the procedure for crystal vibrations is briefly as follows.

It is necessary first to identify the primitive cell of the crystal structure; this is the smallest repeating unit from which the periodic lattice can be constructed. The number of distinct vibrational modes for any given wavevector is equal to the number $3n$ of the degrees of freedom of the n atoms in the primitive cell. The symmetries of the zero-wavevector vibrations are obtained by constructing the $3n$-dimensional representation of the crystal point group based on three orthogonal unit vectors located on each atom in the primitive cell, in the manner described by Tinkham (1964). In performing this calculation, no distinction is made between corresponding atoms in different primitive cells. The final step is the reduction of the $3n$-dimensional representation into a sum of irreducible representations of the point group. The number and dimensions of the resulting irreducible representations give the number of different vibrational frequencies and the degeneracies of the corresponding modes of vibration.

The symmetries of the zero-wavevector lattice vibrations for a selection of crystals of interest in light-scattering experiments are shown in Table 3.1. The more compact irreducible-representation notation of Bethe is used (see Table 1.2 for conversion to the Mulliken notation). Three of the degrees of freedom of the primitive cell at zero wavevector in every crystal structure correspond to rigid displacements of the cell along the three coordinate axes. A displacement of the crystal as a whole is clearly not subject to any internal restoring force, and these degrees of freedom are associated with zero vibrational frequency. However, for a small but nonzero wavevector q, different primitive cells suffer different displacements in a manner similar to the deformation produced by the passage of a sound wave through a continuous medium. The resulting restoring forces cause the vibrational frequencies to rise linearly from zero with increasing q. These three lattice vibrations are the acoustic modes; their $q=0$ symmetries shown in the third column of the table are always the irreducible representations associated with a polar vector in the appropriate point group.

Table 3.1 Vibrational Symmetries of Common Crystals in the Notation of Koster et al. (1963). Bracketed representations are degenerate

Crystal	Class		Vibrational mode symmetries		Other Examples
		Acoustic	Raman-Active Optic	Raman-Inactive Optic	
Sodium nitrite NaNO$_2$	$mm2$	C_{2v}	$\Gamma_1 + \Gamma_2 + \Gamma_4$	$3\Gamma_1 + 3\Gamma_2 + \Gamma_3 + 2\Gamma_4$	
Calcium tungstate CaWO$_4$	$4/m$	C_{4h}	$\Gamma_1^- + (\Gamma_3^- + \Gamma_4^-)$	$2\Gamma_1^+ + 5\Gamma_2^+ + 5(\Gamma_3^+ + \Gamma_4^+)$	$4\Gamma_1^- + 3\Gamma_2^- + 4(\Gamma_3^- + \Gamma_4^-)$	CaMoO$_4$, SrMoO$_4$
Barium titanate BaTiO$_3$	$4mm$	C_{4v}	$\Gamma_1 + \Gamma_5$	$3\Gamma_1 + \Gamma_3 + 4\Gamma_5$		
Strontium titanate SrTiO$_3$	$4/mmm$	D_{4h}	$\Gamma_2^- + \Gamma_5^-$	$\Gamma_1^+ + 2\Gamma_3^+ + \Gamma_4^+ + 3\Gamma_5^+$	$2\Gamma_2^+ + \Gamma_1^- + 3\Gamma_2^- + \Gamma_4^- + 5\Gamma_5^-$	
Rutile TiO$_2$	$4/mmm$	D_{4h}	$\Gamma_2^- + \Gamma_5^-$	$\Gamma_1^+ + \Gamma_3^+ + \Gamma_4^+ + \Gamma_5^+$	$\Gamma_2^+ + \Gamma_2^- + 2\Gamma_3^- + 3\Gamma_5^-$	MnF$_2$, FeF$_2$, CoF$_2$
α-Quartz SiO$_2$	32	D_3	$\Gamma_2 + \Gamma_3$	$4\Gamma_1 + 8\Gamma_3$	$4\Gamma_2$	
Bismuth Bi	$\bar{3}m$	D_{3d}	$\Gamma_2^- + \Gamma_3^-$	$\Gamma_2^+ + \Gamma_3^+$		As, Sb
Lithium iodate LiIO$_3$	6	C_6	$\Gamma_1 + (\Gamma_5 + \Gamma_6)$	$4\Gamma_1 + 5(\Gamma_2 + \Gamma_3) + 4(\Gamma_5 + \Gamma_6)$		
Wurtzite ZnS	$6mm$	C_{6v}	$\Gamma_1 + \Gamma_5$	$\Gamma_1 + \Gamma_5 + 2\Gamma_6$	$5\Gamma_4$	ZnO, CdS, BeO
Hexagonal close packed	$6/mmm$	D_{6h}	$\Gamma_2^- + \Gamma_5^-$	Γ_6^+	$2\Gamma_3^-$	Be, Mg, Cd, Zn
				Γ_3^+		
Zinc blende ZnS	$\bar{4}3m$	T_d	Γ_5	Γ_5		GaAs, GaP
Cubic perovskite CaTiO$_3$	$m3m$	O_h	Γ_4^-		$3\Gamma_4^- + \Gamma_5^-$	BaTiO$_3$, SrTiO$_3$
Fluorite CaF$_2$	$m3m$	O_h	Γ_4^-	Γ_5^+	Γ_4^-	SrF$_2$, BaF$_2$, AuAl$_2$
Diamond C	$m3m$	O_h	Γ_4^-	Γ_5^+		Si, Ge
Rocksalt NaCl	$m3m$	O_h	Γ_4^-		Γ_4^-	KBr, NaI
Caesium chloride CsCl	$m3m$	O_h	Γ_4^-		Γ_4^-	
Simple cubic, bcc, fcc	$m3m$	O_h	Γ_4^-			

First-Order Scattering

The remaining vibrational modes at $q=0$ with nonzero frequencies are the optic modes. They are divided into two categories in Table 3.1, in accordance with the selection rules for light scattering. The allowed excitation symmetries are listed in Table 1.2, but it is necessary to take account of the additional symmetry restriction (2.37) for vibrational scattering. An important general feature, illustrated by the examples of the cubic perovskite, caesium chloride, rocksalt, and basic cubic structures, is the presence of only negative-parity vibrational symmetries for crystals in which every ion lies at a center of inversion symmetry. There are thus no Raman-active modes for such crystal structures.

The properties of most optic modes at the small wavevectors observed in light-scattering experiments are indistinguishable from their $q=0$ properties, but exceptions to this rule occur for polar lattice vibrations. These are optic modes whose symmetries are the same as those of a polar vector in the lattice point group. They appear in the table with irreducible representations in common with acoustic modes of the same crystal; for example the Γ_1, Γ_2, and Γ_4 optic modes of sodium nitrite and the Γ_5 optic mode of zinc blende are polar. It is shown in Chapter 4 that polar optic modes generate an electric polarization in the crystal whose associated electric field produces rapid changes in the vibrational frequencies at wavevectors within the range covered by light scattering. The scattering by such modes is excluded from the considerations of the present chapter.

The acoustic modes form another category of vibrations for which it is not adequate to set the small but finite wavevector q equal to zero. The light scattering by acoustic modes, known as Brillouin scattering (see Figure 1.2), is treated in Chapter 8.

3.1.2. Lattice Dynamics

Although the symmetry properties of molecular or crystal-lattice vibrations are obtained from a brief calculation, the theoretical derivation of their frequencies is usually a lengthy task. The discussion here is restricted to an outline of some of the formal results required for the theory of the scattering cross section. Comprehensive accounts of the dynamical properties of crystal lattices are given by Born and Huang (1954) or Cochran and Cowley (1967); more elementary treatments are given by Cochran (1973) or Reissland (1973). The theory of the zero-wavevector optic modes of a crystal is very similar to the theory of molecular vibrations, very clearly described by Woodward (1972).

Consider a crystal lattice with n atoms in the primitive cell, where the mass of the αth atom is denoted M_α. Atom α in every primitive cell is displaced from its equilibrium position by the same vector $\mathbf{U}_\alpha$ in the presence of a zero-wavevector lattice vibration. The harmonic approximation is made by expanding the potential energy V of the displaced lattice as far as the term of second order in the displacements, namely

$$\tfrac{1}{2} \sum_{\alpha,\beta} A_{\alpha\beta}^{ij} U_\alpha^i U_\beta^j, \tag{3.1}$$

where the force constants are given by

$$A_{\alpha\beta}^{ij} = \left(\frac{\partial^2 V}{\partial U_\alpha^i \partial U_\beta^j} \right)_O, \tag{3.2}$$

i and j run over the three Cartesian components with repeated indices assumed summed, and α and β run over the atoms in the primitive cell. The classical equation of motion of the atoms α is therefore

$$M_\alpha \ddot{U}_\alpha^i = - \sum_\beta A_{\alpha\beta}^{ij} U_\beta^j. \tag{3.3}$$

The motion of the lattice is described by $3n$ of these coupled harmonic oscillator equations. Equations of exactly the same form describe the vibrations of an n-atom molecule.

The equations of motion are simplified by a transformation to the *mass-weighted* displacement coordinates defined as

$$\mathbf{w}_\alpha = M_\alpha^{1/2} \mathbf{U}_\alpha. \tag{3.4}$$

Then (3.3) becomes

$$\ddot{w}_\alpha^i = - \sum_\beta \overline{A}_{\alpha\beta}^{ij} w_\beta^j, \tag{3.5}$$

where

$$\overline{A}_{\alpha\beta}^{ij} = A_{\alpha\beta}^{ij} (M_\alpha M_\beta)^{-1/2} \tag{3.6}$$

is a mass-weighted force constant.

The system of $3n$ simultaneous equations represented by (3.5) is formally solved by a further transformation of the displacements to the *normal coordinates* W_σ. These are linear combinations of displacements which diagonalize the equations of motion,

$$W_\sigma = \sum_\alpha c_{\alpha\sigma}^i w_\alpha^i = \sum_\alpha M_\alpha^{1/2} c_{\alpha\sigma}^i U_\alpha^i, \tag{3.7}$$

First-Order Scattering

where $\sigma\ (=1,2,\ldots,3n)$ is a label, called the branch index, that distinguishes different solutions of the equations. The transformation coefficients can be chosen to satisfy the orthonormality relations

$$\sum_\sigma c_{\alpha\sigma}^{i*} c_{\beta\sigma}^{j} = \delta_{\alpha\beta}\delta^{ij} \tag{3.8}$$

$$\sum_\alpha c_{\alpha\sigma}^{i*} c_{\alpha\sigma'}^{i} = \delta_{\sigma\sigma'}. \tag{3.9}$$

The normal coordinate transformation decouples the motions of the atoms into $3n$ noninteracting collective oscillations called *normal modes*. The decoupled equations of motion have the form

$$\ddot{W}_\sigma + \omega_\sigma^2 W_\sigma = 0, \tag{3.10}$$

where the natural vibrational frequency ω_σ is determined by the force constants and the transformation coefficients. The time dependence of the normal coordinate has the form

$$W_\sigma(t) = W_\sigma \exp(-i\omega_\sigma t) + W_\sigma^* \exp(i\omega_\sigma t), \tag{3.11}$$

where the complex quantity W_σ determines the amplitude and phase of vibration. The inverse of (3.7),

$$w_\alpha^i = \sum_\sigma c_{\alpha\sigma}^{i*} W_\sigma \tag{3.12}$$

expresses the displacements of the various atoms in the primitive cell in terms of the normal coordinates.

The symmetry of the crystal lattice produces relations between the magnitudes of some of the force constants with the consequence that the mode degeneracies are the same as those predicted by the group-theory method described in Section 3.1.1. It is not difficult to calculate the mode frequencies and transformation coefficients if the force constants are known. However, even though the number of independent force constants is restricted by the lattice symmetry, their determination is the main hurdle of lattice dynamics theory. The lattice-dynamics references cited above describe the various models used to reduce the range of choice of force constants to manageable proportions.

Three of the zero-wavevector normal modes are easily determined for any crystal lattice. The forces represented in (3.3) arise from interactions between the various atoms in the primitive cell. According to Newton's third law of motion, the total force in any coordinate direction must vanish when summed over all atoms,

$$\sum_\alpha M_\alpha \ddot{U}^i_\alpha = 0. \tag{3.13}$$

This equation can be written with the help of (3.4) as

$$\ddot{W}_1 = 0, \tag{3.14}$$

where

$$W_1 = \sum_\alpha \left(\frac{M_\alpha}{M}\right)^{1/2} w^i_\alpha \tag{3.15}$$

and

$$M = \sum_\alpha M_\alpha \tag{3.16}$$

is the total mass of the primitive cell. The equation of motion (3.14) is similar to the general form (3.10) but with frequency

$$\omega_1 = 0. \tag{3.17}$$

There are two further zero-frequency normal coordinates W_2 and W_3 corresponding to displacements in the other coordinate directions. These are the three acoustic modes. Note that the coefficients in the normal coordinate transformation (3.15) satisfy the normalization condition (3.9) and that there is no summation over i in (3.15), each acoustic mode corresponding to a single coordinate direction.

The anharmonic terms that occur in the expansion of the potential energy beyond the second-order part (3.1) destroy the independence of the normal-mode vibrations calculated in the harmonic approximation. They lead to a progressive redistribution of energy initially present in one normal mode among other modes, generally at all wavevectors $\mathbf{q}$ in the Brillouin zone. The decay with time of a normal-mode amplitude is approximately represented by insertion of a damping term in (3.10) to produce the damped harmonic oscillator equation (1.84). The damping parameters Γ_σ determined experimentally are usually much smaller than the mode frequencies ω_σ and the harmonic approximation is normally a good one.

3.1.3 Internal and External Vibrations

Calculations of the vibrational frequencies are usually feasible only for isolated molecules, or for crystal lattices where the primitive cell contains a small number of atoms. It is normally very difficult to collect sufficient information on the magnitudes of the large number of independent force constants in crystals with many atoms in the primitive cell.

It is however often possible in such crystals to identify molecular groups of atoms for which the forces between atoms in the same molecule are stronger than those between atoms in different molecules. Those lattice vibrations in which the molecules suffer internal distortions (called *internal* vibrations) then have higher frequencies than vibrations in which different molecules are displaced relative to one another without internal distortion (called *external* vibrations). Thus ideally the normal modes of the lattice have two distinct groups of frequencies corresponding to internal and external vibrations, and this separation can greatly help the interpretation of infrared and Raman spectra of complex crystals. In less ideal cases, the higher-frequency part of the external-vibration spectrum overlaps the lower-frequency part of the internal-vibration spectrum leading to ambiguities in interpretation, but it can still be useful to divide the vibrations into two categories.

The internal vibrations of a molecular group are often relatively insensitive to the crystal environment and it is then possible to treat the vibrations of the isolated molecule as a first step in the lattice vibration problem. The same molecule may occur in several different crystals, allowing an experimental study of the influence of the crystalline environment on the internal vibrations. In some cases it is also possible to study the spectrum of the free molecule in solution.

In the present section we illustrate the usefulness of the division into internal and external vibrations for the interpretation of symmetries of lattice vibrations. It is convenient to consider a specific example and we take the WO_4^{--} molecular anion, first as a free molecule in solution and then as a constituent of the calcium tungstate $CaWO_4$ crystal (Russell and Loudon 1965). The free molecule is tetrahedral with symmetry group $\bar{4}3m$ or T_d, and the standard procedure described in the previous section shows that there are four distinct molecular vibration frequencies. Their symmetry characters are entered on the left-hand side of Figure 3.1 ($A_1 + E + 2T_2$ in the Mulliken notation) with the values for their frequencies determined by Krebs and Müller (1967).

The primitive cell of $CaWO_4$ contains two WO_4 molecules on equivalent sites. Equivalent sites are those produced from an arbitrarily chosen site by

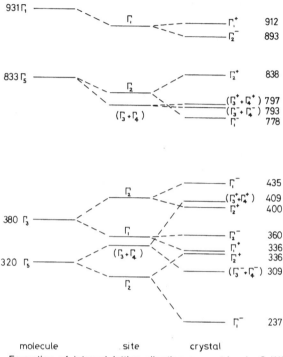

Figure 3.1 Formation of internal lattice-vibration symmetries in CaWO$_4$ from the vibrations of WO$_4$ molecules. The connections between molecular, site, and crystal point groups are explained in the text, where credits for the experimental frequencies (in cm^{-1}) are given.

applying all the operations of the crystal space group to it. Each site in a crystal lattice is characterized by a particular configuration of surrounding atoms whose point symmetry defines the *site group* of the lattice position. Equivalent sites have similar surroundings and similar site groups, but generally with different orientations of the rotation axes and reflection planes. The two equivalent sites in CaWO$_4$ have $\bar{4}$ or S_4 site groups. The lower-than-tetragonal symmetry at these sites produces shifts and splittings of the free-molecule vibrational frequencies. The splittings are determined by the $T_d \rightarrow S_4$ compatibility (Table 3.2) in the usual way (Tinkham 1964, Cracknell 1968), and the results for a single WO$_4$ molecule are represented schematically in the central column of Figure 3.1. The splittings should be much smaller than the molecular frequencies if the concept of internal vibrations is valid.

First-Order Scattering

Table 3.2 Compatibility Between Molecular and Site Symmetry Groups for WO_4 in $CaWO_4$

$\bar{4}3m - T_d$	Γ_1	Γ_2	Γ_3	Γ_4	Γ_5
$\bar{4} - S_4$	Γ_1	Γ_2	$\Gamma_1 + \Gamma_2$	$\Gamma_1 + (\Gamma_3 + \Gamma_4)$	$\Gamma_2 + (\Gamma_3 + \Gamma_4)$

The vibrations of the crystal as a whole must have symmetries appropriate to the space group $I4_1/a$ or C_{4h}^6 of the lattice. The zero-wavevector vibrations considered here form a simple special case describable by irreducible representations of the crystal point group $4/m$ or C_{4h}. The normal modes of lattice vibration are formed from combinations of the vibrations of molecules on equivalent sites. The combinations must be chosen to diagonalize the relatively small couplings between the vibrations of different molecular groups. These couplings lead in general to further shifts and splittings of the vibrational frequencies as shown in the right-hand column of Figure 3.1 representing the mode frequencies of the complete crystal. The method of calculation of the lattice-vibration symmetries that stem from the vibrations of a molecule on a single site is outlined below for the example of $CaWO_4$. Proofs of the steps and more extensive discussion of the method are provided by McClure (1959), Davydov (1962), and Poulet and Mathieu (1976).

The number of equivalent sites in a primitive cell is determined by the relative sizes of the crystal point group and the site group,

$$\text{number of equivalent sites} = \frac{\text{order of crystal point group}}{\text{order of site group}}. \tag{3.18}$$

The orders of the two groups are 8 and 4, respectively, for WO_4 sites in $CaWO_4$. The group of each site in the primitive cell must of course be a subgroup of the crystal point group (sites that have no equivalents in the primitive cell have site groups that are the same as the crystal point group). The compatibility between the irreducible representations of the two groups is shown in Table 3.3

Table 3.3 Compatibility Between Crystal Point Group and the Site Group for WO_4 in $CaWO_4$

$4/m - C_{4h}$	Γ_1^+	Γ_2^+	$(\Gamma_3^+ + \Gamma_4^+)$	Γ_1^-	Γ_2^-	$(\Gamma_3^- + \Gamma_4^-)$
$\bar{4} - S_4$	Γ_1	Γ_2	$(\Gamma_3 + \Gamma_4)$	Γ_2	Γ_1	$(\Gamma_4 + \Gamma_3)$

The main step in the determination of the lattice vibration symmetries is the construction of the *correlation table* (Table 3.4) by the following procedure. The first row of the table lists the irreducible representations of the site group, and beneath each of these are certain representations of the crystal point group. The entry below a site-group representation Γ_i in Table 3.4 is the collection of all the crystal point-group representations that occur *above* Γ_i in the compatibility Table 3.3. It is thus a trivial matter to construct Table 3.4 from Table 3.3. The physical significance of the correlation table is that a molecular vibration of site symmetry Γ_i forms lattice vibrations of the crystal point-group symmetries given below Γ_i when the vibrational amplitudes on the two equivalent sites are appropriately combined. The symmetry properties of the light-scattering cross sections are determined by the crystal point-group symmetries of the vibrations and not by their site-group symmetries.

The right-hand column of Figure 3.1 is straightforwardly constructed from the information given in the correlation table. It is seen that each vibration of a molecule at a single site gives rise to a pair of lattice vibrations. This doubling of the number of vibrations because of the presence of two equivalent sites in the primitive cell is known as Davydov splitting. Each molecular vibration gives rise to one even-parity and one odd-parity lattice vibration in the case of $CaWO_4$. Note that the pairs of representations with subscripts 3 and 4 are degenerate by time reversal in both the site group and the crystal point group. The crystal point-group representations Γ_1^+, Γ_2^+, and $(\Gamma_3^+ + \Gamma_4^+)$ are equivalent to A_g, B_g, and E_g in the Mulliken notation, with a similar relation for the odd-parity representations.

A calculation of the kind outlined above is called a factor-group analysis. The method can be applied in general to any crystal structure, however many equivalent sites may be contained in the primitive cell. In every case each molecular vibration generates as many lattice vibrations as there are equivalent sites in the primitive cell. These do not always have Davydov splittings and thus distinct frequencies as in the case of $CaWO_4$. In other examples where the overall lattice symmetry is higher, some of the lattice vibrations may have greater degeneracies than the parent molecular vibrations at a single site. Factor-group analyses can also be applied to other

Table 3.4 Correlation for WO_4 in $CaWO_4$

$\bar{4} - S_4$	Γ_1	Γ_2	$(\Gamma_3 + \Gamma_4)$
$4/m - C_{4h}$	Γ_1^+	Γ_2^+	$(\Gamma_3^+ + \Gamma_4^+)$
	Γ_2^-	Γ_1^-	$(\Gamma_4^- + \Gamma_3^-)$

First-Order Scattering

kinds of crystal excitation constructed from contributions well-localized on individual sites, and the case of electronic excitations is discussed in Section 7.1.

The even-parity lattice vibrations in $CaWO_4$ are all active in light scattering (see Table 1.2), whereas those of symmetries Γ_1^- and $(\Gamma_3^- + \Gamma_4^-)$ are active in infrared absorption. All the active vibrations associated with internal motion of the WO_4 ions have been observed by light scattering (Porto and Scott 1967) or absorption spectroscopy (Barker 1964), and the experimental frequencies are entered on Figure 3.1. These internal vibration frequencies have been carefully considered by Scott (1968a,b) who deduced the frequencies of the inactive Γ_2^- symmetry modes. It is seen that the splittings of the higher-frequency groups of internal vibrations are indeed small compared to the frequency of the parent vibration of the free molecule, but this statement does not hold for the lower-frequency groups. Porto and Scott (1967) also measured the light-scattering spectrum of $SrWO_4$, finding internal vibration frequencies not very different from those in $CaWO_4$.

Once the internal vibrations of a complex crystal have been identified by their symmetry properties and perhaps also by information on the molecular frequencies obtained by calculations or independent measurements, the remaining oscillations of the lattice fall into three categories: (a) librations that involve rotational motion of the molecules, (b) the three acoustic modes whose frequencies tend to zero at zero wavevector, (c) vibrations in which the molecules as a whole move as rigid units. The oscillations in the third category are the external vibrations; they can be treated to a first approximation by lattice dynamics calculations in which the molecules are effectively replaced by single ions. This area of lattice-dynamical theory is reviewed by Venkataraman and Sahni (1970).

The particular example of $CaWO_4$ is discussed by these last authors and also by Kanamori et al. (1974a,b). A full list of the lattice vibration symmetries is given in Table 3.1. The librational and external frequencies are determined in the experimental references cited above.

3.1.4 The Scattering Cross Section

The macroscopic theory of light scattering given in Section 1.4 applies directly to scattering by nonpolar vibrations, and it is only necessary to make a few comments on this particular application of the general theory. It has been common since the early work of Born and his collaborators (Born and Bradburn 1947, Smith 1948, Born and Huang 1954) to replace the second-order susceptibility, which describes the vibrational modulation of the linear susceptibility, by a susceptibility derivative as in (1.47). It is convenient to adopt a simple notation for the susceptibility derivative, and we define

$$a_\sigma^{ij} \equiv \epsilon_0 \chi_\sigma^{ij}(\omega_I, 0) = \epsilon_0 \frac{\partial \chi^{ij}(\omega_I)}{\partial W_\sigma^*}. \tag{3.19}$$

The validity of the susceptibility-derivative approximation for light scattering by lattice vibrations is examined in Section 4.2.2 on the basis of a quantum-mechanical expression for the second-order susceptibility.

The power spectrum of the vibrational fluctuations is determined by the fluctuation-dissipation theorem of Section 1.4.4. Suppose that fictitious forces $F_\alpha^i(t)$ of frequency ω are applied in coordinate directions i to the atoms α in the primitive cell. The energy of interaction is

$$H = -N \sum_\alpha U_\alpha^i F_\alpha^i(t) = -\sum_\sigma W_\sigma F_\sigma(t), \tag{3.20}$$

where N is the number of primitive cells in the crystal sample and

$$F_\sigma(t) = N \sum_\alpha \frac{c_{\alpha\sigma}^{i*} F_\alpha^i(t)}{M_\alpha^{1/2}}. \tag{3.21}$$

In the second step of (3.20), (3.4) and (3.12) are used to express the atomic displacement components in terms of the normal coordinates, and the interaction energy then takes the form of a sum of contributions like (1.74), one for each normal mode.

In a similar way, it is not difficult to show that the presence of the applied forces modifies the normal-mode equation of motion to a form like (1.87), where the effective force f is equal to F_σ/N. The final result of the macroscopic theory is thus a cross section given by (1.93), except that we now use the abbreviated form (3.19) of susceptibility derivative. The total cross section of all the modes in a molecule or crystal is obtained by simple summation of (1.93) over the normal-mode index σ.

The symmetry properties of cross sections are discussed in Section 1.6. However, in addition to the requirements of spatial symmetry embodied in the matrices of Table 1.2, there is a further approximate symmetry property for vibrational scattering. The linear susceptibility tensor of a nonmagnetic medium has the property (Nye 1957)

$$\chi^{ij}(\omega_I) = \chi^{ji}(\omega_I). \tag{3.22}$$

It follows that the susceptibility derivative (3.19) has the same property, in

First-Order Scattering

agreement with (1.47), leading to the symmetric requirement (2.37) for the second-order susceptibility in vibrational scattering. This additional symmetry is approximate because of the approximate nature of the relation between susceptibility derivatives and second-order susceptibilities. However, it is shown in Section 4.2.2 that the approximation is valid except when ω_I or ω_S lies close to a strong absorption line of the scattering medium. With the same exception, the susceptibility and its derivatives are real quantities.

The macroscopic theory itself produces no basis for assessing the validity of the susceptibility-derivative approximation or for predicting the magnitudes and frequency variations of the susceptibility derivatives. For these aspects it is necessary to set up a microscopic theory of the cross section, using a quantum-mechanical description of the vibrational and electronic states of the scatterer. The microscopic calculation provides an expression for the cross section in terms of matrix elements and eigenvalues of the Hamiltonians of the interacting light and scattering medium, as described in Section 1.5. The explicit frequency dependence of the microscopic expression is particularly useful in interpreting the rapid variations of the cross section in resonance-scattering conditions where the frequencies of the light approach allowed electronic-transition frequencies of the crystal.

It is convenient to defer a detailed account of the microscopic theory to Chapter 4, where polar-mode scattering can be included in a unified treatment. However, we introduce the microscopic viewpoint here by considering the basic mechanisms by which energy is transferred from the electromagnetic radiation to the crystal in the scattering process. The discussion is restricted to insulating or semiconducting crystals (see Mills et al. 1968, 1970 for consideration of the metallic case). The crystal is assumed to be in its electronic ground state before and after each scattering event. The only real change in the event being the excitation of a vibrational quantum or phonon. This account is explicitly concerned with scattering by crystals, but similar remarks apply to molecular scattering.

Figure 3.2 shows diagrammatic representations of four microscopic mechanisms by which the scattering of light can be accomplished. The conventions are the same as in Figure 1.9 with the addition of the continuous line to represent the phonon. Figure 3.2a shows a mechanism in which the radiation causes a virtual excitation of an electron from a valence band to a conduction band of the crystal by the part (1.111) of the electron-radiation interaction. The excited electron is represented by a dashed line with its arrow directed to the left, and the missing electron or hole in the valence band is represented by a dashed line with its arrow directed toward the right. The virtual electron-hole pair subsequently

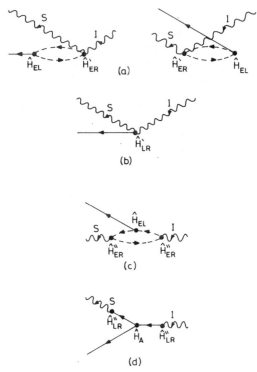

Figure 3.2 Diagrammatic representations of four mechanisms for the first-order scattering of light by lattice vibrations. As discussed in the text, the process of diagram (c) makes the largest contribution to the cross section.

recombines as the electron returns to the valence band, with emission of a phonon via the electron-lattice interaction. Figure 3.2a (right) shows an analogous process with the order of interactions reversed. These diagrams make negligible contributions to the cross section because, as discussed in Section 1.5.1, the A^2 electron-radiation Hamiltonian has very small matrix elements between different electronic states.

Analogous to the A^2 interaction of radiation with electrons, there is a similar coupling of the radiation to the charged ions in a polar lattice. The Hamiltonian $\hat{H}'_{LR}$ is the same as (1.111) but with the ionic charges, masses, and positions substituted. This interaction gives the very direct scattering mechanism shown in Figure 3.2b, but it produces a negligible effect owing

First-Order Scattering

to the large ionic mass that appears squared in the denominator of the cross section.

Figure 3.2c represents another scattering mechanism in which the electrons participate. There are three virtual electronic transitions accompanied, respectively, by absorption of an incident photon, creation of an optic phonon, and emission of a scattered photon. The three interactions can occur in any order, giving a total of six similar diagrams. In each of these, the first interaction excites an electron-hole pair, the second interaction causes a transition of the electron or hole to a different state, and the final interaction is accompanied by recombination of the electron and hole. This scattering mechanism normally makes by far the largest contribution to the cross section, and the detailed derivations of Section 4.2 are restricted to this case.

Figure 3.2d shows another mechanism in which the radiation couples directly to a polar crystal lattice, this time by the lattice analog $\hat{H}''_{LR}$ of the $\mathbf{A} \cdot \mathbf{p}$ electron-radiation Hamiltonian (1.112). The coupling scheme is completed by the third-order anharmonic lattice Hamiltonian $\hat{H}_A$, which arises from the term subsequent to (3.1) in the expansion of the lattice potential energy V in powers of the atomic displacements U^i_α. This mechanism normally makes a small contribution to the cross section, but it has been studied by Maradudin and Wallis (1970), Wallis and Maradudin (1971), and Humphreys (1972) who show that it can become important for experiments with incident light of frequency comparable to the lattice-vibration frequencies.

Other mechanisms for light scattering by lattice vibrations can be concocted, but they normally make very small contributions to the cross section. Exceptions occur in a few special cases and an example is the scattering by certain semiconductors that show magnetic ordering below a transition temperature. The additional microscopic mechanisms that arise in these crystals are discussed by Sokoloff (1972) and Susuki and Kamimura (1973). They give rise to a strong dependence of the vibrational cross section on the degree of magnetic ordering, as observed in $CdCr_2Se_4$ and $CdCr_2S_4$ (Steigmeier and Harbeke 1970), and a breakdown of the usual selection rules, as observed in EuO, EuSe, and related rocksalt-structure crystals (Tsang et al. 1974a,b, Safran et al. 1976). Other exceptions can occur for lattice vibrations that are coupled to different kinds of crystal excitation, as in the Jahn-Teller interaction described in Section 5.2.4. However, the development of the microscopic theory in Section 4.2 is aimed at the great majority of crystals where the dominant scattering mechanism is that of Figure 3.2c.

3.1.5 Randomly Oriented Scatterers

The cross sections considered so far refer to scattering by a molecule or crystal sample whose symmetry axes have fixed and known orientations. However, for scattering by free molecules in a fluid or by a powdered crystalline sample, the symmetry axes of the molecules or crystallites are distributed at random over all directions in space. For a sufficiently large number of scatterers, the scattering is determined by a continuous angular average of the theoretical cross section.

Let a^{ij} be a susceptibility-derivative component of the fluid or powder as a whole in a coordinate system

$$i,j = x,y,z, \qquad (3.23)$$

which is fixed in the laboratory relative to the scattering experiment. The normal-mode label is dropped for simplicity. On the other hand, let a^{IJ} be a susceptibility-derivative component of N oriented particles of the scatterer in a coordinate system

$$I,J = X,Y,Z, \qquad (3.24)$$

which is fixed relative to the symmetry axes of an individual molecule or crystallite. The interpretation of an experiment requires a knowledge of the a^{ij}, but it is the properties of the a^{IJ}, which are determined by the symmetry of the scatterer, as in the matrices of Table 1.2.

The two kinds of susceptibility derivative are related by

$$a^{ij} = a^{IJ}\cos(iI)\cos(jJ) \qquad (3.25)$$

for a given orientation of the scatterer, where (iI) denotes the angle between axes i and I, and repeated indices are summed over the appropriate Cartesian coordinates. The cross section (1.93) is determined by products of susceptibility-derivative components that must now be averaged over all orientations of the scatterers. With the use of a bar to represent the angular averaging, the required quantities have the form

$$\overline{a^{ij}a^{kl}} = \overline{a^{IJ}\cos(iI)\cos(jJ)a^{KL}\cos(kK)\cos(lL)}, \qquad (3.26)$$

where the susceptibility derivatives are assumed real.

It is somewhat tedious, but not very difficult, to evaluate the various angular averages required (fuller details are given for example by Landau and Lifshitz 1960, Chantry 1971, Woodward 1972). The average in (3.26) vanishes unless the axes (i,j,k,l) are identical in pairs and similarly for the

First-Order Scattering

axes (I, J, K, L). With these constraints there are only four basic angular averages to be evaluated. The results obtained by a combination of integration and the use of various orthogonality and normalization properties of the two sets of Cartesian axes (see Chapter 19 of Woodward 1972) are

$$\overline{\cos^4(iI)} = 1/5, \tag{3.27}$$

$$\overline{\cos^2(iI)\cos^2(iJ)} = 1/15 \tag{3.28}$$

$$\overline{\cos^2(iI)\cos^2(jJ)} = 2/15 \tag{3.29}$$

$$\overline{\cos(iI)\cos(jI)\cos(iJ)\cos(jJ)} = -1/30. \tag{3.30}$$

Substitution of these angular averages into (3.26) now produces results for the three different kinds of contribution that can occur in the cross section (1.93)

$$(a^{ii})^2 = \left\{ 3\left[(a^{XX})^2 + (a^{YY})^2 + (a^{ZZ})^2 \right] + 2\left[a^{XX}a^{YY} + a^{YY}a^{ZZ} + a^{ZZ}a^{XX} \right] \right.$$
$$\left. + 4\left[(a^{XY})^2 + (a^{YZ})^2 + (a^{ZX})^2 \right] \right\}/15 \tag{3.31}$$

$$a^{ii}a^{jj} = \left\{ (a^{XX})^2 + (a^{YY})^2 + (a^{ZZ})^2 + 4\left[a^{XX}a^{YY} + a^{YY}a^{ZZ} + a^{ZZ}a^{XX} \right] \right.$$
$$\left. - 2\left[(a^{XY})^2 + (a^{YZ})^2 + (a^{ZX})^2 \right] \right\}/15 \tag{3.32}$$

$$(a^{ij})^2 = \left\{ (a^{XX})^2 + (a^{YY})^2 + (a^{ZZ})^2 - \left[a^{XX}a^{YY} + a^{YY}a^{ZZ} + a^{ZZ}a^{XX} \right] \right.$$
$$\left. + 3\left[(a^{XY})^2 + (a^{YZ})^2 + (a^{ZX})^2 \right] \right\}/15, \tag{3.33}$$

where i and j are assumed to be different in the last two results.

It is seen that much of the detailed information on the susceptibility derivatives provided by the molecular or crystalline symmetry is lost in the orientation averaging. Indeed, inspection of the last three equations shows that

$$(a^{ii})^2 - a^{ii}a^{jj} = 2(a^{ij})^2, \tag{3.34}$$

and the form of the cross section is governed by no more than two independent quantities. Since one of these controls the absolute magnitude of the scattering, there remains only a single number to be determined by measurements of the polarization and angular dependences of the scattering by each mode of vibration.

The degree of depolarization or the depolarization ratio ρ is one way of expressing this single quantity. Consider the experimental arrangement of Figure 2.22 where polarized incident light is scattered through 90°. If the components of scattered intensity polarized parallel and perpendicular to the incident polarization are denoted I_S^z and I_S^y, respectively, then the degree of depolarization is defined by

$$\rho = \frac{I_S^y}{I_S^z} = \frac{(a^{yz})^2}{(a^{zz})^2}. \tag{3.35}$$

A little rearrangement of the results obtained from (3.31) and (3.33) gives

$$\rho = \frac{3\{(a^{XX}-a^{YY})^2 + (a^{YY}-a^{ZZ})^2 + (a^{ZZ}-a^{XX})^2 + 6[(a^{XY})^2 + (a^{YZ})^2 + (a^{ZX})^2]\}}{10(a^{XX}+a^{YY}+a^{ZZ})^2 + 4\{(a^{XX}-a^{YY})^2 + (a^{YY}-a^{ZZ})^2 + (a^{ZZ}-a^{XX})^2 + 6[(a^{XY})^2 + (a^{YZ})^2 + (a^{ZX})^2]\}} \tag{3.36}$$

An alternative definition of the degree of depolarization for an experiment with unpolarized incident light is sometimes used but we do not consider it here.

The maximum value of ρ occurs when the sum of the diagonal susceptibility derivatives is zero, as happens for all vibrational symmetries except A (or A', A_1, A_g, A_{1g}) or Γ_1 (or Γ_1^+) (see Table 1.2),

$$\rho_{max} = \tfrac{3}{4} \tag{3.37}$$

for all vibrations except A or Γ_1. The scattered light is least polarized in this case. The minimum value of ρ occurs when the numerator of (3.36) is zero, that is, when the diagonal susceptibility derivatives are all equal and the off-diagonal derivatives are zero. It is seen from Table 1.2 that these conditions are satisfied only for the A (or A_1, A_g, A_{1g}) or Γ_1 (or Γ_1^+) symmetry vibrations in the isotropic or cubic symmetry groups,

$$\rho_{min} = 0 \tag{3.38}$$

for cubic A or Γ_1 vibrations. The scattered light in this case is linearly polarized parallel to the incident polarization. The remaining types of vibration, of A or Γ_1 symmetry in the biaxial or uniaxial symmetry groups, have intermediate values of ρ,

$$0 \leqslant \rho \leqslant \tfrac{3}{4} \tag{3.39}$$

for noncubic A or Γ_1 vibrations.

It is clear that a measurement of the degree of depolarization can in principle distinguish A or Γ_1 from other symmetries of vibration, but cannot produce unambiguous identifications of the other vibrational symmetries. This contrasts with the situation for scattering by oriented single crystals where, as described in chapter 2 and the following section, the vibrational symmetry for any line in the Raman spectrum can in principle be determined if the crystal symmetry is known.

The properties of light scattering by fluids are very nicely illustrated by a series of experiments performed by Porto (1966). The experimental arrangement is shown in Figure 3.3 and the total scattered intensity was determined as a function of the scattering angle ϕ for incident light polarized parallel to the z axis or the x axis. The angular dependence of the cross section is determined by

$$\begin{aligned}\left|\varepsilon_S^i \varepsilon_I^j a^{ij}\right|^2 &= (\varepsilon_I^x)^2 \left\{ (a^{xx})^2 \cos^2\phi + (a^{yx})^2 \sin^2\phi + (a^{zx})^2 \right\} \\ &\quad + (\varepsilon_I^z)^2 \left\{ (a^{xz})^2 \cos^2\phi + (a^{yz})^2 \sin^2\phi + (a^{zz})^2 \right\} \\ &= (a^{ii})^2 \left\{ (\varepsilon_I^x)^2 \left[(1-\rho)\cos^2\phi + 2\rho \right] + (\varepsilon_I^z)^2 [1+\rho] \right\}, \end{aligned} \tag{3.40}$$

where (3.35) has been used. Figure 3.4 shows experimental results corresponding to the three cases of depolarization ratio given in (3.37), (3.38),

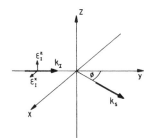

Figure 3.3 Experimental scattering geometry for the results shown in Figure 3.4.

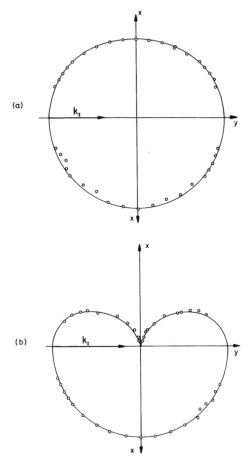

Figure 3.4 Polar plots of the angular dependence of scattering by fluids for the geometry of Figure 3.3. The upper (lower) part of each plot refers to incident polarization parallel to $x(z)$. The theoretical curves are constructed from (3.40) with ρ as given below (after Porto 1966).

	Molecule	Symmetry	Vibration	Frequency	ρ
(a)	CCl_4	$\bar{4}3m$—T_d	T_2—Γ_5	318	0.75
(b)	CCl_4	$\bar{4}3m$—T_d	A_1—Γ_1	458	0.005
(c)	C_6H_6	$6/mmm$—D_{6h}	A_{1g}—Γ_1^+	992	0.065

First-Order Scattering

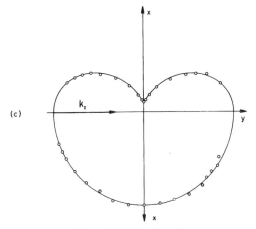

(c)

Figure 3.4 (*continued*)

and (3.39). It is seen that the measured degrees of depolarization and the angular variations (3.40) constructed therefrom give excellent agreement with the experimental points. Skinner and Nilsen (1968) have also measured the angular dependence for the 992 cm^{-1} vibration of benzene.

An additional angular dependence occurs in the scattering of light by molecules in a gas, where the molecular motion causes a Doppler broadening of the Raman lines that varies with scattering angle (Weber 1973). Almost Doppler-free lines are obtained for forward scattering ($\phi=0$).

The analysis of the present section assumes that the vibrational frequency is independent of the orientations of the individual scatterers relative to the incident- and scattered-light wavevectors. This condition does not hold for polar vibrations in crystallites (see Section 4.1.7).

3.1.6 Experiments on Crystals

A very large number of crystals have been studied by Raman spectroscopy (for a comprehensive review and listing of results, see Wilkinson 1973). We illustrate the nature and quality of the results obtained by discussing the diamond and rutile structure semiconductors or insulators and the hexagonal close-packed metals.

Diamond-Structure Crystals

The space group is $Fd3m$ or O_h^7 and the crystal point group is $m3m$ or O_h. There are two carbon atoms in the primitive cell, giving a single threefold-degenerate optic mode at zero wavevector (see Table 3.1). The polarization dependence of its cross section obtained from Table 1.2 is

$$\Gamma_5^+ \text{ or } T_{2g}: |d|^2 \{(\varepsilon_S^y \varepsilon_I^z + \varepsilon_S^z \varepsilon_I^y)^2 + (\varepsilon_S^z \varepsilon_I^x + \varepsilon_S^x \varepsilon_I^z)^2 + (\varepsilon_S^x \varepsilon_I^y + \varepsilon_S^y \varepsilon_I^x)^2\}. \quad (3.41)$$

The Raman line of diamond itself has been studied since the early days of light-scattering spectroscopy. References to much of the earlier work, extending back to 1930, are given by Solin and Ramdas (1970). The low-temperature frequency of the vibration found by these authors is 1333.3 ± 0.5 cm^{-1} and the spectral linewidth is 1.48 ± 0.02 cm^{-1}. The above polarization properties are verified by these measurements, which also determined the temperature dependence of the vibrational frequency and linewidth.

McQuillan et al. (1970) verified that the Raman line of diamond has a Lorentzian shape in accordance with (1.90) and (1.93). They determined an absolute value for the cross section, which corresponds to a susceptibility derivative

$$a_\sigma^{yz} = a_\sigma^{zx} = a_\sigma^{xy} = 4 \cdot 6 \times 10^{12} \text{ Fm}^{-2}\text{kg}^{-1/2}. \quad (3.42)$$

The resulting differential cross section for an experiment with the incident frequency (1.23), and where the polarization factor in (3.41) is unity, is

$$\frac{d\sigma}{d\Omega} \approx 4 \times 10^{-5} \text{ m}^2. \quad (3.43)$$

The optic vibration in diamond therefore scatters one photon in 25,000 into unit solid angle during a pathlength in the crystal of 1m. An independent experimental verification of the value (3.42) for the susceptibility derivative is described in Section 4.1.3.

Silicon has the same crystal structure as diamond, but is opaque to visible light. The first observation of the Raman line in silicon was made by Russell (1965) using backscattering of visible HeNe laser light from the crystal surface (see Section 2.2.1); this was the first measurement of a light-scattering spectrum in reflection. Temple and Hathaway (1973) have since made more detailed measurements of the reflection spectrum with light from an argon-ion laser. They find the Raman line to have a Lorentzian shape with its peak at 519 ± 1 cm^{-1} and a width of 1.45 ± 0.05 cm^{-1} at low temperatures. Ralston and Chang (1970) have made measurements with light from a Nd:YAG laser to which silicon is transparent. The absolute cross section at an incident frequency $\omega_I/2\pi$ of about 3×10^{14} Hz with the polarization factor in (3.41) again equal to unity is found to be

$$\frac{d\sigma}{d\Omega} \approx 3 \times 10^{-4} \text{m}^2 \quad (3.44)$$

for a cubic meter of crystal.

A further example of the same crystal structure is germanium, where Parker et al. (1967) have measured the reflection Raman spectrum for argon-ion laser light. They find the Raman line to have a peak frequency of 300.7 ± 0.5 cm^{-1} and a width of 5.3 cm^{-1} at 300 K.

Rutile-Structure Crystals

The structure (illustrated in Figure 6.7) has space group $P4_2/mnm$ or D_{4h}^{14} and the crystal point group is $4/mmm$ or D_{4h}. Rutile itself has two TiO$_2$ units in the primitive cell and as shown in Table 3.1 there are four zero-wavevector Raman-active optic vibrations. Three of these are singlets and the fourth is a doublet. Table 1.2 gives the polarization dependences of their cross sections as

$$\Gamma_1^+ \text{ or } A_{1g} : |a(\varepsilon_S^x \varepsilon_I^x + \varepsilon_S^y \varepsilon_I^y) + b\varepsilon_S^z \varepsilon_I^z|^2 \tag{3.45}$$

$$\Gamma_3^+ \text{ or } B_{1g} : |d|^2 (\varepsilon_S^x \varepsilon_I^x - \varepsilon_S^y \varepsilon_I^y)^2 \tag{3.46}$$

$$\Gamma_4^+ \text{ or } B_{2g} : |e|^2 (\varepsilon_S^x \varepsilon_I^y + \varepsilon_S^y \varepsilon_I^x)^2 \tag{3.47}$$

$$\Gamma_5^+ \text{ or } E_g : |f|^2 \{ (\varepsilon_S^x \varepsilon_I^z + \varepsilon_S^z \varepsilon_I^x)^2 + (\varepsilon_S^y \varepsilon_I^z + \varepsilon_S^z \varepsilon_I^y)^2 \}, \tag{3.48}$$

where the additional symmetry (2.37) is used in the last result.

Porto et al. (1967) have made careful measurements of the light-scattering spectra of rutile itself, and of four other crystals of the same structure, with the results shown in Table 3.5. Figure 3.5 shows the measured spectra for MnF$_2$ for various combinations of incident and scattered polarization directions. It is clear that the four spectra shown provide sufficient information for an unambiguous assignment of vibrational symmetries to be made on the basis of (3.45) to (3.48). The measurements were made at room temperature; MnF$_2$ and FeF$_2$ show additional Raman lines associated with their antiferromagnetic ordering at lower temperatures, as described in Sections 6.2 and 6.3.

Table 3.5 Vibrational Frequencies in cm^{-1} for some Rutile-Structure Crystals*

	TiO$_2$	MgF$_2$	ZnF$_2$	FeF$_2$	MnF$_2$
Γ_3^+ or B_{1g}	143	92	70	73	61
Γ_5^+ or E_g	447	295	253	257	247
Γ_1^+ or A_{1g}	612	410	350	340	341
Γ_4^+ or B_{2g}	826	515	522	496	476

*From Porto et al. (1967).

Nonpolar Vibrational Scattering

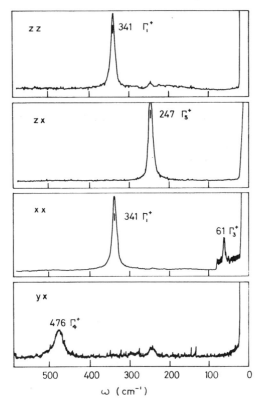

Figure 3.5 Raman spectra of MnF$_2$ for the incident and scattered polarizations indicated (after Porto et al. 1967).

Hexagonal Close-Packed Metals

Many of the common metals crystallize in monatomic structures that have no optic lattice vibrations. However, the hexagonal close-packed structure with space group $P6_3/mmc$ or D_{6h}^4 and crystal point group $6/mmm$ or D_{6h} has two atoms in the primitive cell. As shown in Table 3.1, there is a single Raman-active optic lattice vibration and the polarization dependence of its cross section from Table 1.2 is

$$\Gamma_6^+ \text{ or } E_{2g} : |f|^2 \{ (\varepsilon_S^x \varepsilon_I^x - \varepsilon_S^y \varepsilon_I^y)^2 + (\varepsilon_S^x \varepsilon_I^y + \varepsilon_S^y \varepsilon_I^x)^2 \}. \quad (3.49)$$

Beryllium and magnesium are examples of hcp metals, whereas cadmium and zinc have a slightly modified form of the same structure with larger spacings between hcp-type layers of atoms.

Measurements of light-scattering spectra of metals can only be made in reflection because of their strong absorption of light. The first observations by Feldman et al. (1968) followed their earlier work on Ge. They found the Γ_6^+ vibration of Be at 455 cm^{-1} at room temperature. Subsequent work included Mg (120 cm^{-1}) and Zn (70 cm^{-1}) (Parker et al. 1969), and the fluorite-structure metal AuAl$_2$ and bismuth were also studied (see Table 3.1 for details of Raman-active modes). Fraas et al. (1970) verified the predicted scattering symmetry (3.49) in the case of Be. More recently Grant et al. (1973) (this paper lists most previous references on metals) and Schultz and Hüfner (1976) have made careful measurements of the temperature dependences of the linewidths in the hpc metals and Bi, interpreting their results in terms of anharmonic and electron-phonon interactions.

3.2 SECOND-ORDER SCATTERING

3.2.1 Density of States and Selection Rules

The second-order Raman effect is a scattering process in which two phonons participate. They may both be created (giving a Stokes component in the scattered light) or one may be created and the other destroyed (giving a Stokes or anti-Stokes component) or both may be destroyed (giving an anti-Stokes component). Most of the scattering at low temperature arises from the two-phonon creation process and we concentrate on this case. The analogous process of second-order scattering by magnons is discussed in Section 6.3.1.

If the two created phonons have branch indices σ and σ', and their wavevectors are $\mathbf{q}$ and $\mathbf{q}'$, then energy and momentum conservation give

$$\omega_{\sigma\mathbf{q}} + \omega_{\sigma'\mathbf{q}'} = \omega_I - \omega_S \tag{3.50}$$

$$\mathbf{q} + \mathbf{q}' = \mathbf{k}_I - \mathbf{k}_S. \tag{3.51}$$

The branch indices and wavevectors can take all values consistent with these restrictions. Thus the wavevectors need not be small, as in first-order scattering, but can take on values throughout the Brillouin zone. The phonon wavevectors are typically three orders of magnitude larger than the light wavevectors over the major part of the zone, and (3.51) can be replaced by

$$\mathbf{q} + \mathbf{q}' = 0 \tag{3.52}$$

to a very good approximation. The vibrational frequencies are the same at wavevectors $+\mathbf{q}$ and $-\mathbf{q}$, and the energy conservation condition can thus be written

$$\omega_{\sigma\mathbf{q}} + \omega_{\sigma'\mathbf{q}} = \omega_I - \omega_S. \tag{3.53}$$

The theory of second-order scattering is described later but it is evident without any detailed analysis that the intensity of scattering at frequency shift ω is controlled by the number of pairs of vibrational frequencies whose sum is ω. More formally, the important quantity is the combined density of states of pairs of phonons

$$\rho_2(\omega) = \sum_{\sigma,\sigma'} \sum_{\mathbf{q}} \delta(\omega - \omega_{\sigma\mathbf{q}} - \omega_{\sigma'\mathbf{q}}), \tag{3.54}$$

where σ and σ' run over all branches of the vibrational spectrum and $\mathbf{q}$ runs over all wavevectors in the Brillouin zone.

It is clear that the combined density of states is a continuous function of frequency for crystals of macroscopic size, and the second-order spectrum is a continuous one, in contrast to the line spectrum of first-order scattering. The nature of the second-order spectrum was the subject of a celebrated controversy between Born and Raman in the 1940s, the latter contending that the vibrational frequencies of a crystal lie not in the continuous branches of the Born theory of lattice dynamics, but are restricted to a small number of discrete values (for references, see Krishnan 1971). Raman's theory, which did not receive any substantial acceptance outside India, leads to the prediction of a discrete second-order spectrum, in conflict with observations.

Any density of states for excitations in a crystal shows features known as Van Hove singularities. These usually take the form of infinities in the frequency-derivative of the density of states, and Figure 6.12 shows examples in the density of states for the single magnon branch in MnF_2. Each branch of the phonon dispersion curves makes a qualitatively similar contribution to the phonon density of states, leading to a function of the frequency with many slope discontinuities. The Van Hove singularities in the single-phonon density of states occur at frequencies where

$$\nabla_\mathbf{q} \omega_{\sigma\mathbf{q}} = 0. \tag{3.55}$$

These positions of extrema in the dependence of frequency on wavevector are known as critical points on the dispersion curve of phonon branch σ.

The various types of critical point and the rules that govern their occurrence are reviewed by Johnson and Loudon (1964). Most of the

Second-Order Scattering

critical points occur on the zone boundaries where the periodic properties of the crystal structure often require zero wavevector-derivatives of the vibrational frequencies. The corresponding properties of the single-magnon critical points are illustrated by Figures 6.8 and 6.12. It is also seen in these figures that most of the strength of the density of states function lies in the vicinity of the zone-boundary magnon frequencies, with very little strength in the region of the zone-center frequency. This is generally true for all kinds of excitation and results from the geometrical consideration that only roughly $\frac{1}{8}$ of the volume of the Brillouin zone lies at wavevectors smaller than $\frac{1}{2}$ the maximum wavevectors q_M in the zone.

These properties of the single-excitation density of states carry over to the combined two-phonon density of states (3.54). A Van Hove singularity in the combined density of states occurs at any frequency where

$$\nabla_{\mathbf{q}}(\omega_{\sigma\mathbf{q}} + \omega_{\sigma'\mathbf{q}}) = 0. \tag{3.56}$$

Such singularities occur whenever both branches σ and σ' have critical points at wavevector $\mathbf{q}$ in accordance with (3.55), but there are additional singularities corresponding to pairs of branches that have equal and opposite slopes at the same wavevector $\mathbf{q}$. These possibilities usually give rise to a large number of Van Hove singularities in the combined density of states, even for crystals with only a small number of atoms in the primitive cell.

To a first approximation then, the second-order light-scattering spectrum of the phonons reflects the density of states (3.54), with peaks at shifts close to combinations of zone-boundary phonon frequencies and sharp features associated with the Van Hove singularities. A better approximation must take account of the variation of the coupling of the light to the phonons with the branch indices and wavevector. This information is partly contained in the selection rules for particular pairs $\sigma, \mathbf{q}$ and $\sigma', -\mathbf{q}$ of phonons in the second-order scattering. The selection rules are obtained by an extension of the method of Section 1.6 in which the excitation symmetry Γ_X is obtained by taking a direct product of the irreducible representations of the participating phonons

$$\Gamma_X = \Gamma_{\sigma,\mathbf{q}} \times \Gamma_{\sigma',-\mathbf{q}}. \tag{3.57}$$

The excitation symmetry corresponds to a zero-wavevector representation of the crystal space group, equivalent to a representation of the crystal point group. However, the nonzero-wavevector phonons require the use of irreducible representations of the space group itself on the right-hand side of (3.57). Birman (1974) gives an exhaustive account of the methods of calculation.

It can be shown (Loudon 1965) that there are no selection rules on the participation in second-order light scattering of phonons at general wavevectors in the Brillouin zone since the excitation symmetry Γ_X contains all Raman-active $q=0$ symmetries in this case. Consideration of specific crystal structures shows, however, that restrictions do sometimes occur for the wavevectors of symmetric positions in the Brillouin zone. Some of the Van Hove singularities in the combined density of states may thus be suppressed in the second-order Raman spectrum since many critical points usually occur at high-symmetry positions in the zone. Crystal structures, for which the second-order selection rules have been derived, include diamond (Birman 1963, Johnson and Loudon 1964), zinc blende (Birman 1963), and rocksalt (Burstein et al. 1965, Chen et al. 1968).

Armed with selection rules and with some independent knowledge of the phonon dispersion curves from neutron-scattering experiments, it is in principle possible to identify features in the experimental second-order spectra with particular critical points in the combined dispersion curves of pairs of phonons. Analyses of this kind have been attempted (see Birman 1974 for a review) but they are made difficult by the large numbers of Van Hove singularities, many of which escape resolution in even the most careful of experiments.

3.2.2 Experiments and Calculations

Rocksalt-Structure Crystals

Alkali halide crystals have either the rocksalt structure ($Fm3m$ or O_h^5) or, less commonly, the caesium chloride structure ($Pm3m$ or O_h^1). Rocksalt-structure crystals have dominated experiments and theories on second-order scattering. Rocksalt itself was the first crystal to have its second-order spectrum recorded (Rasetti 1931). Figure 3.6 shows more recent spectra of NaCl (Krauzman 1968) and Figure 3.7 shows similar results for KBr (Krauzman 1967). Both structures of alkali halide crystal have two ions in the primitive cell, but in neither case is the zero-wavevector optic mode Raman active (see Table 3.1). The second-order spectra of these crystals are not therefore obscured by any first-order lines, as is sometimes the case in other materials.

Some evidence of Van Hove singularities can be seen in Figures 3.6 and 3.7, but the more striking properties are the overall shapes of the intensity distributions and their dependence on the polarizations of the incident and scattered light. The excitation symmetries detected for the three experimental polarizations are shown in Table 3.6. It is convenient in second-order scattering to regard the coefficients in the matrices of Table 1.2 as

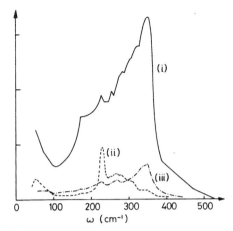

Figure 3.6 Second-order Raman spectra of NaCl for the polarizations shown in Table 3.6 (after Krauzman 1968). By permission of Gauthier-Villars.

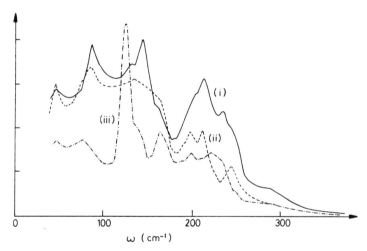

Figure 3.7 Second-order Raman spectra of KBr for the polarizations shown in Table 3.6. The scattering for polarization (i) is 10 times more intense than plotted (after Krauzman 1967). By permission of Gauthier-Villars.

Table 3.6 Polarizations and Scattering Symmetries for the Experimental Traces of Figures 3.6 and 3.7.

Trace	ε_S	ε_I	Symmetry	Cross Section
(i)	$1,0,0$	$1,0,0$	$\Gamma_1^+ + \Gamma_3^+$	$a^2 + 4b^2$
(ii)	$-2^{-1/2}, 2^{-1/2}, 0$	$2^{-1/2}, 2^{-1/2}, 0$	Γ_3^+	$3b^2$
(iii)	$1,0,0$	$0,1,0$	Γ_5^+	d^2

continuous functions of the frequency, and Table 3.6 also shows the functions that determine the frequency dependences of the cross sections for each of the three spectra. The frequency dependence of a^2 alone can of course be obtained by appropriate subtraction of the first two spectra. The frequency spreads of the spectra for a given crystal are roughly the same for different polarizations, and this is a consequence of the production of all Raman-active $q = 0$ excitation symmetries by pairs of phonons at general wavevectors $\pm \mathbf{q}$ in the Brillouin zone. The antisymmetric Γ_4^+ matrices of the crystal point group $m3m$ or O_h are excluded from second-order vibrational scattering for the same reason as their exclusion in the first-order case.

The strong dependences of the integrated scattered intensity and its frequency distribution on the incident and scattered polarizations in principle provide information on the interionic interactions in alkali halide crystals. The main aim of theoretical analyses of the spectra is the extraction of such information. Any successful calculation of second-order spectra must clearly be based on a lattice-dynamical theory capable of providing accurate frequencies for all vibrational branches throughout the Brillouin zone.

The theory of the second-order cross section can be obtained from either the macroscopic or microscopic approaches used in the theory of first-order scattering. The macroscopic approach begins with an extension of the susceptibility-derivative expansion of Sections 1.4 or 3.1.4 to include contributions of the form

$$P_S^i = \epsilon_O \sum_{\sigma,\sigma'} \sum_{\mathbf{q}} \chi_{\sigma\sigma'\mathbf{q}}^{ij}(\omega_I, 0) W_{\sigma,\mathbf{q}}^* W_{\sigma',-\mathbf{q}}^* E_I^j, \qquad (3.58)$$

where

$$\chi_{\sigma\sigma'\mathbf{q}}^{ij}(\omega_I, 0) = \frac{\partial^2 \chi^{ij}(\omega_I)}{\partial W_{\sigma,\mathbf{q}}^* \partial W_{\sigma',-\mathbf{q}}^*}, \qquad (3.59)$$

and we include only the creation parts of the normal coordinates.

Second-Order Scattering

The second-order cross section obtained from (1.70) now contains a higher-order power spectrum of the form

$$\langle W_{\sigma'',\mathbf{q}'} W_{\sigma''',-\mathbf{q}'} W^*_{\sigma,\mathbf{q}} W^*_{\sigma',-\mathbf{q}} \rangle_\omega. \tag{3.60}$$

In thermal equilibrium, with independent random fluctuations in the normal-mode amplitudes, the power spectrum vanishes unless the subscripts on the starred and unstarred amplitudes are equal in pairs. Thus only power spectra of the form

$$\langle W_{\sigma',-\mathbf{q}} W_{\sigma,\mathbf{q}} W^*_{\sigma,\mathbf{q}} W^*_{\sigma',-\mathbf{q}} \rangle_\omega \tag{3.61}$$

are needed and these can be factorized into products of contributions from the phonons $\sigma, \mathbf{q}$ and $\sigma', -\mathbf{q}$. In the limit of zero damping where the Lorentzians in the power spectra (1.92) reduce to delta functions, the factored form of (3.61) gives

$$\int d\omega' \langle W_{\sigma,\mathbf{q}} W^*_{\sigma,\mathbf{q}} \rangle_{\omega'} \langle W_{\sigma',-\mathbf{q}} W^*_{\sigma',-\mathbf{q}} \rangle_{\omega-\omega'}$$

$$= \frac{\hbar^2}{4N^2 \omega_{\sigma\mathbf{q}} \omega_{\sigma'\mathbf{q}}} \{n(\omega_{\sigma\mathbf{q}}) + 1\} \{n(\omega_{\sigma'\mathbf{q}}) + 1\} \delta(\omega - \omega_{\sigma\mathbf{q}} - \omega_{\sigma'\mathbf{q}}). \tag{3.62}$$

The cross section obtained from (1.70) is therefore

$$\frac{d^2\sigma}{d\Omega d\omega_S} = \frac{\omega_I \omega_S^3 v V \eta_S}{(4\pi\epsilon_0)^2 c^4 \eta_I} \frac{\hbar^2}{4N^2} \sum_{\sigma,\sigma'} \sum_{\mathbf{q}} |\epsilon_O \epsilon_S^i \epsilon_I^j \chi^{ij}_{\sigma\sigma'\mathbf{q}}(\omega_I, 0)|^2 (\omega_{\sigma\mathbf{q}} \omega_{\sigma'\mathbf{q}})^{-1}$$

$$\times \{n(\omega_{\sigma\mathbf{q}}) + 1\} \{n(\omega_{\sigma'\mathbf{q}}) + 1\} \delta(\omega - \omega_{\sigma\mathbf{q}} - \omega_{\sigma'\mathbf{q}}). \tag{3.63}$$

The Bose-Einstein thermal factors are modified in the manner of Table 2.2 for the other contributions in which phonons are destroyed in the scattering process. The remaining part of the summation in (3.63) is the same as the combined density of states (3.54) but weighted by branch and wavevector-dependent factors. The ordinary results of a lattice-dynamics calculation must be supplemented by a knowledge of the susceptibility derivatives (3.59) to obtain the theoretical second-order spectrum.

The normal coordinates are linear combinations of the displacements of the ions in the crystal lattice, and the Stokes polarization (3.58) is accordingly quadratic in the ionic displacements. The magnitudes of the susceptibility derivatives are determined physically by the ways in which interactions between displacements of the same or neighboring ions contribute to the susceptibility at the incident frequency. It is necessary to find some relatively simple model for the description of these interactions, consistent with the ionic model assumed in the underlying lattice-dynamics calculation.

The first calculation of a second-order spectrum by Born and Bradburn (1947) was based on a rigid-ion model of the NaCl lattice. It was assumed that only the displacements of nearest-neighbor ions contribute to the Stokes polarization. The shell model was introduced into the theory of second-order scattering by Cowley (1964), who treated NaI and KBr, and most subsequent work uses some form of the shell model. Representative crystals treated in this way include KBr (Bruce and Cowley 1971, 1972), NaCl (Bruce 1972), and NaF, KF, RbF, and CsF (Cunningham et al. 1974). Other kinds of rocksalt-structure crystal have been studied by similar methods; for example Pasternak et al. (1974) treat MgO in addition to KBr and NaCl, and Buchanan et al. (1974) calculate second-order spectra of CaO to compare with the measurements of Rieder et al. (1973). The number of independent contributions to the susceptibility derivatives in these calculations is minimized by use of the symmetry of the lattice and the assumption that only displacements on the same or nearest-neighbor ions couple to produce a Stokes polarization. The unknown quantities that remain are determined by fits to the experimental spectra.

The main microscopic mechanisms of second-order scattering are represented in the diagrams of Figure 3.8, which follow the same conventions as the first-order diagrams of Figure 3.2. The processes are indeed extensions to a higher order of the dominant kind of first-order process shown in Figure 3.2c. The two second-order processes differ only in whether the two phonons are created separately in a pair of first-order electron-lattice interactions, as in Figure 3.8a, or together in a single second-order electron-lattice interaction, as in Figure 3.8b. Coupling of the vibrations to the incident and scattered light occurs via the intermediary of the electrons.

The shell-model approach to calculations of second-order cross sections closely mirrors the main features of the microscopic scattering

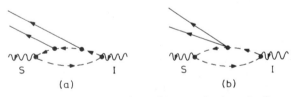

Figure 3.8 Diagrammatic representations of two mechanisms for the second-order scattering of light by lattice vibrations.

mechanisms. Thus the electron-lattice interactions take the form of anharmonic couplings between the motions of the electron shells alone and the core-shell motions that represent the normal modes of ionic vibration. The vibrations are coupled to the light via the electron shells. Each stage in the microscopic processes of Figure 3.8 has an appropriate representation in the shell model, and the resulting expression for the second-order cross section depends only on the harmonic and anharmonic parameters of the model (Cowley 1964, Bruce and Cowley 1971, 1972).

The calculation of second-order spectra provides a sensitive test of lattice-dynamical theories since it involves not only the normal-mode frequencies but also the normal coordinates and the anharmonic couplings. Much progress has been made since the work of Born and Bradburn, but the second-order alkali halide spectra present a continuing challenge to the validity of lattice-dynamical models.

Diamond-Structure Crystals

Crystals of the diamond structure ($Fd3m$ or O_h^7) show a single first-order line discussed in Section 3.1.6. The second-order scattering by diamond itself was first observed by Krishnan in 1944 (for a reproduction of the results, see Krishnan 1971). Figure 3.9 shows some more recent room-temperature spectra in which the different symmetries of contribution are separated by appropriate experimental polarizations (Solin and Ramdas 1970). The spectra show some additional features of the Van Hove type at lower temperatures, but the main interest of the spectra again lies in their overall shapes and relative intensities.

The same general comments on the calculation of second-order spectra apply to diamond as were made above for the alkali halides. Smith (1948) calculated the diamond spectrum by the Born and Bradburn method, while subsequent work has applied the shell model to diamond-structure crystals. However, the most successful calculation of the diamond spectra, by Go et al. (1975) uses the bond-charge model for the lattice dynamics of covalent crystals (Weber 1974). The results of this calculation included in Figure 3.9 achieve a good representation of the main features of the experimental spectra with only one adjustable anharmonic parameter.

The distributions of intensity in the experimental diamond spectra extend up to a common cut-off in the region of 2690 cm^{-1}. An intriguing feature of the Γ_1^+ spectrum, and to a lesser extent of the Γ_3^+ spectrum, is the rather sharp line just below the cut-off. Its frequency of 2667 cm^{-1} is close to twice the frequency of the first-order line in diamond, and it is natural to assume that it results in some way from scattering by a pair of zero-wavevector optic phonons. However, the second-order spectrum is

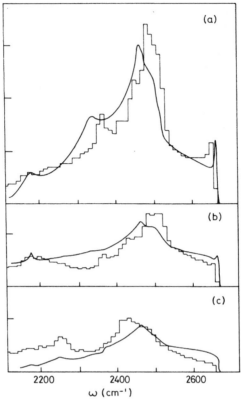

Figure 3.9 Second-order Raman spectra of diamond measured by Solin and Ramdas (1970) (continuous lines) and calculated by Go et al. (1975) (histograms). The symmetries of the contributions are (a) Γ_1^+, (b) Γ_3^+ (vertical scale multiplied by 600), and (c) Γ_5^+ (after Go et al. 1975).

controlled by the combined density of states (3.54) and is not normally expected to have sharp peaks at particular two-phonon frequencies. Some special mechanism for the 2667 cm^{-1} line in diamond has therefore been sought. It could in principle result from two successive first-order scatterings of the same incident photon, but this process is too weak to account for the experimental line. Another proposal is the emergence of a two-phonon bound state slightly above the highest frequency of the combined density of states, caused by anharmonic forces in the lattice (Cohen and Ruvalds 1969, Ngai et al. 1974). However, it is seen in Figure 3.9 that the calculations of Go et al. (1975) include the second-order peak at 2667 cm^{-1} without any necessity for two-phonon bound states. It therefore seems that the peak may be an intrinsic feature of the combined density of states in diamond.

The second-order spectrum of silicon was first measured by Parker et al. (1967) and more detailed observations and interpretations have been made by Temple and Hathaway (1973) and Uchinokura et al. (1974). The theory of the silicon spectra follows the same lines as diamond and the bond-charge model also provides a good account of the experimental results (Go et al. 1975). Resonance effects in the second-order scattering have been studied by Klein et al. (1974); in general terms these are similar to the resonance effects in first-order scattering (see Section 4.2.3) and the microscopic theory is useful in interpreting the results, particularly the identification of the relative contributions to the cross section from the two kinds of diagram in Figure 3.8.

The second-order Raman spectrum of germanium has been measured by Weinstein and Cardona (1973a), and resonance scattering effects have been investigated by Renucci et al. (1974).

Zinc-Blende-Structure Crystals

The other main crystal structure investigated by second-order Raman scattering is zinc blende, $F\bar{4}3m$ or T_d^2. The long-wavelength optic vibrations are polar and give rise to the double-peaked first-order spectrum described in Section 4.1.4. We mention here only the example of GaP, which has been the subject of most of the work. Its second-order spectra have been measured by Hoff and Irwin (1973), and resonant effects have been studied by Weinstein and Cardona (1973b) and by Klein et al. (1974). Zeyher (1974) has made a thorough theoretical investigation of the variation of the spectrum with the frequency of the incident light.

As a final comment on second-order spectra, it should be emphasized that the calculations mentioned above assume that the two created phonons behave independently. However, with short-range anharmonic forces as the source of the second-order Stokes polarization, the two phonons are created in close proximity in the lattice, and their interaction may distort the spectrum. Similar effects in second-order magnetic spectra are described in Section 6.3.2.

3.3 DEFECT-INDUCED SCATTERING

Raman scattering is an effective tool for the study of vibrational excitations in imperfect crystals, being complementary to infrared techniques in many cases. Both these methods are much more sensitive than neutron scattering for investigating crystal defects present in small concentrations ($\lesssim 1$ mole%).

The introduction of point defects into crystals activates Raman scattering and infrared absorption by a variety of modes (for reviews of the background theory, see Maradudin 1966a,b, Elliott 1966, Maradudin et al. 1971) as follows:

1. Localized modes arising from vibrations of light impurities: This type of vibration occurs at frequencies above the vibrational bands of the perfect crystal and gives rise to sharp lines. The amplitude of a localized mode is large near the defect and dies away rapidly with distance from the defect.
2. Gap modes in crystals with a frequency gap in the single-phonon density of states of the perfect crystal: Gaps often occur between acoustic and optic modes in diatomic crystals with a large mass difference between the constituent ions, for example KI. These modes may be activated by all kinds of defect, giving localized excitations with sharp spectral lines.
3. Resonance modes within the vibrational bands of the perfect crystal: The amplitudes of resonance modes are enhanced near the defect but these modes may be transmitted through the lattice and closely resemble perfect-crystal modes at large distances from the defect. They give rise to broader peaks and may be activated by all kinds of defect but are generally associated with heavy impurities.
4. Band modes of the host crystal activated by the removal of translational invariance, and hence of wavevector conservation, by the presence of the defect: The band modes are heavily perturbed, perhaps showing resonances, if the point defect is very different from the host atom it replaces.

As the concentration of defects in a crystal increases, small clusters begin to form, that is pairs and triplets, etc., but light-scattering studies in this regime have been limited (for a review of small clusters and molecular impurities, see Barker and Sievers 1975). Eventually, when impurity ions are substituted, we arrive with increasing concentration in the alloy or mixed-crystal regime. Mixed crystals represent a type of disorder arising from random distribution of different atomic species on a well-defined lattice. By contrast amorphous solids have a type of disorder that is spatial rather than configurational in nature. Dean (1972) reviews numerical methods for calculating the vibrations of mixed crystals and amorphous solids.

Defect-Induced Scattering

We are concerned with point defects in Section 3.3.1 and with mixed crystals and amorphous solids in Section 3.3.2. The relevant literature is very extensive, and we shall have to be content with a few illustrative examples.

3.3.1 Light Scattering by Point Defects

Many kinds of point defect have been studied by spectroscopic methods (for reviews, see Newman 1969, Barker and Sievers 1975). We consider here the hydride ion in simple ionic solids. Hydrogen dissolves in alkali halides and alkaline earth fluorides as H^-, replacing a lattice anion. Because of its very light mass, it gives rise to localized vibrational modes at frequencies higher than the maximum band frequency by factors of two or three. The first measurements were made by Schaefer (1960) on H^- in KCl using infrared absorption, and this investigation triggered detailed studies of H^- in this and other crystals.

In view of the high local-mode frequencies of H^-, Elliott et al. (1965) suggested that an adiabatic approximation could be used to describe the local oscillator. They assumed that the H^- ion vibrates in a static potential having the symmetry of the host lattice at the impurity site. The fluorine-site symmetry in the alkaline-earth fluorides is $\overline{4}3m$ or T_d and the local-oscillator Hamiltonian has the form

$$H = Ar^2 + Bxyz + C_1(x^4 + y^4 + z^4) + C_2(x^2y^2 + y^2z^2 + z^2x^2). \quad (3.64)$$

This is a Taylor expansion of the energy to fourth order in the displacement of the H^- ion. For alkali halides all lattice sites have $m3m$ or O_h symmetry and the additional center of inversion removes the cubic term from (3.64).

The energy-level structures of the anharmonic oscillator for both T_d and O_h symmetries are shown in Figure 3.10. For a site of T_d symmetry, electric-dipole transitions are allowed from the singlet A_1 ground state only to the T_2 triply degenerate excited states (see Table 1.2 for conversion to the Bethe notation). The four allowed transitions were observed in fluorides by Elliott et al. (1965), enabling them to determine the four constants in (3.64). Transitions to A_1, E, and T_2 levels are allowed in Raman scattering, and some of these were observed by Harrington et al. (1970) as shown in Figure 3.10. The energy levels not observed directly are calculated from (3.64).

In the case of O_h-site symmetry, electric-dipole transitions are allowed from the singlet A_{1g} ground state to the T_{1u} triply degenerate excited states. Only the T_{1u} state in the harmonic $n = 1$ level has been observed by

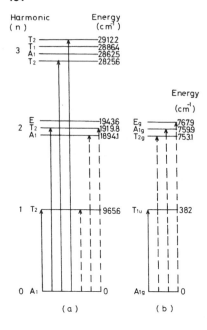

Figure 3.10 Schematic energy-level diagrams for the H^- ion in (a) CaF_2 (T_d site symmetry) and (b) KI (O_h site symmetry). The solid lines indicate observed infrared transitions and the dashed lines observed Raman transitions.

infrared absorption. Although this fundamental excitation is Raman forbidden, the three states of the $n=2$ level are allowed in light scattering and have been observed in KI by Montgomery et al. (1972), enabling them to determine the constants in (3.64).

Equation (3.64) is an approximation to the extent that it neglects motion of the lattice ions. In fact, anharmonic coupling of the local mode to the lattice phonons gives rise to pronounced temperature-dependent variations of the local-mode linewidths and peak positions that have been studied in considerable detail (for reviews, see Newman 1969, Barker and Sievers 1975).

Gap modes have been studied by Raman scattering almost exclusively in alkali halides. Five of these materials, LiCl, NaBr, NaI, KBr, and KI have gaps between the acoustic and optic modes, and gap modes have been found in all except the first. The gap in KI extends from 70 to 96 cm^{-1} and there is an A_{1g}-symmetry gap mode of H^- at 93 cm^{-1}, observed by Montgomery et al. (1972) using light scattering. A gap mode with T_{1u} symmetry caused by F centers (electrons bound to anion vacancies, first seen in light scattering by Worlock and Porto 1965) in KI was found in the infrared absorption at about 82 cm^{-1} by Bauerle and Fritz (1968). A gap mode of E_g symmetry was found in the same system at 78 cm^{-1} by

Defect-Induced Scattering

Buisson et al. (1975, 1976) using Raman spectroscopy, and there is also a resonance mode of predominantly A_{1g} character centered on 96.4 cm^{-1}, as shown in Figure 3.11.

The calculation of the defect-activated spectrum of the vibrational modes of the host lattice normally requires a lattice-dynamical model (e.g., shell model) to fit to phonon dispersion curves measured by neutron diffraction. The defect itself does not move in the even-symmetry lattice modes centered on the defect, so that its mass difference from the atom it replaces is not important. In its simplest form the calculation for an F center in an alkali halide involves computing the projected densities of states of A_{1g}, E_g, and T_{2g} symmetry for the unperturbed crystal on to the motion of the nearest-neighbor ions to the F center. On a lower level of sophistication it is possible by a simple group-theory calculation to determine which critical points on the phonon dispersion curves are activated in the various symmetries of light scattering (Loudon 1964, Kristofel 1966). To predict the frequencies of gap modes and resonance modes, it is necessary to take account of the change in force constant between the defect and its neighbors. In the case of F centers for example, the central

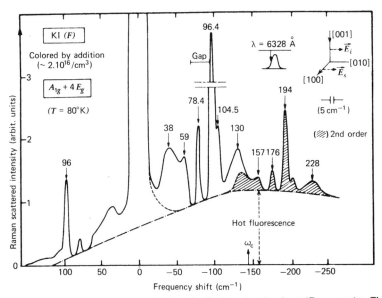

Figure 3.11 Raman spectrum of KI doped with F centres for $A_{1g}+4E_g$ symmetry. The hot fluorescence indicated is probably due to emission from excited vibrational states of the excited state of the F centre (after Buisson et al. 1975). By permission of Flammarion.

force constant between the defect and the nearest cations is about 70% smaller than the corresponding anion-cation force constant in the perfect crystal, and a reduction of this magnitude in KI accounts quite well for the peak positions of the observed resonance and gap modes (Buisson et al. 1975).

An early study of defect-activated lattice scattering was carried out by Harley et al. (1971) on alkali halides doped with the Tl^+ ion, which contrasts with the F center in having force-constant changes smaller than 10%. They observed Raman spectra of E_g and T_{2g} symmetry, but the A_{1g} spectrum was not detectable. Their model calculations of the spectra assumed that the Tl^+ electronic wavefunctions are modulated by vibrations of the nearest anions only. This approximation is consistent with the observation that the most intense Raman bands in KI and KBr lie in the acoustic region of the spectrum, since the anions are the heavier in these crystals. In RbCl, where the nearest neighbors are the lighter ions, the stronger band is in the optic phonon region, whereas in KCl the acoustic and optic bands are almost equally intense. It is generally true that the calculation of defect-activated lattice scattering in simple ionic solids with a nearest-neighbor model gives good agreement with experiment only if the masses of cation and anion are comparable. The next-nearest neighbors should also be considered when there is a sizeable difference between the masses (Chase et al. 1973, Robbins and Page 1976).

Light scattering induced in SrF_2 and BaF_2 by Eu^{2+} has been studied by Chase et al. (1973) and in SrF_2 containing F centers by Glynn et al. (1977). It is essential for agreement between theory and experiment to include second neighbors, because of the mass-difference between cations and anions in these crystals. It is interesting that the scattering induced by the F center in SrF_2 is about four orders of magnitude more intense than the scattering induced by Eu^{2+}. This is primarily caused by the stronger coupling of the F center wavefunction to the lattice phonons.

In concluding this section, we point out that Raman scattering is also a useful tool for studying low-lying vibronic energy levels, that is, levels of mixed vibrational and electronic character, of impurity ions with degenerate electronic ground states. For example, the Cu^{2+} ion substituted for Ca^{2+} in CaO has site symmetry $m3m$ or O_h and a ^2E_g electronic ground state. The twofold orbital degeneracy can be lifted in first order by E_g symmetry distortions of the surrounding ions, thereby reducing the total energy of the interacting system. This is an example of the Jahn-Teller effect, discussed at greater length in Section 5.2.4. The energy minima correspond to tetragonal distortions of the complex consisting of the Cu^{2+} ion and the six nearest oxygen ions; the lowest energy levels correspond to hindered rotations of the distorted complex between the three equivalent

Defect-Induced Scattering

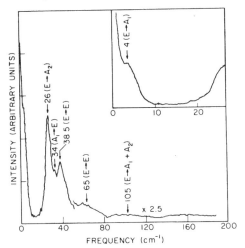

Figure 3.12 The E_g symmetry light-scattering spectrum of the hindered rotational levels of CaO:Cu^{2+} (after Guha and Chase 1974).

minima (O'Brien 1964). These vibronic levels transform as the A_{1g}, A_{2g}, and E_g representations of the cubic site group. For the example of E_g symmetry scattering shown in Figure 3.12 (Guha and Chase 1974, 1975), the allowed Raman transitions are $A_{1g} \rightarrow E_g$, $A_{2g} \rightarrow E_g$, and $E_g \rightarrow E_g$. Chase and Hao (1975) have carried out a similar study of the Jahn-Teller ion Ni^{3+} in Al_2O_3, and Chase et al. (1977) of the Mn^{3+} ion in Al_2O_3.

3.3.2 Mixed Crystals and Amorphous Solids

The properties of mixed crystals are reviewed by Chang and Mitra (1971) and by Elliott et al. (1974). The most interesting systems are those available with a complete range of concentrations, for example the mixed crystal $Ca_{1-x}Sr_xF_2$ shows solid solubility over the range $x=0$ to 1. True mixed crystals have a structure identical to that of the end members, the lattice constant varying approximately linearly with x. In many cases it is possible to see local modes and gap modes become more intense and broaden continuously into lattice bands as x is varied.

Some crystals such as $Ca_{1-x}Sr_xF_2$ show a single Raman-active mode or a single reststrahl peak whose frequency in general varies approximately linearly with x. Other mixed crystals such as Ge_xSi_{1-x} show two separate modes simultaneously, one characteristic of Ge and the other of Si. These distinct types of behavior are referred to as one-mode and two-mode, and more complex kinds of behavior occur in more complex systems. One-mode behavior can be described in terms of the virtual-crystal approximation in which all masses, force constants, and effective charges are treated as averages weighted by the mixed-crystal composition. A variety of other models have been tried for both one- and two-mode behaviors, including linear chain and cluster models (Verleur and Barker 1966). Green-function techniques have been used by Taylor (1967) for Ge_xSi_{1-x} and by Beserman and Balkanski (1970) for $CdS_{1-x}Se_x$.

Detailed studies of the one-mode systems $Ca_{1-x}Sr_xF_2$ and $Sr_{1-x}Ba_xF_2$ have been carried out by Chang et al. (1966) and Lacina and Pershan (1970) using Raman scattering, and by Verleur and Barker (1967) using infrared techniques. The Raman and infrared-active modes in the pure crystals belong to different irreducible representations of the crystal point group (see Table 3.1) and hence are independent (see Figure 3.13). The Raman-active mode (symmetry Γ_5^+ or T_{2g}) shown in Figure 3.13a has the

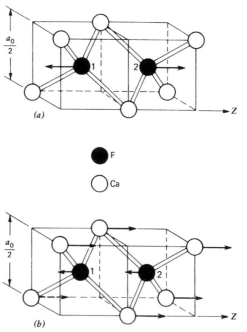

Figure 3.13 Primitive cell of the fluorite structure showing the ionic displacements in (a) the Raman-active Γ_5^+ or T_{2g} mode, and (b) the infrared-active Γ_4^- or T_{1u} mode (after Barker and Sievers 1975).

two fluorine ions in the primitive cell moving in antiphase parallel to a crystal cube edge, is threefold degenerate, and does not involve any cation motion. The Raman frequency varies linearly with x, which is perhaps not surprising in view of the absence of cation motion. The Raman mode doubles its linewidth on going from $x=0$ to $x=0.5$ and then falls again.

A search of the infrared spectrum for the Raman-mode frequency showed it to be absent in spite of the disorder. The infrared-active mode (symmetry Γ_4^- or T_{1u}) is shown in Figure 3.13b. It gives rise to a peak that shows some structure in the middle of the concentration range. Verleur and Barker compare their infrared results with a virtual-crystal model and a cluster model, while Lancina and Pershan also use a virtual-crystal approach, calculating an average Green function to first order in x. The cluster model of Verleur and Barker emphasizes the short-range order by tabulating the possible configurations of a fluorine ion and the nearest mixed cations. Different force constants are adopted for each configuration and preferential clustering of cations of a particular species is allowed for. The various approaches work well for the Raman mode but the infrared modes are more complicated because of the involvement of the cations. Verleur and Barker conclude that the structure observed in the reststrahl peak is a consequence of the sizeable variation in frequency from one cluster configuration to another.

Amorphous materials are less well defined than mixed crystals (for a general review, see Stuke and Brenig 1973 and, for a review of Raman scattering in amorphous semiconductors, Brodsky 1975). They are generally covalent in character and valence requirements are satisfied locally; for example each Ge atom in amorphous germanium has four nearest neighbors at approximately the same distance and bond angle as in crystalline germanium. However, atom spacings and bond angles between a particular atom and second or third neighbors tend to be different. In contrast with pure and mixed crystals, the x-ray pattern of an amorphous material shows broad halos rather than sharp rings and spots. It should be emphasized that a given material may have no unique amorphous state, different preparation techniques yielding different forms of the same solid. Although much of the interest in amorphous materials has centered on Si and Ge, there is a growing interest in Raman scattering in other amorphous systems (see, e.g., the work of Nemanich et al. 1977b on amorphous $GeSe_2$ and of Nemanich et al. 1977a on glass).

The absence of periodicity in amorphous systems allows many vibrational modes to be both Raman and infrared active. Figure 3.14 shows a Raman spectrum of amorphous silicon taken by Smith et al. (1971) using an argon-ion laser and backscattering geometry. Following Shuker and Gammon (1970), they assume that the Stokes intensity at frequency shift ω

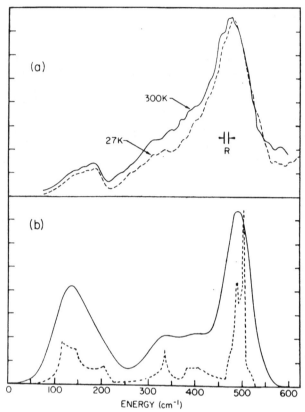

Figure 3.14 (a) Raman spectra of amorphous Si at 27 and 300 K reduced by the factor $\{n(\omega)+1\}/\omega$. (b) Single-phonon density of states $\rho_1(\omega)$ of crystalline Si (dashed line) and the same density of states after broadening (solid line) (after Smith et al. 1971).

satisfies [compare also (1.93)]

$$\bar{I}_S \propto \{n(\omega)+1\} \frac{\rho_1(\omega)}{\omega}, \tag{3.65}$$

where $\rho_1(\omega)$ is the single-phonon density of states defined analogously to the combined density of states (3.54). Figure 3.14 shows a comparison of the scaled scattered intensity with an arbitrarily broadened single-phonon density of states of crystalline silicon. It is apparent that the major features of the reduced Raman spectrum are in reasonable agreement with the

Defect-Induced Scattering

major features of the density of states, and this reflects the similarity in the short-range order in the crystalline and amorphous phases (for discussion, see Alben et al. 1975). However, it would be surprising if (3.65) held over the entire frequency range of the spectrum and indeed Lannin (1973) finds that the Raman intensity of amorphous silicon satisfies

$$I_S \propto \omega \{n(\omega)+1\}\rho_1(\omega) \qquad (3.66)$$

over the frequency range 20 to 70 cm^{-1}. Martin and Brenig (1974) discuss the possible origin of the extra factor of ω^2.

There is now a variety of experimental evidence to suggest that many amorphous materials contain anomalous low-frequency modes in addition to the long-wavelength acoustic modes. Measurement of the specific heat of fused quartz for example shows a large non-Debye-like contribution below 2K, which varies linearly with temperature (for references, see Stephens 1973). The microscopic nature of the additional states is uncertain, but Raman scattering measurements by Winterling (1975) provide information about the states in vitreous silica in the region 2 to 20 cm^{-1}

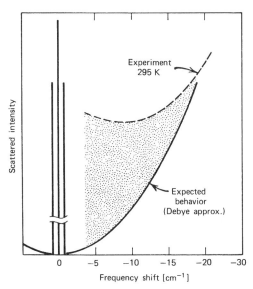

Figure 3.15 Representation of the expected dependence of the scattered intensity of vitreous silica on frequency shift at small shifts. The dashed curve represents the experimental results and the shaded area marks the excess scattering (after Winterling 1975). By permission of Flammarion.

(for the region 20 to 200 cm^{-1}, see Hass 1970, Stolen 1970). Equation (3.66) predicts an ω^2 dependence of the low-frequency scattering at room temperature, but the measured intensity in Figure 3.15 actually increases for $\omega < 10$ cm^{-1}, and is an order of magnitude larger than in crystalline quartz in the range 4 to 20 cm^{-1}. The excess scattering is assigned by Winterling to the low-frequency tails of strongly damped acoustic modes of higher frequency and correspondingly higher density. In a microscopic description the damping is associated with relaxation of the sound waves by mechanical coupling to structural defects (Anderson et al. 1972). The defects give rise to a weak central mode in the Raman spectrum whose width is determined by the inverse of the relaxation rate (see Chapter 5 for a more detailed discussion of central modes).

There is considerable interest in the microscopic structure of amorphous materials, particularly silicon and germanium, where the continuous-random-network model and the microcrystallite model have received considerable attention. The latter model assumes a composition of small crystallites (~ 15 Å in size) randomly oriented and joined by irregular bonds. Raman scattering provides information that helps to distinguish between the models (Solin and Kobliska 1973), although the scattering of light by small crystallites is itself a somewhat complicated phenomenon (see Section 4.1.7)

REFERENCES

Alben R., Weaire D., Smith J. E., and Brodsky M. H. (1975), *Phys. Rev.* **B11**, 2271.

Anderson P. W., Halperin B. I., and Varma C. M. (1972), *Philos. Mag.* **25**, 1.

Barker A. S. (1964), *Phys. Rev.* **135**, A742.

Barker A. S. and Sievers A. J. (1975), *Rev. Mod. Phys.* Suppl. 2.

Bauerle D. and Fritz B. (1968), *Solid State Commun.* **6**, 453.

Beserman R. and Balkanski M. (1970), *Phys. Rev.* **B1**, 608.

Birman J. L. (1963), *Phys. Rev.* **131**, 1489.

Birman J. L. (1974), *Handbuch der Physik* Vol. XXV/2b (Berlin: Springer).

Born M. and Bradburn M. (1947), *Proc. R. Soc.* **A188**, 161.

Born M. and Huang K. (1954), *Dynamical Theory of Crystal Lattices*, (Oxford: Clarendon Press).

Brodsky M. H. (1975), in M. Cardona, Ed., *Light Scattering in Solids* (Berlin: Springer), p. 208.

Bruce A. D. (1972), *J. Phys.* **C5**, 2909.

Bruce A. D. and Cowley R. A. (1971), *Ind. J. Pure Appl. Phys.* **9**, 877.

Bruce A. D. and Cowley R. A. (1972), *J. Phys.* **C5**, 595.

Buchanan M., Haberkorn R., and Bilz H. (1974), *J. Phys.* **C7**, 439.

References

Buisson J. P., Lefrant S., Sadoc A., Taurel L., and Billardon M. (1976), *Phys. Stat. Sol. (b)* **78**, 779.

Buisson J. P., Sadoc A., Taurel L., and Billardon M. (1975), in M. Balkanski, R. C. C. Leite, and S. P. S. Porto, Eds., *Light Scattering in Solids* (Paris: Flammarion), p. 587.

Burstein E., Johnson F. A., and Loudon R. (1965), *Phys. Rev.* **139**, A1239.

Chang I. F. and Mitra S. S. (1971), *Adv. Phys.* **20**, 359.

Chang R. K., Lacina W. B., and Pershan P. S. (1966), *Phys. Rev. Lett.* **17**, 755.

Chantry G. W. (1971), in A. Anderson, Ed., *The Raman Effect*, Vol. 1 (New York: Marcel Dekker), p. 49.

Chase L. L. and Hao C. H. (1975), *Phys. Rev.* **B12**, 5990.

Chase L. L., Hayes W., and Rushworth A. J. (1977), *J. Phys.* **C10**, L575.

Chase L. L., Kuhner D., and Bron W. E. (1973), *Phys. Rev.* **B7**, 3892.

Chen L. C., Berenson R., and Birman J. L. (1968), *Phys. Rev.* **170**, 639.

Cochran W. (1973), *The Dynamics of Atoms in Crystals* (London: Edward Arnold).

Cochran W. and Cowley R. A. (1967), *Handbuch der Physik*, Vol. XXV/2a (Berlin: Springer), p. 59.

Cohen M. H. and Ruvalds J. (1969), *Phys. Rev. Lett.* **23**, 1378.

Cowley R. A. (1964), *Proc. Phys. Soc.* **84**, 281.

Cracknell A. P. (1968), *Applied Group Theory* (Oxford: Pergamon).

Cunningham S. L., Sharma T. P., Jaswal S. S., Hass M., and Hardy J. R. (1974), *Phys. Rev.* **B10**, 3500.

Davydov A. S. (1962), *Theory of Molecular Excitons* (New York: McGraw-Hill).

Dean P. (1972), *Rev. Mod. Phys.* **44**, 127.

Elliott R. J. (1966), in R. W. H. Steverson, Ed., *Phonons in Perfect Lattices and in Lattices with Point Imperfections* (London: Oliver and Boyd), p. 377.

Elliott R. J., Hayes W., Jones G. D., Macdonald H. F., and Sennett C. T. (1965), *Proc. R. Soc.* **A289**, 1.

Elliott R. J., Krumhansl J. A., and Leath P. L. (1974), *Rev. Mod. Phys.* **46**, 465.

Feldman D. W., Parker J. H., and Ashkin M. (1968), *Phys. Rev. Lett.* **21**, 607.

Fraas L. M., Porto S. P. S., and Loh E. (1970), *Solid State Commun.* **8**, 803.

Glynn T. J., Hayes W., and Wiltshire M. C. K. (1977), *J. Phys.* **C10**, 137.

Go S., Bilz H., and Cardona M. (1975), *Phys. Rev. Lett.* **34**, 580.

Grant W. B., Schulz H., Hüfner S., and Pelzl J. (1973), *Phys. Stat. Sol. (b)* **60**, 331

Guha S. and Chase L. L. (1974), *Phys. Rev. Lett.* **32**, 869.

Guha S. and Chase L. L. (1975), *Phys. Rev.* **B12**, 1658.

Harley R. T., Page J. B., and Walker C. T. (1971), *Phys. Rev.* **B3**, 1365.

Harrington J. A., Harley R. T., and Walker C. T. (1970), *Solid State Commun.* **8**, 407.

Hass M. (1970), *J. Phys. Chem. Solids* **31**, 415.

Hoff R. M. and Irwin J. C. (1973), *Can. J. Phys.* **51**, 63.

Humphreys L. B. (1972), *Phys. Rev.* **B6**, 3886.

Johnson F. A. and Loudon R. (1964), *Proc. R. Soc.* **A281**, 274.

Kanamori H., Hayashi S., and Ikeda Y. (1974a), *J. Phys. Soc. Jap.* **36**, 511.

Kanamori H., Hayashi S., and Ikeda Y. (1974b), *J. Phys. Soc. Jap.* **37**, 1375.

Klein P. B., Masui H., Song J. J., and Chang R. K. (1974), *Solid State Commun.* **14**, 1163.
Koster G. F., Dimmock J. O., Wheeler R. G., and Statz H. (1963), *Properties of the Thirty-Two Point Groups* (Cambridge Mass.: MIT Press).
Krauzman M. (1967), *Comptes Rendus* **265B**, 1029.
Krauzman M. (1968), *Comptes Rendus* **266B**, 186.
Krebs B. and Müller A. (1967), *J. Mol. Spect.* **22**, 290.
Krishnan R. S. (1971), in A. Anderson, Ed., *The Raman Effect*, Vol. 1 (New York: Marcel Dekker), p. 1.
Kristofel' N. N. (1966) *Sov. Phys.-Solid State* **7**, 2027.
Lacina W. B. and Pershan P. S. (1970), *Phys. Rev.* **B1**, 1765.
Landau L. D. and Lifshitz E. M. (1960), *Electrodynamics of Continuous Media* (Oxford: Pergamon).
Lannin J. S. (1973), *Solid State Commun.* **12**, 947.
Loudon R. (1964), *Proc. Phys. Soc.* **84**, 379.
Loudon R. (1965), *Phys. Rev.* **137**, A1784.
Maradudin A. A. (1966a), *Solid State Phys.* **18**, 273.
Maradudin A. A. (1966b), *Solid State Phys.* **19**, 1.
Maradudin A. A., Montroll E. W., Weiss C. H., and Ipatova I. P. (1971), *Solid State Physics*, suppl. 3.
Maradudin A. A. and Wallis R. F. (1970), *Phys. Rev.* **B2**, 4294.
Martin A. and Brenig W. (1974), *Phys. Stat. Sol. (b)* **64**, 163.
McClure D. S. (1959) *Electronic Spectra of Molecules and Ions in Crystals* (New York: Academic Press).
McQuillan A. K., Clements W. R. L., and Stoicheff B. P. (1970), *Phys. Rev.* **A1**, 628.
Mills D. L., Maradudin A. A., and Burstein E. (1968), *Phys. Rev. Lett.* **21**, 1178.
Mills D. L., Maradudin A. A., and Burstein E. (1970), *Ann. Phys. (N.Y.)* **56**, 504.
Montgomery G. P., Fenner W. R., Klein M. V., and Timusk T. (1972), *Phys. Rev.* **B5**, 3343.
Nemanich R. J., Gorman M., and Solin S. A. (1977a), *Solid State Commun.* **21**, 277.
Nemanich R. J., Solin S. A., and Lucovsky G. (1977b), *Solid State Commun.* **21**, 273.
Newman R. C. (1969), *Adv. Phys.* **18**, 545.
Ngai K. L., Ganguly A. K., and Ruvalds J. (1974), *Phys. Rev.* **B10**, 3280.
Nye J. F. (1957), *Physical Properties of Crystals* (Oxford: Clarendon Press).
O'Brien M. C. M. (1964), *Proc. R. Soc.* **A281**, 323.
Parker J. H., Feldman D. W., and Ashkin M. (1967), *Phys. Rev.* **155**, 712.
Parker J. H., Feldman D. W., and Ashkin M. (1969), in G. B. Wright, Ed., *Light Scattering Spectra of Solids* (New York: Springer), p. 389.
Pasternak A., Cohen E., and Gilat G. (1974), *Phys. Rev.* **B9**, 4584.
Porto S. P. S. (1966), *J. Opt. Soc. Am.* **56**, 1585.
Porto S. P. S., Fleury P. A., and Damen T. C. (1967), *Phys. Rev.* **154**, 522.
Porto S. P. S. and Scott J. F. (1967), *Phys. Rev.* **157**, 716.
Poulet H. and Mathieu J. P. (1976), *Vibration Spectra and Symmetry of Crystals* (Paris: Gordon and Breach).

References

Ralston J. M. and Chang R. K. (1970), *Phys. Rev.* **B2**, 1858.
Rasetti E. (1931), *Nature* **127**, 626.
Reissland J. A. (1973), *The Physics of Phonons* (London: Wiley).
Renucci M. A., Renucci J. B., Zeyher R., and Cardona M. (1974), *Phys. Rev.* **B10**, 4309.
Rieder K. H., Weinstein B. A., Cardona M., and Bilz H. (1973), *Phys. Rev.* **B8**, 4780.
Robbins D. and Page J. B. (1976), *Phys. Rev.* **B13**, 3604.
Russell J. P. (1965), *Appl. Phys. Lett.* **6**, 223.
Russell J. P. and Loudon R. (1965), *Proc. Phys. Soc.* **85**, 1029.
Safran S. A., Dresselhaus G., and Lax B. (1976), *Solid State Commun.* **19**, 1217.
Schaefer G. (1960), *J. Phys. Chem. Solids* **12**, 233.
Schulz H. and Hüfner S. (1976), *Solid State Commun.* **20**, 827.
Scott J. F. (1968a), *J. Chem. Phys.* **48**, 874.
Scott J. F. (1968b), *J. Chem. Phys.* **49**, 98.
Shuker R. and Gammon R. W. (1970), *Phys. Rev. Lett.* **25**, 222.
Skinner J. G. and Nilsen W. G. (1968), *J. Opt. Soc. Am.* **58**, 113.
Smith H. M. J. (1948), *Philos. Trans. R. Soc.* **A241**, 105.
Smith J. E., Brodsky M. H., Crowder B. L., Nathan M. I., and Pinczuk A. (1971), *Phys. Rev. Lett.* **26**, 642.
Sokoloff J. B. (1972), *J. Phys.* **C5**, 2482.
Solin S. A. and Kobliska R. J. (1973), see Stuke and Brenig (1973), p. 1251.
Solin S. A. and Ramdas A. K. (1970), *Phys. Rev.* **B1**, 1687.
Steigmeier E. F. and Harbeke G. (1970), *Phys. Kondens. Mat.* **12**, 1.
Stephens R. B. (1973), *Phys. Rev.* **B8**, 2896.
Stolen R. H. (1970), *Phys. Chem. Glasses* **11**, 83.
Stuke J. and Brenig W. (1973), *Amorphous and Liquid Semiconductors*, (London: Taylor and Francis).
Suzuki N. and Kamimura H. (1973), *J. Phys. Soc. Japan* **35**, 985.
Taylor D. W. (1967), *Phys. Rev.* **156**, 1017.
Temple P. A. and Hathaway C. E. (1973), *Phys. Rev.* **B7**, 3685.
Tinkham M. (1964), *Group Theory and Quantum Mechanics* (New York: McGraw-Hill).
Tsang J. C., Dresselhaus M. S., Aggarwal R. L., and Reed T. B. (1974a), *Phys. Rev.* **B9**, 984.
Tsang J. C., Dresselhaus M. S., Aggarwal R. L., and Reed T. B. (1974b), *Phys. Rev.* **B9**, 997.
Uchinokura K., Sekine T., and Matsuura E. (1974), *J. Phys. Chem. Solids* **35**, 171.
Venkataraman G. and Sahni V. C. (1970), *Rev. Mod. Phys.* **42**, 409.
Verleur H. W. and Barker A. S. (1966), *Phys. Rev.* **149**, 715.
Verleur H. W. and Barker A. S. (1967), *Phys. Rev.* **164**, 1169.
Wallis R. F. and Maradudin A. A. (1971), *Phys. Rev.* **B3**, 2063.
Weber A. (1973), in A. Anderson, Ed., *The Raman Effect*, Vol. 2, (New York: Marcel Dekker), p. 543.
Weber W. (1974), *Phys. Rev. Lett.* **33**, 371.
Weinstein B. A. and Cardona M. (1973a), *Phys. Rev.* **B7**, 2545.

Weinstein B. A. and Cardona M. (1973b), *Phys. Rev.* **B8**, 2795.

Wilkinson G. R. (1973), in A. Anderson, Ed., *The Raman Effect*, Vol. 2 (New York: Marcel Dekker), p. 811.

Winterling G. (1975), in M. Balkanski, R. C. C. Leite, and S. P. S. Porto, Eds., *Light Scattering in Solids* (Paris: Flammarion), p. 663.

Woodward L. A. (1972), *Introduction to the Theory of Molecular Vibrations and Vibrational Spectroscopy* (Oxford: Clarendon Press).

Worlock J. M. and Porto S. P. S. (1965), *Phys. Rev. Lett.* **15**, 697.

Zeyher R. (1974), *Phys. Rev.* **B9**, 4439.

CHAPTER FOUR
Polar Vibrational Scattering

4.1 Macroscopic Theory _____ 148
    4.1.1 Lattice Dynamics of Polar Modes
    4.1.2 The Scattering Cross Section
    4.1.3 Properties of the Susceptibility Derivatives
    4.1.4 Polar-Mode Scattering in Cubic Crystals
    4.1.5 Polar-Mode Scattering in Uniaxial Crystals
    4.1.6 Polar-Mode Scattering in Biaxial Crystals
    4.1.7 Scattering by Powdered Crystals

4.2 Microscopic Theory _____ 179
    4.2.1 Electrons and Phonons in Crystals
    4.2.2 The Scattering Cross Section
    4.2.3 Resonance Scattering

4.3 Light Scattering by Polaritons _____ 192

Lattice vibrations that carry an electric-dipole moment have radically different long-wavelength properties from nonpolar vibrations. The dipole moment generates an electromagnetic field that accompanies the polar vibration, and it is necessary to solve the coupled equations of motion of the lattice and the electromagnetic field to obtain the properties of the polar modes. As a consequence of the electromagnetic-field effects, the polar-mode frequencies depend in general on the magnitude and orientation of the mode wavevector. The degeneracies of polar modes are often smaller than group theory predicts, and the angular anisotropies of their scattering cross sections differ from those of nonpolar vibrations of the same symmetries.

The polar modes in cubic crystals were first comprehensively treated by Huang (1951, see also Born and Huang 1954), and the nature of their light-scattering spectra was first understood as a result of the work of Poulet (1955). The light-scattering theory is most easily approached from the macroscopic treatment of polar crystals. We give here a somewhat general theory that applies to most symmetries of crystal and then look at the simpler special cases of cubic and uniaxial crystals. The microscopic theory presented subsequently is restricted to cubic crystals. Typical examples of nonpolar and polar modes are provided by Figures 3.13a and 3.13b, respectively, although the latter is not Raman active in this case. The theory of polar-mode scattering is considerably more complicated than that of scattering by nonpolar vibrations.

4.1 MACROSCOPIC THEORY

4.1.1 Lattice Dynamics of Polar Modes

The dielectric properties of a polar crystal are described by its linear susceptibility, which determines the polarization according to

$$P^i = \epsilon_O \chi^{ij}(\omega) E^j, \qquad (4.1)$$

similar to (1.36). The susceptibility tensor is symmetric, as in (3.22), and it can be diagonalized by a choice of suitable principal axes (Nye 1957), leaving in general three independent components. However, the different components of the susceptibility generally have different frequency dependences and the principal-axis directions vary with the frequency in crystals of sufficiently low symmetry. Thus two of the principal axes in monoclinic crystals and three of the axes in triclinic crystals point in varying directions not fixed by the crystal symmetry. Table 4.1 shows the

Macroscopic Theory

Table 4.1 Nonzero Susceptibility Components for Fixed Cartesian Axes

Optical Classification	System	Nonzero Tensor Components					
Biaxial	Triclinic	xx	yy	zz	yz	zx	xy
	Monoclinic	xx	yy	zz	zx		
	Orthorhombic	xx	yy	zz			
Uniaxial	Trigonal Tetragonal Hexagonal	xx	=	yy	zz		
Isotropic	Cubic	xx	=	yy	=	zz	

minimum numbers of nonzero susceptibility components for *fixed* Cartesian axes. The theory given below is restricted to crystals of orthorhombic or higher symmetry where the susceptibility is diagonal. The three components of the relative permittivity from (1.73) are

$$\kappa^i = 1 + \chi^{ii}(\omega), \tag{4.2}$$

where a single superscript is sufficient in the diagonal case. The relative permittivity plays a key role in determining the polar-mode frequencies.

The total polarization in the crystal can be expressed as a sum of contributions from the displacements of the charged ions and from the displacements of electrons relative to their ionic nuclei

$$P^i = P^i_{\text{ion}} + P^i_{\text{electron}}. \tag{4.3}$$

With the notation of Section 3.1.2 for displacements, the ionic contribution is

$$P^i_{\text{ion}} = \frac{N}{V} \sum_\alpha e_\alpha U^i_\alpha, \tag{4.4}$$

where N is the number of primitive cells in the crystal volume V and e_α is the charge on ions of label α, all ions of the same label suffering the same displacement.

It is convenient to express the ionic polarization in terms of the normal coordinates (3.7). The polarization is a vector, and the vibrational modes that contribute to the polarization are limited to those with the same symmetry character as a polar vector. These polar modes can be chosen so that their contributions to the polarization are parallel to principal axes of the susceptibility tensor. Let $\boldsymbol{\xi}_\sigma$ be a unit vector parallel to the polarization contributed by normal mode W_σ. Then one of the Cartesian components ξ_σ^i is unity and the other two are zero. The ionic polarization (4.4) can be written

$$P_{\text{ion}}^i = \frac{N}{V} \sum_\sigma Z_\sigma \xi_\sigma^i W_\sigma, \qquad (4.5)$$

where Z_σ, defined by

$$Z_\sigma \xi_\sigma^i = \sum_\alpha \frac{e_\alpha c_{\alpha\sigma}^{i*}}{M_\alpha^{1/2}}, \qquad (4.6)$$

is the *effective charge* of normal mode σ. It is easy to show with the help of (3.7) and (3.8) that (4.5) is equivalent to (4.4). The effective charge defined in this way is zero for all nonpolar modes.

The presence of an electric field **E** modifies the harmonic oscillator equation (1.84) of the normal mode to

$$\ddot{W}_\sigma + \Gamma_\sigma \dot{W}_\sigma + \omega_\sigma^2 W_\sigma = Z_\sigma \xi_\sigma^i E^i. \qquad (4.7)$$

An applied electric field of frequency ω therefore produces a steady-state normal-mode amplitude

$$W_\sigma = \frac{Z_\sigma \xi_\sigma^i E^i}{\omega_\sigma^2 - \omega^2 - i\omega\Gamma_\sigma}, \qquad (4.8)$$

and the ionic polarization (4.5) can be written in a form proportional to the electric-field components. We assume that ω is in the vicinity of the vibrational frequencies and well below the frequencies of all electronic transitions. The relative permittivity obtained from (4.1), (4.2), (4.3), (4.5), and (4.8) is then

$$\kappa^i = \kappa_\infty^i + \sum_\sigma \frac{N Z_\sigma^2 \xi_\sigma^{i2}/\epsilon_0 V}{\omega_\sigma^2 - \omega^2 - i\omega\Gamma_\sigma}, \qquad (4.9)$$

where κ_∞^i is the constant electronic contribution to the relative permittivity, so that

$$P_{\text{electron}}^i = \epsilon_0 (\kappa_\infty^i - 1) E^i \qquad \text{(no } i \text{ summation)}. \qquad (4.10)$$

The well-known Lorentz form (4.9) for the relative permittivity provides a good basic description of the optical properties of crystals for frequencies in the lattice vibration region. Measurements of lattice absorption spectra provide values of the frequencies, damping parameters, polarizations, and effective charges of the polar vibrations. However, some caution is necessary in interpreting the physical meaning of the mode frequencies that appear in the relative permittivity (4.9). It is seen from (4.7) that the ω_o are the frequencies of free vibration of the lattice in the absence of damping and of any electric field **E**.

Electric fields are, however, generally associated with polar vibrational modes in crystals since the above discussion shows that the modes produce a lattice polarization, and the polarization in turn generates a macroscopic electric field. With **E** henceforth taken to represent the macroscopic electric field of the polar modes, the equation of motion (4.7) continues to hold, and the term on its right-hand side changes the free vibrational frequency to be different from ω_o.

The electric field and polarization of any electromagnetic wave of frequency ω and wavevector **q** are required by Maxwell's equations to satisfy

$$-\epsilon_O c^2 \mathbf{q}(\mathbf{q} \cdot \mathbf{E}) + \epsilon_O (c^2 q^2 - \omega^2) \mathbf{E} = \omega^2 \mathbf{P}, \tag{4.11}$$

independent of any properties of the medium that supports the wave. Removal of the polarization with the help of (4.1) and (4.2) gives

$$-c^2 q^i (\mathbf{q} \cdot \mathbf{E}) + (c^2 q^2 - \kappa^i \omega^2) E^i = 0 \qquad \text{(no } i \text{ summation)}. \tag{4.12}$$

There are three simultaneous equations of this form for the electric-field components, and their elimination leads to Fresnel's equation

$$c^4 q^2 \left(\kappa^x q^{x^2} + \kappa^y q^{y^2} + \kappa^z q^{z^2} \right) - c^2 \left\{ \kappa^x (\kappa^y + \kappa^z) q^{x^2} \right.$$
$$\left. + \kappa^y (\kappa^z + \kappa^x) q^{y^2} + \kappa^z (\kappa^x + \kappa^y) q^{z^2} \right\} \omega^2 - \kappa^x \kappa^y \kappa^z \omega^4 = 0. \tag{4.13}$$

This is a formal solution to the problem of determining the frequency-wavevector relation for the polar modes of vibration.

Fresnel's equation is unfortunately very complicated in its most general form, but simplifications can be made for most applications. The most striking simplification occurs when the frequency and wavevector satisfy (1.29), so that the entire second line of Fresnel's equation can be neglected to give

$$\kappa^x q^{x^2} + \kappa^y q^{y^2} + \kappa^z q^{z^2} = 0 \qquad (cq \gg \omega). \tag{4.14}$$

This approximation holds except for scattering close to the forward direction and is therefore valid in the vast majority of experiments. The wavevectors of interest lie in the accessible range (1.27) for light scattering, thus ensuring that the basic vibrational frequencies ω_σ have negligible wavevector dependence.

The real mode frequencies for a given wavevector $\mathbf{q}$ are obtained by solution of (4.13) or (4.14) with the damping parameters Γ_σ set equal to zero to give real relative permittivity components (4.9). In the more common case of large wavevectors where (4.14) is valid, the frequencies generally depend on the orientation of $\mathbf{q}$ but not on its magnitude, and the remainder of the present section is devoted to this case. For near-forward scattering where the full form (4.13) must be used, the frequencies vary with both the orientation and the magnitude of the wavevector; the polar excitations in this region are called *polaritons* and their light scattering is discussed in Section 4.3.

Fresnel's equation is obtained by treating the polar excitations as electromagnetic waves with the mechanical nature of the vibrations, expressed in the normal coordinates W_σ, eliminated in the derivation of the relative permittivity. It is also possible to emphasize the mechanical aspects of the excitations by eliminating the electric field from the equation of motion (4.7). Maxwell's equation for the divergence of the electrical displacement gives

$$\mathbf{q} \cdot (\epsilon_O \mathbf{E} + \mathbf{P}) = 0, \tag{4.15}$$

which combines with (4.11) to give

$$\mathbf{E} = -\frac{c^2 \mathbf{q}(\mathbf{q} \cdot \mathbf{P}) - \omega^2 \mathbf{P}}{\epsilon_O (c^2 q^2 - \omega^2)}, \tag{4.16}$$

becoming

$$\mathbf{E} = -\frac{\mathbf{q}(\mathbf{q} \cdot \mathbf{P})}{\epsilon_O q^2} \qquad (cq \gg \omega) \tag{4.17}$$

in the large-wavevector limit. The electric field of the vibrations is therefore parallel to the wavevector, and it can be reexpressed with the help of (4.3) and (4.10) as

$$\mathbf{E} = -\frac{\mathbf{q}(\mathbf{q} \cdot \mathbf{P}_{\text{ion}})}{\epsilon_O \left(\kappa_\infty^x q^{x^2} + \kappa_\infty^y q^{y^2} + \kappa_\infty^z q^{z^2} \right)}. \tag{4.18}$$

Thus with the ionic polarization taken from (4.5) and the damping re-

Macroscopic Theory

moved, the equation of motion (4.7) becomes

$$\ddot{W}_\sigma + \omega_\sigma^2 W_\sigma = -\frac{NZ_\sigma(\mathbf{q}\cdot\boldsymbol{\xi}_\sigma)\sum_\tau Z_\tau(\mathbf{q}\cdot\boldsymbol{\xi}_\tau)W_\tau}{\epsilon_O V\left(\kappa_\infty^x q^{x^2} + \kappa_\infty^y q^{y^2} + \kappa_\infty^z q^{z^2}\right)}, \tag{4.19}$$

where τ is summed over all the polar modes.

These equations show that nonpolar modes ($Z_\sigma = 0$) and transverse modes ($\mathbf{q}\cdot\boldsymbol{\xi}_\sigma = 0$) have no associated macroscopic electric field; the right-hand side of (4.19) is *zero* and the ω_σ give the correct vibrational frequencies. The right-hand side of (4.19) is *nonzero* for polar modes ($Z_\sigma \neq 0$) that are not transverse ($\mathbf{q}\cdot\boldsymbol{\xi}_\sigma \neq 0$); the macroscopic electric field associated with such modes couples together the equations of motion for their normal coordinates. It is a simple matter to eliminate the normal coordinates from the coupled equations (4.19) and arrive once more, by a different route, at Fresnel's equation (4.14) for the excitation frequencies in terms of the wavevector. These frequencies are in general different from the ω_σ.

For a general wavevector direction, the term on the right of (4.19) couples modes whose polarization vectors point along all three of the principal axes. Such modes span more than one irreducible representation in uniaxial or biaxial crystals, and the macroscopic-field coupling produces excitations of mixed symmetry character. The group-theoretical polar-mode degeneracies are in general lifted by the electric-field coupling. These properties are illustrated for the different crystal symmetries in later sections.

The theory outlined in the present section parallels work of Le Corre (1963). Shapiro and Axe (1972) give a more general theory that embraces monoclinic and triclinic crystals.

4.1.2 The Scattering Cross Section

The electric field that accompanies a polar vibration provides an additional source of modulation of the linear susceptibility of the crystal. The total Stokes polarization accordingly becomes

$$P_S^i = \sum_\sigma a_\sigma^{ij} E_I^j W_\sigma^* + b^{ijh} E_I^j E^{h*}, \tag{4.20}$$

where the susceptibility derivatives a_σ^{ij} are defined in (3.19), and we have introduced a new kind of derivative

$$b^{ijh} = \epsilon_O \frac{\partial \chi^{ij}(\omega_I)}{\partial E^{h*}} \tag{4.21}$$

to describe the modulation of the susceptibility by the macroscopic field. The mixing of modes by the macroscopic field makes it convenient to treat the scattering by all polar modes in a single calculation. Poulet (1955) first realized that the Stokes polarization for polar modes includes the electric-field contribution, and the extension of the susceptibility-derivative method to cover the scattering by polar modes was made by Burstein (1967) for the case of cubic crystals.

A straightforward extension of (1.70) for a Stokes polarization with more than one term [as in (4.20)] produces the polar-mode spectral differential cross section

$$\frac{d^2\sigma}{d\Omega d\omega_S} = \frac{\omega_I \omega_S^3 \text{v} V \eta_S \left\langle \left| \varepsilon_S^i \varepsilon_I^j \left(\sum_\sigma a_\sigma^{ij} W_\sigma^* + b^{ijh} E^{h*} \right) \right|^2 \right\rangle_\omega}{(4\pi\epsilon_0)^2 c^4 \eta_I}. \quad (4.22)$$

The cross section depends on the power spectra of the fluctuations in the electric-field components and on the cross products of field components and normal coordinates. These are additional to the normal-coordinate power spectra which alone provide the spectral distribution of the scattering by nonpolar vibrations. The required power spectra are again obtained with the help of the fluctuation-dissipation theorem. Consider first the electric-field power spectra and let $\mathbf{P}_{\text{ext}}(t)$ be a fictitious polarization of frequency ω and wavevector $\mathbf{q}$ imposed on the crystal by some external influence. The energy of interaction of the polarization with the electric field in the crystal is

$$H = -\int \mathbf{E} \cdot \mathbf{P}_{\text{ext}}(t) dV = -VE^i P_{\text{ext}}^i(t). \quad (4.23)$$

This has the form (1.74) with a generalized force component

$$F^i(t) = VP_{\text{ext}}^i(t). \quad (4.24)$$

The response of the crystal to the external polarization takes the form of an electric field $\overline{\mathbf{E}}$ and some additional polarization $\overline{\mathbf{P}}$ related by

$$\overline{P}^i = \epsilon_0(\kappa^i - 1)\overline{E}^i \quad (4.25)$$

in accordance with (4.1) and (4.2). The total polarization satisfies (4.17),

$$\overline{\mathbf{E}} = -\frac{\mathbf{q}\left\{ \mathbf{q} \cdot (\overline{\mathbf{P}} + \mathbf{P}_{\text{ext}}) \right\}}{\epsilon_0 q^2}. \quad (4.26)$$

Macroscopic Theory

It is not difficult to remove $\bar{\mathbf{P}}$ from these equations, the algebraic steps being identical to those used in the removal of the electronic polarization to obtain (4.18). The result in the present case can be written

$$\bar{E}^i = T^{ij}(\mathbf{q},\omega) F^j, \qquad (4.27)$$

similar to (1.76), where the field and force have frequency ω and wavevector $\mathbf{q}$, and

$$T^{ij}(\mathbf{q},\omega) = -\frac{q^i q^j}{\epsilon_0 V \left(\kappa^x q^{x^2} + \kappa^y q^{y^2} + \kappa^z q^{z^2} \right)}. \qquad (4.28)$$

This is a linear response function that generalizes the definition (1.76) to a situation where several variables (the three components of the electric field) are coupled so that they all respond to the application of a generalized force to one of the variables. The corresponding generalization of the fluctuation-dissipation theorem (1.82) is

$$\langle \hat{E}^i \hat{E}^{j\dagger} \rangle_\omega = \langle \hat{E}^j \hat{E}^{i\dagger} \rangle_\omega = \frac{\hbar}{\pi} \{ n(\omega) + 1 \} \operatorname{Im} T^{ij}(\mathbf{q},\omega), \qquad (4.29)$$

and explicit expressions for the power spectra of the electric-field components are readily obtained by substitution of the response function (4.28). Note that the denominator of the response function is the same as the left-hand side of Fresnel's equation (4.14).

The other kinds of power spectrum required for the cross section (4.22) are derived by similar methods (for details, see Barker and Loudon 1972), and they all involve the linear response function (4.28). The results for the cross products of field components and normal coordinates are

$$\langle \hat{W}_\sigma \hat{E}^{j\dagger} \rangle_\omega = \langle \hat{E}^j \hat{W}_\sigma^\dagger \rangle_\omega = \frac{\hbar}{\pi} \{ n(\omega) + 1 \} \operatorname{Im} \{ \beta_\sigma \xi_\sigma^i T^{ij}(\mathbf{q},\omega) \} \qquad (4.30)$$

where the β factor is a convenient shorthand,

$$\beta_\sigma = \frac{Z_\sigma}{\omega_\sigma^2 - \omega^2 - i\omega\Gamma_\sigma}. \qquad (4.31)$$

Finally, the normal coordinate power spectra are

$$\langle \hat{W}_\tau \hat{W}_\sigma^\dagger \rangle_\omega = \langle \hat{W}_\sigma \hat{W}_\tau^\dagger \rangle_\omega$$

$$= \frac{\hbar}{\pi} \{ n(\omega) + 1 \} \operatorname{Im} \left\{ \beta_\tau \beta_\sigma \xi_\tau^i T^{ij}(\mathbf{q},\omega) \xi_\sigma^j + \frac{\beta_\sigma}{Z_\sigma N} \delta_{\sigma\tau} \right\}. \qquad (4.32)$$

This result is a generalization to polar modes of the power spectrum (1.92) used to obtain the scattering cross section of the nonpolar modes considered in Chapter 3. The nonpolar power spectrum is retrieved from (4.32) when the effective charges are all set equal to zero and only the final term in the large bracket remains.

The results (4.29), (4.30), and (4.32) for the power spectra enable the cross section (4.22) to be written

$$\frac{d^2\sigma}{d\Omega \, d\omega_S} = \frac{\hbar\omega_I \omega_S^3 \mathrm{v} V \eta_S \epsilon_S^i \epsilon_I^j \epsilon_S^{i'} \epsilon_I^{j'} \{n(\omega)+1\}}{(4\pi\epsilon_O)^2 \pi c^4 \eta_I}$$

$$\times \mathrm{Im}\left\{ -\frac{\left(\sum_\sigma a_\sigma^{ij}\beta_\sigma \xi_\sigma^h + b^{ijh}\right)\left(\sum_\tau a_\tau^{i'j'*}\beta_\tau \xi_\tau^{h'} + b^{i'j'h'*}\right) q^h q^{h'}}{\epsilon_O V (\kappa^x q^{x^2} + \kappa^y q^{y^2} + \kappa^z q^{z^2})} + \sum_\sigma \frac{a_\sigma^{ij} a_\sigma^{i'j'*} \beta_\sigma}{Z_\sigma N} \right\},$$

(4.33)

where (4.28) has been used. The second term in the large bracket gives rise to peaks in the cross section at the normal-mode frequencies ω_σ, and with the help of (3.19) and (4.31) it reproduces the nonpolar cross section (1.93) summed over the contributing modes. The first term in the bracket is the additional contribution of the polar modes; it also produces peaks at the normal-mode frequencies ω_σ but in addition it gives peaks at frequencies where its denominator is small. In the limit of zero damping these are just the polar-mode frequencies obtained by solution of the large-wavevector Fresnel equation (4.14). The first term in the bracket cannot in general be written as a sum of contributions from the individual normal modes, reflecting the polar-mode coupling described in Section 4.1.1.

The general scattering cross section (4.33) is applied to the special case of scattering by cubic crystals in Section 4.1.4, where considerable simplifications occur. Shapiro and Axe (1972) have also derived an expression for the polar-mode differential cross section.

4.1.3 Properties of the Susceptibility Derivatives

The polar-mode scattering cross section has a magnitude and symmetry determined by the susceptibility derivatives defined in (3.19) and (4.21). Consider first the symmetry of the scattering. Polar modes can in principle exist for crystals of any point-group symmetry, but, in crystals that are invariant under spatial inversion, they have odd parity and do not contribute to Raman scattering. We are therefore concerned only with crystals

Macroscopic Theory

that lack a center of inversion symmetry, that is *noncentrosymmetric* or *piezoelectric* crystals. The 21 noncentrosymmetric crystal point groups are included in Table 1.2, but some additional information on the susceptibility derivatives is needed for the polar cross section (4.33), namely the Cartesian components h of the mode polarization ξ_σ and macroscopic electric field $\mathbf{E}$ for particular components i and j of the incident and scattered light polarizations. This additional information is listed in Table 4.2, where the conventions are the same as in Table 1.2. The method of derivation of the symmetry properties is also the same as that outlined in Section 1.6, and the vibrational symmetry condition (2.37) is included.

It is shown in Section 4.2.2 that the susceptibility derivatives with respect to the electric field satisfy a further approximate symmetry condition

$$b^{ijh} = b^{hij} = b^{jhi} = b^{hji} = b^{ihj} = b^{jih}, \qquad (4.34)$$

when the incident frequency is much smaller than all the electronic-excitation frequencies of the crystal. This symmetry was first discussed by Kleinman (1962), and it leads to greater restrictions on the susceptibility derivatives, whose resulting forms are tabulated by Shapiro and Axe (1972). There is no analogous additional symmetry for the a_σ^{ij}.

The magnitudes of the susceptibility derivatives can be determined by measurements of the cross section, but there exist several independent methods of determination. The remainder of the present section is devoted to these alternative methods, which provide useful predictions or verifications of cross-section measurements.

Consider first the case of a *nonpolar* crystal and suppose that a *static* field $\mathbf{E}_I$ is applied to it. According to the first term on the right of (4.20), the field produces a second-order polarization

$$P^i = \sum_\sigma a_\sigma^{ij} E_I^j W_\sigma^*, \qquad (4.35)$$

where the susceptibility derivative is evaluated at zero frequency. The polarization oscillates at the mode frequencies ω_σ and the applied field thus induces an ionic polarization similar to that given by (4.5), which ordinarily exists for the polar modes in a polar crystal. Comparison of (4.5) with (4.35) shows that the resulting field-dependent relative permittivity analogous to (4.9) is

$$\kappa^i = \kappa_\infty^i + \sum_\sigma \frac{V\left(a_\sigma^{ij} E_I^j\right)^2 / \epsilon_O N}{\omega_\sigma^2 - \omega^2 - i\omega\Gamma_\sigma}. \qquad (4.36)$$

Table 4.2 Symmetry Properties of the Polar-Mode Susceptibility Derivatives $a_\sigma^{ij}\xi_\sigma^h$ and b^{ijh}*

			x	y	z
Biaxial					
Triclinic			$\begin{bmatrix} a & d & e \\ d & b & f \\ e & f & c \end{bmatrix}$	$\begin{bmatrix} g & j & k \\ j & h & l \\ k & l & i \end{bmatrix}$	$\begin{bmatrix} m & p & q \\ p & n & r \\ q & r & o \end{bmatrix}$
	1	C_1	$A\quad\Gamma_1$	$A\quad\Gamma_1$	$A\quad\Gamma_1$
Monoclinic			$\begin{bmatrix} & & a \\ & & b \\ a & b & \end{bmatrix}$	$\begin{bmatrix} & & c \\ & & d \\ c & d & \end{bmatrix}$	$\begin{bmatrix} e & h & \\ h & f & \\ & & g \end{bmatrix}$
	2	C_2	$B\quad\Gamma_2$	$B\quad\Gamma_2$	$A\quad\Gamma_1$
			$\begin{bmatrix} a & d & \\ d & b & \\ & & c \end{bmatrix}$	$\begin{bmatrix} e & h & \\ h & f & \\ & & g \end{bmatrix}$	$\begin{bmatrix} & & i \\ & & j \\ i & j & \end{bmatrix}$
	m	C_s	$A'\quad\Gamma_1$	$A'\quad\Gamma_1$	$A''\quad\Gamma_2$
Orthorhombic			$\begin{bmatrix} & & \\ & & a \\ & a & \end{bmatrix}$	$\begin{bmatrix} & & b \\ & & \\ b & & \end{bmatrix}$	$\begin{bmatrix} & c & \\ c & & \\ & & \end{bmatrix}$
	222	D_2	$B_3\quad\Gamma_4$	$B_2\quad\Gamma_2$	$B_1\quad\Gamma_3$
			$\begin{bmatrix} & & a \\ & & \\ a & & \end{bmatrix}$	$\begin{bmatrix} & & \\ & & b \\ & b & \end{bmatrix}$	$\begin{bmatrix} c & & \\ & d & \\ & & e \end{bmatrix}$
	$mm2$	C_{2v}	$B_1\quad\Gamma_2$	$B_2\quad\Gamma_4$	$A_1\quad\Gamma_1$
Uniaxial					
Tetragonal			$\begin{bmatrix} & & a \\ & & b \\ a & b & \end{bmatrix}$	$\begin{bmatrix} & & -b \\ & & a \\ -b & a & \end{bmatrix}$	$\begin{bmatrix} c & & \\ & c & \\ & & d \end{bmatrix}$
	4	C_4	E	$\Gamma_3+\Gamma_4$	$A\quad\Gamma_1$
			$\begin{bmatrix} & & a \\ & & b \\ a & b & \end{bmatrix}$	$\begin{bmatrix} & & b \\ & & -a \\ b & -a & \end{bmatrix}$	$\begin{bmatrix} c & d & \\ d & -c & \\ & & \end{bmatrix}$
	$\bar{4}$	S_4	E	$\Gamma_3+\Gamma_4$	$B\quad\Gamma_2$
			$\begin{bmatrix} & & \\ & & a \\ & a & \end{bmatrix}$	$\begin{bmatrix} & & -a \\ & & \\ -a & & \end{bmatrix}$	$\begin{bmatrix} & & \\ & & \\ & & \end{bmatrix}$
	422	D_4	E	Γ_5	
			$\begin{bmatrix} & & a \\ & & \\ a & & \end{bmatrix}$	$\begin{bmatrix} & & \\ & & a \\ & a & \end{bmatrix}$	$\begin{bmatrix} b & & \\ & b & \\ & & c \end{bmatrix}$
	$4mm$	C_{4v}	E	Γ_5	$A_1\quad\Gamma_1$
			$\begin{bmatrix} & & \\ & & a \\ & a & \end{bmatrix}$	$\begin{bmatrix} & & a \\ & & \\ a & & \end{bmatrix}$	$\begin{bmatrix} & b & \\ b & & \\ & & \end{bmatrix}$
	$\bar{4}2m$	D_{2d}	E	Γ_5	$B_2\quad\Gamma_4$

System				$h=x$	$h=y$	$h=z$
Trigonal	3	C_3		$\begin{bmatrix} a & b & c \\ b & -a & d \\ c & d & \end{bmatrix}$ E	$\begin{bmatrix} b & -a & -d \\ -a & -b & c \\ -d & c & \end{bmatrix}$ $\Gamma_2+\Gamma_3$	$\begin{bmatrix} e & & \\ & e & \\ & & f \end{bmatrix}$ $A\ \Gamma_1$
	32	D_3		$\begin{bmatrix} & a & \\ a & & b \\ & b & \end{bmatrix}$ E	$\begin{bmatrix} a & & -b \\ & -a & \\ -b & & \end{bmatrix}$ Γ_3	
	$3m$	C_{3v}		$\begin{bmatrix} a & & b \\ & -a & \\ b & & \end{bmatrix}$ E	$\begin{bmatrix} & -a & \\ -a & & b \\ & b & \end{bmatrix}$ Γ_3	$\begin{bmatrix} c & & \\ & c & \\ & & d \end{bmatrix}$ $A_1\ \Gamma_1$
Hexagonal	6	C_6		$\begin{bmatrix} & & a \\ & & b \\ a & b & \end{bmatrix}$ E_1	$\begin{bmatrix} & & -b \\ & & a \\ -b & a & \end{bmatrix}$ $\Gamma_5+\Gamma_6$	$\begin{bmatrix} c & & \\ & c & \\ & & d \end{bmatrix}$ $A\ \Gamma_1$
	$\bar{6}$	C_{3h}		$\begin{bmatrix} a & b & \\ b & -a & \\ & & \end{bmatrix}$ E'	$\begin{bmatrix} b & -a & \\ -a & -b & \\ & & \end{bmatrix}$ $\Gamma_2+\Gamma_3$	
	622	D_6		$\begin{bmatrix} & & \\ & & a \\ & a & \end{bmatrix}$ E_1	$\begin{bmatrix} & & -a \\ & & \\ -a & & \end{bmatrix}$ Γ_5	
	$6mm$	C_{6v}		$\begin{bmatrix} & & a \\ & & \\ a & & \end{bmatrix}$ E_1	$\begin{bmatrix} & & \\ & & a \\ & a & \end{bmatrix}$ Γ_5	$\begin{bmatrix} b & & \\ & b & \\ & & c \end{bmatrix}$ $A_1\ \Gamma_1$
	$\bar{6}m2$	D_{3h}		$\begin{bmatrix} & a & \\ a & & \\ & & \end{bmatrix}$ E'	$\begin{bmatrix} a & & \\ & -a & \\ & & \end{bmatrix}$ Γ_6	
Isotropic Cubic	23	T		$\begin{bmatrix} & & \\ & & a \\ & a & \end{bmatrix}$ T	$\begin{bmatrix} & & a \\ & & \\ a & & \end{bmatrix}$ T	$\begin{bmatrix} & a & \\ a & & \\ & & \end{bmatrix}$ Γ_4
	$\bar{4}3m$	T_d			T_2	Γ_5
	432	O	All components vanish			

*The rows and columns of the matrices define the i and j indices, and the different matrices correspond to $h=x, y$, and z.

The most striking effect of the applied field is the production of infrared absorption lines centered on the frequencies ω_σ with strengths proportional to the square of the field. Large fields are needed for observation of the effect, but Anastassakis et al. (1966) and Angress et al. (1968) were able to measure the susceptibility derivative of diamond by this technique, and Figures 4.1 and 4.2 show some of the experimental results. The measured value is close to that in (3.42) obtained from the observed cross section, but predates it by a few years. The two kinds of measurement differ in that one determines the susceptibility derivative at zero frequency and the other at a visible-light frequency. The frequency dependence of the susceptibility derivative is discussed in Section 4.2.2.

The above method cannot be used for polar modes since the change in the existing infrared absorption produced by an applied field is normally undetectable. There are however other nonlinear effects that depend upon the susceptibility derivatives. Suppose that an electric field **E** of frequency ω is applied to the polar crystal. The field excites the various polar modes in accordance with (4.7)

$$W_\sigma = \beta_\sigma \xi_\sigma^h E^h, \tag{4.37}$$

where the β factor is defined in (4.31). If a field $\mathbf{E}_I$ of frequency ω_I is also present, the polarization (4.20) produced by coupling of the two fields is

$$P^i = \left\{ \sum_\sigma a_\sigma^{ij} \beta_\sigma^* \xi_\sigma^h + b^{ijh} \right\} E_I^j E^{h*}. \tag{4.38}$$

There are two limiting cases to consider.

Suppose first that ω is much smaller than the mode frequencies ω_σ. The polarization can then be viewed as the result of a field-induced change in

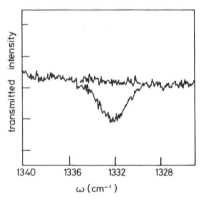

Figure 4.1 Transmitted infrared intensity in diamond in the vicinity of the optic frequency in the presence (lower trace) and absence (upper trace) of a static electric field parallel to [110] (after Angress et al. 1968).

Macroscopic Theory

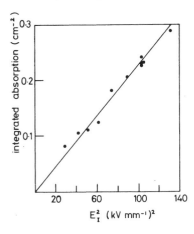

Figure 4.2 The integrated field-induced absorption in diamond as a function of the square of the static applied field (after Angress et al. 1968).

the relative permittivity at frequency ω_I and written

$$P^i = \epsilon_0 \Delta\kappa^{ij} E_I^j. \tag{4.39}$$

This phenomenon is the electrooptic effect and its magnitude is conventionally expressed in terms of the electrooptic coefficients r^{ijh} of the crystal (Nye 1957) according to

$$\Delta\kappa^{ij} = -\kappa^i \kappa^j r^{ijh} E^{h*} \quad \text{(no } i \text{ or } j \text{ summation)}. \tag{4.40}$$

Comparison of (4.38) and (4.39) with (4.40) leads to an expression for the electrooptic coefficients in terms of susceptibility derivatives

$$\epsilon_0 \kappa^i \kappa^j r^{ijh} = -\sum_\sigma a_\sigma^{ij} \frac{Z_\sigma}{\omega_\sigma^2} \xi_\sigma^h - b^{ijh} \quad \text{(no } i \text{ or } j \text{ summation)}, \tag{4.41}$$

where the low-frequency limit of the β-factor is taken from (4.31) and the relative permittivities on the left are evaluated at frequency ω_I. Measured values of electrooptic coefficients are available for a wide range of crystals, and they provide some information on the magnitudes of the susceptibility derivatives but not enough to calculate the cross section (4.33).

Suppose now, however, that the frequency ω is much larger than the mode frequencies ω_σ. The β factors (4.31) are very small in this limit and (4.38) reduces approximately to

$$P^i = b^{ijh} E_I^j E^{h*}. \tag{4.42}$$

This equation describes the nonlinear optical process of difference-frequency generation (for example, see Bloembergen 1965), where two beams of light interact to produce a polarization at the difference of their frequencies. Essentially the same coefficients b^{ijh} control sum-frequency and second-harmonic generation, and these processes can be used for independent measurements of the coefficients. Some caution is necessary since the susceptibility derivatives depend on frequency and do not in general have the same magnitudes for measurements of the light-scattering cross section, electrooptic effect, and second-harmonic generation. It is, however, shown in Section 4.2.2 that the frequency variation should be small for experiments where all of the optical frequencies are well below the electronic transitions of the crystal.

The relations between the coefficients measured in these three kinds of experiment have been studied for various crystals. Kaminow and Johnston (1967) use (4.41) to calculate the electrooptic coefficients of $LiNbO_3$ and $LiTaO_3$, with values of a_σ^{ij} from cross-section measurements, Z_σ and ω_σ from infrared absorption, and b^{ijh} from second-harmonic generation. They compare the computed electrooptic coefficients with direct measurements (but for conflicting measurements on $LiTaO_3$, see Penna et al. 1976). Johnston and Kaminow (1969) obtain the electrooptic and second-harmonic coefficients of GaAs from measured light-scattering cross sections. Johnston (1970) shows that the accuracy of coefficients obtained in this way can be comparable with that attained in direct measurements. Bairamov et al. (1974) have made a similar study of GaP.

4.1.4. Polar-Mode Scattering in Cubic Crystals

The polar-vector representation is threefold degenerate for the cubic symmetry groups. The crystals are optically isotropic and the principal axes are not restricted to lie in any particular directions. The components of the relative permittivity are therefore all equal, as indicated in Table 4.1. For any direction of the wavevector of a threefold polar mode, it is permissible to choose two of the polarization vectors ξ_σ perpendicular to $\mathbf{q}$ and the third ξ_σ parallel to $\mathbf{q}$. In accordance with the discussion of Section 4.1.1, the two transverse polar modes have a frequency ω_σ determined by the standard lattice-dynamics calculation. The longitudinal polar mode of any trio has the associated macroscopic electric field, and its frequency is determined by Fresnel's equation (4.14), which reduces to

$$\kappa = 0 \qquad (4.43)$$

in the cubic case.

The simplest cases to consider first are the cubic crystals that have a single threefold polar mode, for example the zinc blende, fluorite, and rocksalt structures (see Table 3.1). The relative permittivity (4.9) has the isotropic form

$$\kappa = \kappa_\infty + \frac{NZ^2/\epsilon_o V}{\omega_T^2 - \omega^2 - i\omega\Gamma}, \quad (4.44)$$

where the mode frequency ω_σ is replaced by ω_T to emphasize its transverse nature and redundant subscripts and superscripts are omitted. The longitudinal frequency obtained from (4.43) with the damping removed from (4.44) is

$$\omega_L = \left(\frac{\kappa_0}{\kappa_\infty}\right)^{1/2} \omega_T, \quad (4.45)$$

where κ_0 is the zero-frequency value of the relative permittivity

$$\kappa_0 = \kappa_\infty + \frac{NZ^2}{\epsilon_o V \omega_T^2}. \quad (4.46)$$

This expression for the longitudinal frequency is the Lyddane-Sachs-Teller relation. The equation of motion (4.19) for the longitudinal mode takes the form

$$\ddot{W} + \omega_T^2 W = -\frac{NZ^2 W}{\epsilon_o V \kappa_\infty}, \quad (4.47)$$

leading to

$$\omega_L^2 = \omega_T^2 + \frac{NZ^2}{\epsilon_o V \kappa_\infty} \quad (4.48)$$

in agreement with (4.45) and (4.46). The relative permittivity (4.44) can thus be rewritten in the compact form

$$\kappa = \kappa_\infty \left\{ 1 + \frac{\omega_L^2 - \omega_T^2}{\omega_T^2 - \omega^2 - i\omega\Gamma} \right\}. \quad (4.49)$$

The electric field associated with the longitudinal mode is given by (4.5) and (4.18) to be

$$\mathbf{E} = -\frac{NZW\mathbf{q}}{\epsilon_O V \kappa_\infty q};\qquad(4.50)$$

the transverse modes have no associated macroscopic fields.

It is seen that the effect of the macroscopic field in cubic crystals is to lift the group-theoretical degeneracy of polar modes, producing a nondegenerate longitudinal vibration, which lies at a higher frequency than the doubly degenerate transverse vibrations. The modes thus produce two distinct peaks in the Raman spectrum. There has been a great deal of experimental work on crystals of the zinc blende structure, space group $F\bar{4}3m$ or T_d^2, and point group $\bar{4}3m$ or T_d. The single polar optic mode of Γ_5 or T_2 symmetry (see Table 3.1) is Raman-active. The first reported measurements (Hobden and Russell 1964) of a crystal Raman spectrum using incident light from a laser source was on zinc-blende structure GaP. Figure 4.3 shows the spectra of three such crystals measured by Mooradian and Wright (1966) with the longitudinal and transverse Stokes and anti-Stokes peaks very clearly recorded. Other crystals studied include ZnS (Nilsen 1969) and ZnTe (Irwin and LaCombe 1970). The measured values of transverse and longitudinal peak frequencies are shown in Table 4.3.

Similar remarks apply to cubic crystals that have more than one polar-mode triplet, but the algebraic details become more complicated. Each group-theoretical threefold polar vibration splits into a transverse doublet and a longitudinal singlet. The Lyddane-Sachs-Teller relation (4.45) gener-

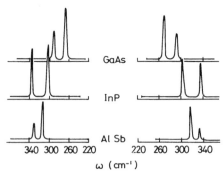

Figure 4.3 Raman spectra of several zinc-blende structure crystals with Stokes peaks on the right and anti-Stokes peaks on the left (after Mooradian and Wright 1966).

Macroscopic Theory

Table 4.3 Transverse and Longitudinal Frequencies in cm^{-1} Determined for Zinc-Blende Structure Crystals by Raman Spectroscopy

	GaP	GaAs	InP	AlSb	ZnS	ZnTe
Longitudinal	403	292	345	340	352	208
Transverse	367	269	304	319	271	177

alizes to (Cochran and Cowley 1962, Barker 1964)

$$\prod_L \omega_L^2 = \frac{\kappa_0}{\kappa_\infty} \prod_T \omega_T^2, \qquad (4.51)$$

all polar modes being included in the products. The relative permittivity in the zero-damping limit can be written

$$\kappa = \kappa_\infty \frac{\prod_L (\omega_L^2 - \omega^2)}{\prod_T (\omega_T^2 - \omega^2)}, \qquad (4.52)$$

a form that shows the association of longitudinal and transverse frequencies with zeros and poles in the undamped relative permittivity.

The polar-mode scattering cross section (4.33) takes on its simplest form for cubic crystals with a single threefold polar vibration, such as zinc blende, and we consider this case in some detail. Poulet (1955) gave the first theoretical treatment of polar-mode scattering by zinc-blende symmetry crystals.

Consider first the peak at the transverse frequency, which results from the β factors in (4.33). With only one polar-mode triplet, and unnecessary subscripts omitted, the β factor (4.31) is related to the relative permittivity (4.44) by

$$\beta = \frac{(\kappa - \kappa_\infty)\epsilon_O V}{NZ}. \qquad (4.53)$$

Both terms in the bracket of (4.33) have contributions that become large for frequencies close to ω_T, and they combine to give

$$\frac{d^2\sigma}{d\Omega d\omega_S} = \frac{\hbar\omega_I \omega_S^3 \mathfrak{v} V \eta_S \{n(\omega)+1\}}{(4\pi\epsilon_O)^2 \pi c^4 \eta_I} \frac{\epsilon_O V}{N^2 Z^2}$$

$$\times \left\{ -\left| \frac{\epsilon_S^i \epsilon_I^j q^h \sum_\sigma a_\sigma^{ij} \xi_\sigma^h}{q} \right|^2 + \sum_\sigma |\epsilon_S^i \epsilon_I^j a_\sigma^{ij}|^2 \right\} \mathrm{Im}\,\kappa. \qquad (4.54)$$

From (4.44), assuming Γ to be much smaller than ω_T,

$$\mathrm{Im}\,\kappa = \frac{\pi N Z^2}{2\epsilon_0 V \omega_T} g_T(\omega), \tag{4.55}$$

where $g_T(\omega)$ is the Lorentzian lineshape (1.90), centered on the transverse vibrational frequency.

Now consider the peak at the longitudinal frequency produced by the denominator of the first term in the bracket of (4.33). With Γ again assumed small, the value of the β factor at the longitudinal frequency is given by (4.31) and (4.48) to be

$$\beta = -\frac{\epsilon_0 V \kappa_\infty}{NZ} \quad \text{at} \quad \omega = \omega_L. \tag{4.56}$$

The longitudinal cross section is therefore

$$\frac{d^2\sigma}{d\Omega\,d\omega_S} = -\frac{\hbar\omega_I \omega_S^3 v V \eta_S \{n(\omega)+1\}}{(4\pi\epsilon_0)^2 \pi c^4 \eta_I} \frac{\epsilon_0 V}{N^2 Z^2}$$

$$\times \left| \epsilon_S^i \epsilon_I^j q^h \frac{\sum_\sigma a_\sigma^{ij} \xi_\sigma^h - (NZ/\epsilon_0 V \kappa_\infty) b^{ijh}}{q} \right|^2 \kappa_\infty^2 \,\mathrm{Im}\,\frac{1}{\kappa}, \tag{4.57}$$

where

$$-\kappa_\infty^2 \,\mathrm{Im}\,\frac{1}{\kappa} = \left(\frac{\pi N Z^2}{2\epsilon_0 V \omega_L}\right) g_L(\omega), \tag{4.58}$$

with a Lorentzian lineshape now centered on the longitudinal vibrational frequency.

These results provide explicit expressions for the transverse and longitudinal spectral differential cross sections. Their proportionality to the imaginary parts of the relative permittivity and its inverse, respectively, is a characteristic feature that occurs in other cross sections (for example, see Section 7.3), and was first considered by Barker (see DiDomenico et al. 1968). The differential cross sections corresponding to (4.54) and (4.57), derived originally by Burstein (1967), are obtained by simple removal of the normalized Lorentzian lineshape functions.

The relative magnitudes of the transverse and longitudinal cross sections depend on the scattering geometry and on the relative sizes and signs of the two kinds of susceptibility derivative. These have opposite signs for many zinc-blende structure crystals, for example

$$\frac{\omega_T^2 b^{xyz}}{Z} = -0.19 a^{xy} \quad \text{for GaP (Faust and Henry 1966)}$$

$$= -0.17 a^{xy} \quad \text{for GaAs (Johnston and Kaminow 1969)} \quad (4.59)$$

but they can also have the same sign in other crystal structures (Scott et al. 1971).

As an example of the angular dependences of the cross sections, consider the scattering geometry shown in Figure 1.6, with incident light polarized perpendicular to the plane of the figure (taken as the y axis)

$$\varepsilon_I = (0, 1, 0). \quad (4.60)$$

If the z axis is taken parallel to $\mathbf{k}_I$ and the experiment records only the component of scattered light polarized in the zx plane, then

$$\varepsilon_S = (\cos\phi, 0, \sin\phi). \quad (4.61)$$

The vibrational wavevector is

$$\mathbf{q} = (q\cos\tfrac{1}{2}\phi, 0, q\sin\tfrac{1}{2}\phi). \quad (4.62)$$

Thus for a crystal of $\bar{4}3m$ symmetry with the susceptibility derivatives given in Table 4.2, the cross sections are proportional to

$$\text{transverse } \Gamma_5 \text{ or } T_2: \ |a|^2 \cos^2\frac{3\phi}{2} \quad (4.63)$$

$$\text{longitudinal } \Gamma_5 \text{ or } T_2: \ |a'|^2 \sin^2\frac{3\phi}{2}, \quad (4.64)$$

where $|a'|$ differs from $|a|$ owing to the inclusion of the b^{ijh} contribution in the longitudinal cross section.

These angular dependences are plotted in the first two parts of Figure 4.4 together with experimental points for ZnS obtained by Dawson (1972). The markedly anisotropic cross sections for the zinc-blende structure contrast with the analogous result for scattering by a nonpolar threefold mode in a cubic crystal, for example the Γ_5^+ or T_{2g} mode in diamond, where (3.41) gives

$$\Gamma_5^+ \text{ or } T_{2g}: \ |d|^2 \quad (4.65)$$

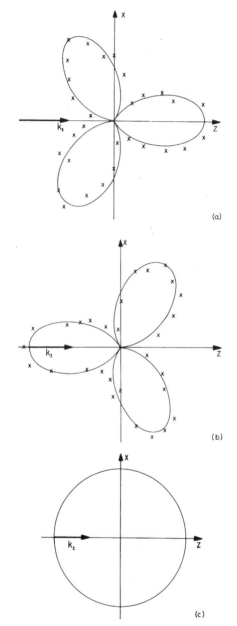

Figure 4.4 Scattering cross sections for threefold modes in cubic crystals for the geometry described in the text: (*a*) transverse mode in zinc blende, (*b*) longitudinal mode in zinc blende, and (*c*) optic mode in diamond (after Dawson 1972).

for the polarization vectors assumed here. This isotropic cross section is shown in the third part of Figure 4.4. The same isotropic result is obtained from the sum of the cross sections (4.63) and (4.64) in the absence of polar effects where $|a'| = |a|$ and the transverse and longitudinal frequencies become degenerate.

The scattering by a cubic crystal that has several threefold polar modes is treated by a suitable extension of the above analysis. The outline of the calculation is the same but the evaluation of the cross sections at individual peaks is more tedious. The general result (4.33) now simplifies only to the extent that the relative permittivity is isotropic and it can be written as a sum of β factors. Peaks associated with the β factors occur at all the transverse frequencies, and peaks from the Fresnel denominator occur at all the longitudinal frequencies. The peaks have Lorentzian shapes, but the spectra as a whole cannot be expressed in forms proportional to the imaginary parts of the relative permittivity and its inverse, since different peaks have intensities determined in part by different susceptibility derivatives. The symmetry of the scattering is not affected by the presence of several polar modes, and for example the angular anisotropies in (4.63) and (4.64) apply to all transverse and longitudinal peaks in crystals of $\bar{4}3m$ symmetry.

4.1.5 Polar-Mode Scattering in Uniaxial Crystals

The properties of polar vibrations in uniaxial crystals were treated first by Merten (1960) and later by Loudon (1964). One principal axis is the crystal c axis and the other two have arbitrary orientations perpendicular to the c axis. Taking these as the z, x, and y axes, respectively, the two independent components of the relative permittivity are as shown in Table 4.1, and it is convenient to use superscripts $\|$ for extraordinary quantities defined parallel to the c axis and $\perp$ for ordinary quantities defined perpendicular to the c axis. With this convention, Fresnel's equation (4.14) becomes

$$\kappa^\perp \sin^2\theta + \kappa^\| \cos^2\theta = 0 \tag{4.66}$$

for an excitation whose wavevector q is inclined at an angle θ to the c axis. The polar-mode frequencies in uniaxial crystals thus depend in general on their directions of propagation.

The simplest case is that of a uniaxial crystal with a single polar mode for each of the principal axes, and we take as an example the wurtzite structure, having space group $P6_3mc$ or C_{6v}^4 and point group $6mm$ or C_{6v}. As shown in Table 3.1 the polar-mode symmetries are $\Gamma_1 + \Gamma_5$ (or $A_1 + E_1$), where Γ_1 is a nondegenerate representation corresponding to motion parallel to the c axis and Γ_5 is twofold degenerate, corresponding to motion perpendicular to the c axis. If $\omega_T^{\parallel}$ and $\omega_T^{\perp}$ are the frequencies of the two kinds of motion, the two components of the relative permittivity (4.9) have the forms

$$\kappa^{\parallel} = \kappa_{\infty}^{\parallel} + \frac{NZ^{\parallel^2}/\epsilon_0 V}{\omega_T^{\parallel 2} - \omega^2 - i\omega\Gamma^{\parallel}} \tag{4.67}$$

$$\kappa^{\perp} = \kappa_{\infty}^{\perp} + \frac{NZ^{\perp^2}/\epsilon_0 V}{\omega_T^{\perp 2} - \omega^2 - i\omega\Gamma^{\perp}}. \tag{4.68}$$

The frequencies of those modes that have a macroscopic electric field are obtained by substitution of the undamped expressions for the relative permittivities into (4.66).

Fresnel's equation is of course easiest to solve when θ is either zero or a right-angle, in which case it reduces essentially to the cubic form (4.43). For **q** parallel to the c axis, the Γ_5 modes are transverse with the common frequency $\omega_T^{\perp}$, whereas the Γ_1 mode is longitudinal with a frequency given by (4.66) as

$$\omega_L^{\parallel} = \left(\frac{\kappa_0^{\parallel}}{\kappa_{\infty}^{\parallel}}\right)^{1/2} \omega_T^{\parallel}, \tag{4.69}$$

where $\kappa_0^{\parallel}$ is the zero-frequency limit of (4.67). Similarly for **q** perpendicular to the c axis, the Γ_1 mode is transverse with frequency $\omega_T^{\parallel}$, one of the Γ_5 modes is transverse with frequency $\omega_T^{\perp}$, and the other Γ_5 mode is longitudinal with a frequency given by (4.66) as

$$\omega_L^{\perp} = \left(\frac{\kappa_0^{\perp}}{\kappa_{\infty}^{\perp}}\right)^{1/2} \omega_T^{\perp}. \tag{4.70}$$

Equations (4.69) and (4.70) are the Lyddane-Sachs-Teller relations for a uniaxial crystal; they enable Fresnel's equation to be written

$$\frac{\omega_L^{\perp 2} - \omega^2}{\omega_T^{\perp 2} - \omega^2} \kappa_{\infty}^{\perp} \sin^2\theta + \frac{\omega_L^{\parallel 2} - \omega^2}{\omega_T^{\parallel 2} - \omega^2} \kappa_{\infty}^{\parallel} \cos^2\theta = 0 \tag{4.71}$$

for a general angle θ.

Figure 4.5 shows the angular variation of the polar-mode frequencies in wurtzite-structure ZnO constructed using numerical data of Arguello et al. (1969). The frequencies at the ends of the angular range are those de-

Macroscopic Theory

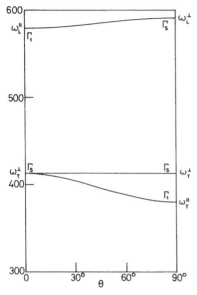

Figure 4.5 Angular variation of the polar-vibrational frequencies of ZnO.

scribed above, and the intermediate frequencies are deduced by solution of (4.71). One of the Γ_5 modes is transverse for any value of θ and this is represented by the flat branch of frequency $\omega_T^\perp$ in the figure. The two remaining branches obtained from the two roots of (4.71) have variable frequencies and no simple polarization except at the ends of the range. The macroscopic electric fields lift the group-theoretical degeneracy of the Γ_5 modes and mix the Γ_1 and Γ_5 symmetries of motion in the two variable-frequency excitations. There are three distinct frequencies at general angle.

It is seen from Figure 4.5 that the differences between the $\parallel$ and $\perp$ frequencies are rather small compared to the splittings between longitudinal and transverse frequencies. This reflects the rather small force-constant anisotropies in the crystal, and indeed the uniaxial wurtzite structure is similar to the cubic zinc-blende structure, being obtained from the latter by rearrangement of planes of atoms perpendicular to the (111) axis. The relatively large splitting produced by the macroscopic electric field in this case enables approximate solutions of (4.71) to be obtained (Loudon 1964),

$$\omega^2 = \omega_L^{\perp^2} \sin^2\theta + \omega_L^{\parallel^2} \cos^2\theta \tag{4.72}$$

$$\omega^2 = \omega_T^{\perp^2} \cos^2\theta + \omega_T^{\parallel^2} \sin^2\theta, \tag{4.73}$$

and the modes approximate to longitudinal and transverse polarization throughout the angular range.

There has been a good deal of experimental work on wurtzite-structure crystals. The polar modes produce three peaks in the light-scattering cross section, of which two frequencies vary with scattering geometry, and there are in addition two nonpolar Raman-active modes of Γ_6 or E_2 symmetry (see Table 3.1). Table 4.4 shows results for the two nonpolar frequencies and the four frequencies associated with the polar modes in five crystals. Note that ZnS and AgI are obtainable in both the zinc-blende and wurtzite structures, but are so isotropic in the latter structure that no angular variation of the polar-mode frequencies could be detected in the light-scattering experiments. All of the remaining crystals are sufficiently isotropic for the approximate relations (4.72) and (4.73) to provide a good description of the angular variations of the mode frequencies. Loose et al. (1976) have made careful measurements of the directional dependence of the polar-mode frequencies in ZnO.

By way of contrast to the ZnO curves of Figure 4.5, we show in Figure 4.6 the different qualitative behavior of the polar modes for crystals in which the force constants for motion parallel and perpendicular to the c axis are very dissimilar. One mode is now approximately polarized parallel to the c axis and the other perpendicular to the c axis for all wavevector directions. The variable-frequency modes change their polarization characters between transverse and longitudinal as θ is varied across its range of values. Poulet (1952, 1955) has derived approximations analogous to (4.72) and (4.73) for this limit.

The discussion so far is restricted to crystals that have a single polar vibration for each of the three principal axes. Most uniaxial crystals have larger numbers of polar modes, but the method of calculation of their frequencies remains the same. The expressions (4.67) and (4.68) must be summed over all the modes polarized parallel and perpendicular to the c axis, and these relative permittivities are then substituted into Fresnel's equation (4.66). As an example, we consider the case of quartz (see also

Table 4.4 Vibrational Frequencies of Wurtzite-Structure Crystals in cm^{-1} Measured by Light Scattering

		BeO	ZnO	CdS	ZnS	AgI
E_1 or Γ_5	$\omega_L^\perp$	1097	591	307	356	124
A_1 or Γ_1	$\omega_L^\parallel$	1081	579	305	356	124
E_1 or Γ_5	$\omega_T^\perp$	722	413	243	280	106
A_1 or Γ_1	$\omega_T^\parallel$	678	380	234	280	106
E_2 or Γ_6		684	444	256	280	112
E_2 or Γ_6		338	101	43	55	17

AgI—Bottger and Damsgard (1972); Others—Arguello et al. (1969).

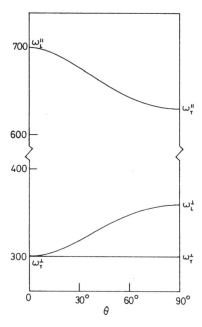

Figure 4.6 Angular variation of the polar-vibrational frequencies for a hypothetical crystal with highly anisotropic force constants.

Section 5.2.1), a crystal much studied by light-scattering spectroscopy. The space group of α quartz is $P3_121$ or D_3^4 and the crystal point group is 32 or D_3. The optic vibrations listed in Table 3.1 have eight doubly-degenerate Raman-active Γ_3 or E polar modes and four nondegenerate inactive Γ_2 or A_2 polar modes. The dashed lines in Figure 4.7 show the angular-dependent frequencies calculated by solution of (4.66), and there are in addition eight constant-frequency transverse branches that join the other branches at the points marked E_T at $\theta=0$. It is seen that some modes have the same qualitative behavior as in Figure 4.5 and others behave more like the branches of Figure 4.6. These behaviors result from the competing influences of the force-constant anisotropy and the macroscopic field, and it is not possible to identify one of these as having an overall dominance in the case of quartz.

Scott and Porto (1967) made the first careful measurement of the quartz Raman spectrum with laser exciting light. They obtained all 16 of the Γ_3 or E symmetry polar-mode frequencies marked on Figure 4.7 at $\theta=0$ and 90°. More recent measurements have been made by She et al. (1971) and Shapiro and Axe (1972). The latter workers studied the angular variation of the mode frequencies, obtaining the good agreement between theory and experiment shown in Figure 4.7. Note the absence of experimental

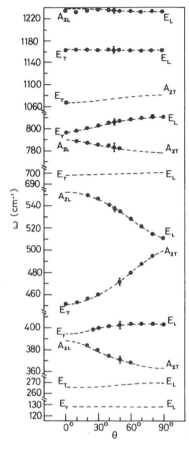

Figure 4.7 Calculated angular variation of the polar-vibrational frequencies of quartz with observed frequencies shown as circles (after Shapiro and Axe 1972).

points at the ends of branches where the mode symmetry is the Raman-forbidden Γ_2 or A_2. Pine and Tannenwald (1969) have measured the temperature dependence of Raman linewidths in quartz, and Pine and Dresselhaus (1969) have studied the small splitting of Γ_3 or E modes proportional to their wavevector, which can be detected in large-angle scattering.

The general expression (4.33) for the scattering cross section can be used to predict the magnitude and anisotropy of the scattering for uniaxial crystals. The calculations are similar to those described in the cubic case, but the details are more complicated, particularly for the polar modes whose frequencies vary with direction. Claus et al. (1975) give a much more complete discussion of polar-mode scattering in uniaxial crystals.

4.1.6 Polar-Mode Scattering in Biaxial Crystals

Polar vibrations in biaxial crystals were first treated by Merten (1968) and subsequently by Mathieu and Krauzman (1971) and Barker and Loudon (1972). The principal axes in orthorhombic crystals are at right-angles to each other in directions determined by the crystal symmetry. The three components of the relative permittivity are all different, and the polar-mode frequencies are obtained from Fresnel's equation in its most general form (4.14). The basic polar modes are all nondegenerate by group theory and the macroscopic electric field cannot produce any additional mode splittings in the biaxial case.

For a general wavevector direction, all the modes have associated electric fields and all their frequencies are obtained by solution of Fresnel's equation. None of the modes possesses simple transverse or longitudinal polarization in the general case. The mode properties are a little simpler when the wavevector lies in a principal plane, for example the xy plane. The modes polarized parallel to z are then purely transverse, and they have fixed frequencies equal to the ω_σ that appear in the expression for κ^z, that is, whose ξ_σ are parallel to z. The frequencies of the remaining modes are obtained from Fresnel's equation; they vary with direction in the xy plane and have mixed polarization.

The mode properties are simplest of all when the wavevector is parallel to a principal axis, for example the x axis. The modes polarized parallel to y and z are then transverse with frequencies given by the ω_σ whose ξ_σ are parallel to y and z. The remaining modes are determined by Fresnel's equation, which reduces to

$$\kappa^x = 0; \tag{4.74}$$

these modes have longitudinal polarization parallel to x. Note that for biaxial crystals, pure longitudinal modes occur only for wavevectors parallel to a principal axis, whereas pure transverse modes can occur for all wavevectors in principal planes.

Sodium nitrite is the most studied biaxial polar crystal (see also Section 5.2.3) and Figure 4.8 shows the polar-mode frequencies for each of the three principal planes. The $NaNO_2$ crystal is piezoelectric below its ferroelectric transition temperature of about 437 K, with orthorhombic space group $Imm2$ or C_{2v}^{20} and point group $mm2$ or C_{2v}. The optic-mode symmetries listed in Table 3.1 show eight polar vibrations, divided into three of symmetry Γ_1 or A_1 (z polarization), three of symmetry Γ_2 or B_1 (x polarization), and two of symmetry Γ_4 or B_2 (y polarization), where the polarization directions are obtained from Table 4.2. The branches in the figure are labelled with the appropriate irreducible representations, showing the pure symmetries of the flat transverse branches and the mixed

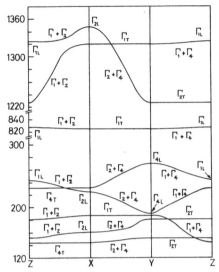

Figure 4.8 Calculated angular variation of the polar-vibrational frequencies of sodium nitrite for wavevectors in the three principal planes (after Anda 1971).

symmetries of the variable-frequency branches. For more general wavevector directions that do not lie in any principal plane, the excitations have mixed $\Gamma_1 + \Gamma_2 + \Gamma_4$ symmetry.

The polar-mode frequencies of sodium nitrite have been measured by light-scattering spectroscopy by Tsuboi et al. (1969), Asawa and Barnoski (1970), and Hartwig et al. (1971, 1972). The experiments particularly study the mode frequencies in the xy plane but there are some results for other principal planes. The experimental frequencies agree well with the calculated angular dependences. Another polar orthorhombic crystal studied by Raman spectroscopy is HIO_3 (Couture et al. 1970). The most complete determinations of the angular variations of polar-mode frequencies in biaxial crystals are those of Quilichini and Poulet (1974) on $BaZnF_4$ and of Quilichini (1975) on $BaMgF_4$. These crystals have the orthorhombic $mm2$ or C_{2v} point symmetry with 27 polar modes, all of whose frequencies were measured in each of the three principal planes. The general expression (4.33) for the cross section applies to biaxial crystals but there has been little quantitative work in this case. Claus et al. (1975) mention other examples and give a general account of polar-mode scattering in biaxial crystals.

Finally, as an example of a very complicated biaxial crystal with only one principal axis fixed by the crystal symmetry, we mention the work of

Canterford and Ninio (1975) on monoclinic $Li_2SO_4H_2O$. The space group is $P2_1$ or C_2^2 and the point group is 2 or C_2. There are 57 optic modes, divided into 29 of Γ_1 or A symmetry and 28 of Γ_2 or B symmetry. All of these modes are polar and have frequencies that vary with their wavevector orientation. The work is of particular interest, because it demonstrates the large amount of information about the lattice vibrations of a complex crystal that can be obtained from careful and detailed measurements of light-scattering spectra for different wavevector and polarization directions, coupled with theoretical work on the symmetries of the internal vibrations of the SO_4 and H_2O molecules and on the angular dependence of the mode frequencies. Canterford and Ninio (1975) were able in this way to observe and interpret all except 4 of the 57 polar optic modes.

4.1.7. Scattering by Powdered Crystals

The calculation of the cross section for a collection of randomly oriented scatterers is described in Section 3.1.5, but the theory needs some extension for the scattering by polar modes in a powdered uniaxial or biaxial crystal, where the mode frequencies now also depend on the orientations of the individual crystallites.

Figure 4.9 shows some experimental results of Gualberto and Arguello (1974) on the spectral differential cross section of $LiIO_3$ powder. The orientation dependence of the single-crystal polar-mode frequencies in the same spectral region (Otaguro et al. 1971) is shown for comparison in Figure 4.10. The modes are labelled with their symmetry characters in the

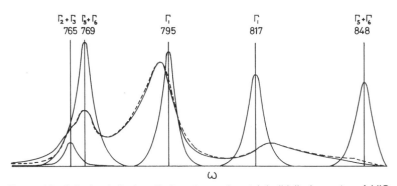

Figure 4.9 Calculated (broken line) and experimental (solid line) spectra of $LiIO_3$ powder. The five isolated peaks are the cross sections of pure-symmetry phonons obtained from single-crystal samples (after Gualberto and Arguello 1974).

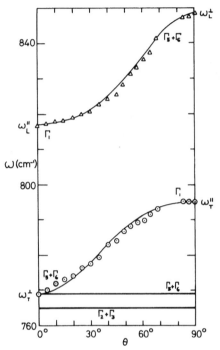

Figure 4.10 Orientation dependence of the single-crystal polar-mode frequencies of LiIO$_3$ in the same spectral range as Figure 4.9 (after Otaguro et al. 1971).

crystal point group 6 or C_6 and all the mode symmetries are listed in Table 3.1. It is seen that the powder spectrum is quite different from the single-crystal spectra; not all the single-crystal peaks appear in the powder spectra, and some that do appear are displaced in frequency. The powder spectrum shows a continuous distribution over the range of the single-crystal polar-mode frequencies.

The theory of polar-mode scattering for powders was considered by Burns and Scott (1970). Let $f(\omega)$ be the fraction of crystallites in a random powder per unit frequency range that have a polar-mode frequency equal to ω. Only those crystallites whose c axes have the appropriate orientation θ with respect to the wavevector transfer $\mathbf{q}$ belong to the fraction $f(\omega)$, and a simple solid-angle calculation gives

$$f(\omega) = \frac{d\theta}{d\omega} \sin\theta \tag{4.75}$$

the mode frequencies being independent of azimuthal angle in a uniaxial crystal. The additional factor $f(\omega)$ that weights the polar-mode scattering cross section for a powder is readily calculated if the angular dependence of the mode frequencies is known. The dashed curve in Figure 4.9 is the result of a calculation of the powder spectrum of $LiIO_3$ (Gualberto and Arguello 1974) using measured parameters and taking account of all the necessary angular averages. The agreement with experiment is very good.

These results provide a warning against too naive an assignment of peak frequencies in powder spectra to polar-mode single-crystal vibrations. This may be a valid procedure in some cases, as in the powder studies of Burns and Scott (1970) and Brya (1971), but the $LiIO_3$ results show that it can lead to incorrect determinations of polar-mode frequencies.

The above remarks assume that the basic polar-mode properties of the single crystal persist in the powdered specimen. This is no longer the case when the crystallite dimension L is sufficiently small for qL to be comparable to or smaller than unity. Then, as mentioned in Section 1.4.2, the wavevector conservation condition no longer limits the modes that contribute to light scattering. The most striking modification of the spectra results, however, from the occurrence of new contributions from crystallite modes whose frequencies generally lie between longitudinal and transverse frequencies of the bulk crystal. The new modes have their largest amplitudes on the crystallite surface and fall off toward its interior. The mode frequencies depend on the size and shape of the crystallite (Ruppin and Englman 1970).

The size of the contribution of these surface modes to the light-scattering cross section of the powder can be calculated for simple crystallite shapes, such as spheres or ellipsoids (Martin and Genzel 1973, Ruppin 1975). For example Ruppin shows that the surface-mode cross section in spheres of GaP exceeds the bulk-mode transverse and longitudinal cross sections for sphere diameters smaller than about 10^{-7} m. The scattering is nevertheless difficult to observe, and the crystallites in real powders are not usually spherical or even ellipsoidal, but have irregularities that broaden the surface-mode peaks. The clearest observation of surface-mode scattering appears to be a peak detected at 296 cm^{-1} in scattering by CdS crystallites (Scott and Damen 1972). Scattering peaks characteristic of MgO microcrystals have also been observed (Schlecht and Böckelmann 1973).

4.2. MICROSCOPIC THEORY

A descriptive outline of the microscopic theory of light scattering is given in Section 3.1.4. The present section provides some of the mathematical details of the microscopic theory, following the approach of Loudon (1963,

1964). The theory is based on the quantum mechanics of electrons and phonons in crystals, and the resulting expressions for cross sections are particularly useful in interpreting the dependence of the scattering on the incident frequency. The scattering crystal is assumed to be polar, diatomic, and cubic, for example, like zinc blende. The results also apply to similar nonpolar crystals, for example diamond, on setting the ionic charge equal to zero.

4.2.1. Electrons and Phonons in Crystals

The evaluation of the cross section resulting from the diagram in Figure 3.2c requires the matrix elements of various operators between electronic wavefunctions. The one-electron wavefunctions have the form

$$\psi_{m\mathbf{k}}(\mathbf{r}) = \exp(i\mathbf{k}\cdot\mathbf{r})u_{m\mathbf{k}}(\mathbf{r}), \qquad (4.76)$$

where m is a band index, $\mathbf{k}$ is a wavevector, and the final function has the full translational symmetry of the lattice. These Bloch functions are eigenstates of the electronic Hamiltonian,

$$\hat{H}_E \psi_{m\mathbf{k}}(\mathbf{r}) = \hbar\omega_{m\mathbf{k}}\psi_{m\mathbf{k}}(\mathbf{r}). \qquad (4.77)$$

The total wavefunction of all the electrons in the crystal is written in zero order as a Slater determinant of the Bloch functions. The electronic ground state $|0\rangle$ of a semiconductor or insulator has all the valence-band states occupied and all the conduction-band states empty. The light-scattering calculation involves only excited states in which a single electron is promoted from a valence-band state $v\mathbf{k}_h$ to a conduction-band state $c\mathbf{k}_e$. The e and h subscripts symbolize the behavior of the state as an electron-hole pair. These excited states are denoted $|\alpha\rangle$, where the single label α is a shorthand for the four symbols c, $\mathbf{k}_e$, v, and $\mathbf{k}_h$ needed to specify the state. The use of α, and sometimes β, to specify electron-hole pair states greatly simplifies some of the equations. The pair excitation-energy is denoted

$$\omega_\alpha = \omega_{c\mathbf{k}_e} - \omega_{v\mathbf{k}_h}. \qquad (4.78)$$

The pair-state wavefunctions in nonmagnetic crystals can be arranged to be real by taking suitable linear combinations of the Bloch functions for equal and opposite wavevectors.

Consider the excitation of a pair state by the interaction (1.112) of the electrons with the incident light beam. With the use of (1.105) and (1.108), the matrix element is

$$\langle \alpha, n_I - 1|\hat{H}''_{ER}|0, n_I\rangle = \frac{e}{m}\left(\frac{\hbar n_I}{2\epsilon_0 \eta_I^2 V \omega_I}\right)^{1/2} \boldsymbol{\varepsilon}_I \cdot \mathbf{p}_{\alpha 0}(\mathbf{k}_I), \quad (4.79)$$

where

$$\mathbf{p}_{\alpha 0}(\mathbf{k}_I) = \int \psi^*_{c\mathbf{k}_e}(\mathbf{r})\exp(i\mathbf{k}_I\cdot\mathbf{r})\hat{\mathbf{p}}\psi_{v\mathbf{k}_h}(\mathbf{r})\,d\mathbf{r}$$

$$= (2\pi)^3 \frac{N}{V}\delta(\mathbf{k}_e - \mathbf{k}_I - \mathbf{k}_h)\int_{\text{unit cell}} u^*_{c\mathbf{k}_e}(\mathbf{r})(\hat{\mathbf{p}} + \hbar\mathbf{k}_h)u_{v\mathbf{k}_h}(\mathbf{r})\,d\mathbf{r}. \quad (4.80)$$

The development in the second line of this equation depends on the translational symmetry of the Bloch functions, and the delta function is a demonstration of the momentum conservation automatically provided by the form (1.110) of electron-radiation interaction.

The photon wavevectors are normally much smaller than the electronic wavevectors, and $\mathbf{k}_e$ can be taken equal to $\mathbf{k}_h$ for most purposes. The notation is conveniently further simplified by denoting the matrix element (4.80) simply $\mathbf{p}_{\alpha 0}$. The electron-radiation interaction matrix elements (4.79) then lead to a first-order susceptibility of the form (for example, see Loudon 1973 for details of the method)

$$\chi^{ij}(\omega_I) = \frac{e^2}{\epsilon_0 \hbar V m^2 \omega_I^2}\sum_\alpha \left\{\frac{p^i_{0\alpha}p^j_{\alpha 0}}{\omega_\alpha - \omega_I} + \frac{p^j_{0\alpha}p^i_{\alpha 0}}{\omega_\alpha + \omega_I}\right\}, \quad (4.81)$$

where η_I is set equal to unity for a crystal in free space. The summation in (4.81) runs over all pairs of valence and conduction bands and over a common wavevector for the two band states.

The vibrational modes of a lattice are represented in quantum mechanics by normal-coordinate operators (see for example Peierls 1955, Reissland 1973)

$$\hat{W}_{\sigma,\mathbf{q}} = i\left(\frac{\hbar}{2N\omega_{\sigma\mathbf{q}}}\right)^{1/2}\left\{\hat{b}_{\sigma\mathbf{q}}\exp(i\mathbf{q}\cdot\mathbf{r}) - \hat{b}^\dagger_{\sigma\mathbf{q}}\exp(-i\mathbf{q}\cdot\mathbf{r})\right\} \quad (4.82)$$

where $\hat{b}^\dagger_{\sigma\mathbf{q}}$ and $\hat{b}_{\sigma\mathbf{q}}$ are creation and destruction operators for the phonon of branch σ and wavevector $\mathbf{q}$. The lattice displacements produced by excitation of a phonon perturb the potential experienced by the electrons and lead to the electron-lattice interactions that play an essential role in the microscopic scattering mechanism of Figure 3.2c. There are two contributions to the electron-lattice interaction in a polar crystal, known as the short-range part and the long-range part.

The short-range part $\hat{H}'_{EL}$ results from the effect of the lattice displacements on the periodic electronic potential energy. Its strength for a long-wavelength optic vibration is proportional to the relative displacement of the two ions in the primitive cell (Bir and Pikus 1961). A brief lattice-dynamics calculation along the lines of Section 3.1.2 gives the relative-displacement operator to be

$$\hat{U}_1 - \hat{U}_2 = i\left(\frac{\hbar}{2\mu N \omega_{\sigma q}}\right)^{1/2} \xi_{\sigma q}\{\hat{b}_{\sigma q}\exp(i\mathbf{q}\cdot\mathbf{r}) - \hat{b}^\dagger_{\sigma q}\exp(-i\mathbf{q}\cdot\mathbf{r})\}, \quad (4.83)$$

where μ is the reduced mass of the two ions. Let ξ be the unit polarization vector of one of the cubic-crystal longitudinal or transverse polar-modes described in Section 4.1.4. The matrix element of the short-range electron-lattice interaction between an electron-hole pair state α with $n(\omega)$ polar phonons of frequency ω and a pair state β with one additional phonon is conventionally written

$$\langle \beta, n(\omega)+1|\hat{H}'_{EL}|\alpha,n(\omega)\rangle = -i\left(\frac{\hbar}{2\mu N\omega}\right)^{1/2}\{n(\omega)+1\}^{1/2}\sum_\sigma \xi\cdot\xi_{\sigma q}\frac{\Xi_{\sigma,\beta\alpha}}{d}.$$

(4.84)

Here d is the lattice constant, introduced to give the *deformation potentials* $\Xi_{\sigma,\beta\alpha}$ the dimensions of energy. These are matrix elements of real operators that include the derivatives of the periodic electronic potential with respect to atomic displacements, and they have built-in momentum conservation.

The long-range part $\hat{H}''_{EL}$ arises from the interaction of the electrons with the macroscopic electric field of the lattice vibrations. Only the longitudinal polar mode has an electric field, as discussed in Section 4.1.4, and the quantized form of its magnitude (4.50) obtained with the use of (4.82) is

$$\hat{\mathbf{E}} = -i\frac{Z}{\epsilon_0 V \kappa_\infty q}\left(\frac{\hbar N}{2\omega_L}\right)^{1/2}\mathbf{q}\{\hat{b}_{Lq}\exp(i\mathbf{q}\cdot\mathbf{r}) - \hat{b}^\dagger_{Lq}\exp(-i\mathbf{q}\cdot\mathbf{r})\}. \quad (4.85)$$

The long-range electron-lattice interaction is now similar to the electron-radiation Hamiltonian (1.112) but with a vector potential related to the electric field (4.85) in the same manner as the potential and field in (1.105) and (1.106). The matrix element of the long-range interaction similar to

(4.84) is accordingly

$$\langle \beta, n(\omega_L)+1|\hat{H}''_{EL}|\alpha, n(\omega_L)\rangle = -\frac{Ze}{\epsilon_O V \kappa_\infty mq}\left(\frac{\hbar N}{2\omega_L^3}\right)^{1/2}\{n(\omega_L)+1\}^{1/2}\mathbf{q}\cdot\mathbf{p}_{\beta\alpha}.$$

(4.86)

The long-range interaction is sometimes named after Fröhlich (1954) who first derived it in a theory of the polaron. It vanishes in a nonpolar crystal like diamond where the effective charge Z is zero.

4.2.2. The Scattering Cross Section

The transition rate for the scattering mechanism represented in Figure 3.2c is given by third-order time-dependent perturbation theory [see (11.57) of Loudon 1973], with one matrix element of the electron-lattice interaction and two matrix elements of the electron-radiation interaction. In addition to the six similar diagrams that must be taken account of, the second interaction can occur in either the hole or electron parts of the pair excitation. These possibilities are illustrated in Figure 4.11, which is an alternative way of representing the scattering transitions.

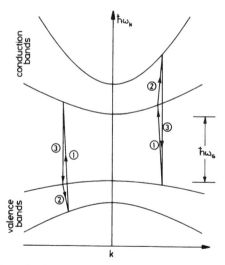

Figure 4.11 Schematic electronic energy bands of a crystal showing the electronic transitions of the third-order light-scattering process. The pair state created in the first transition can make its second transition within the valence bands (left) or conduction bands (right) before recombination in the final transition.

Consider the scattering by the longitudinal polar phonon where both kinds of electron-lattice interaction occur. The matrix elements required for the transition rate are given in (4.79), (4.84), and (4.86). The differential cross section obtained by substitution of the transition rate into (1.103), summation of the scattered wavevector directions within a solid angle $d\Omega$ with the help of (1.64), and conversion of the variable of integration from k_S to ω_S with the help of (1.19), is

$$\frac{d\sigma}{d\Omega} = \frac{\omega_S \mathfrak{v} V \eta_S e^4 \{n(\omega_L)+1\}}{(4\pi\epsilon_O)^2 2c^4 \omega_I \eta_I N \hbar^3 m^4 \omega_L} \left| \frac{\epsilon_S^i \epsilon_I^j q^h}{q} \left\{ \frac{i}{d\mu^{1/2}} \sum_\sigma R_\sigma^{ij}(\omega_S, -\omega_I, \omega_L) \xi_\sigma^h \right. \right.$$

$$\left. \left. + \frac{ZeN}{\epsilon_O V \kappa_\infty m \omega_L} P^{ijh}(\omega_S, -\omega_I, \omega_L) \right\} \right|^2. \qquad (4.87)$$

The mode polarization is parallel to the wavevector for the longitudinal mode.

The two terms in the square modulus of (4.87) are the contributions of the short-range and long-range electron-lattice interactions. With the assumption that the incident and scattered frequencies do not match any electronic transitions of the crystal

$$R_\sigma^{ij}(\omega_S, -\omega_I, \omega) =$$

$$\frac{1}{V} \sum_{\alpha,\beta} \left\{ \frac{p_{0\beta}^i p_{\beta\alpha}^j \Xi_{\sigma,\alpha 0}}{(\omega_I - \omega - \omega_\beta)(-\omega - \omega_\alpha)} + \frac{p_{0\beta}^j p_{\beta\alpha}^i \Xi_{\sigma,\alpha 0}}{(-\omega - \omega_S - \omega_\beta)(-\omega - \omega_\alpha)} \right.$$

$$+ \frac{p_{0\beta}^i \Xi_{\sigma,\beta\alpha} p_{\alpha 0}^j}{(\omega_I - \omega - \omega_\beta)(\omega_I - \omega_\alpha)} + \frac{p_{0\beta}^j \Xi_{\sigma,\beta\alpha} p_{\alpha 0}^i}{(-\omega - \omega_S - \omega_\beta)(-\omega_S - \omega_\alpha)}$$

$$\left. + \frac{\Xi_{\sigma,0\beta} p_{\beta\alpha}^i p_{\alpha 0}^j}{(\omega_I - \omega_S - \omega_\beta)(\omega_I - \omega_\alpha)} + \frac{\Xi_{\sigma,0\beta} p_{\beta\alpha}^j p_{\alpha 0}^i}{(\omega_I - \omega_S - \omega_\beta)(-\omega_S - \omega_\alpha)} \right\} \qquad (4.88)$$

and

$$P^{ijh}(\omega_S, -\omega_I, \omega) = \text{same expression but replace } \Xi_\sigma \text{ by } p^h. \qquad (4.89)$$

These quantities are called the scattering tensors; their notation is such that a negative (positive) frequency denotes destruction (creation) of the corresponding photon or phonon in the scattering. Their sizes can be roughly estimated from a knowledge of typical values of crystal electronic transition frequencies and matrix elements. Both parts of the electron-lattice interaction make contributions of the same orders of magnitude to give a differential cross section

$$\frac{d\sigma}{d\Omega} \approx 10^{-4} \text{ to } 10^{-5} \text{ m}^2, \qquad (4.90)$$

consistent with the measured cross section (3.43) for diamond.

The cross section analogous to (4.87) for the transverse polar phonons is obtained similarly. The result differs from (4.87) in having no contribution from the long-range electron-lattice interaction, and the phonon frequency and polarization are changed to their transverse values. By contrast with the macroscopic results of Section 4.1.4, the microscopic cross sections derived here do not include the spectral lineshapes. The microscopic calculation of the spectral differential cross section is relatively difficult and requires the inclusion in the quantum-mechanical Hamiltonian of terms describing the phonon decay processes.

It is seen that the macroscopic differential cross section from (4.57) is converted into the microscopic cross section (4.87) if the following identifications are made

$$a_\sigma^{ij} \to -\frac{e^2}{\omega_I \omega_S \hbar^2 m^2 d\mu^{1/2}} R_\sigma^{ij}(\omega_S, -\omega_I, \omega_L) \qquad (4.91)$$

$$b^{ijh} \to -i\frac{e^3}{\omega_I \omega_S \omega_L \hbar^2 m^3} P^{ijh}(\omega_S, -\omega_I, \omega_L), \qquad (4.92)$$

and similar replacements bring the transverse cross sections into agreement. The phases in these connections are not determined by the comparison of cross sections, but there is another way of relating macroscopic and microscopic theories (Loudon 1963). The linear susceptibility (4.81), which appears in the definitions (3.19) and (4.21) of the susceptibility derivatives, depends on the electronic wavefunctions and energy levels. These are perturbed by the electron-lattice interactions and the resulting first-order change in the linear susceptibility can be expressed in contributions proportional to the normal-mode amplitude and macroscopic electric field of the phonon. The coefficients of proportionality are the quantum-mechanical expressions for the susceptibility derivatives. The results of such calculations are identical to (4.91) and (4.92) in the limit $\omega_L \to 0$ (the evaluation of this limit requires some care in the case of the latter equation).

The quantum-mechanical expressions (4.88) and (4.89) taken in conjunction with (4.91) and (4.92) enable the frequency dependence of the cross section to be understood. It is seen that the expressions should be very weakly dependent on the phonon frequency in the normal situation where this is much smaller than the frequencies of the incident and scattered light, and where these are far removed from electronic transitions. This confirms the validity of the susceptibility-derivative approximation.

Inspection of the Raman tensors shows that they are proportional to $\omega_I \omega_S$ in the limit of an incident frequency much smaller than the electronic frequencies ω_α and ω_β. The susceptibility derivatives are therefore independent of frequency in the low-frequency limit. Diamond is an example of a crystal where the electronic frequencies are much higher than the incident light frequencies normally used in scattering experiments. The additional symmetry (4.34) of the electric-field susceptibility derivative is readily verified with the help of (4.89) and (4.92) in the low-frequency limit.

The Raman tensors have one symmetry property that holds completely generally for any nonmagnetic crystal with real electronic wavefunctions, where

$$\mathbf{p}_{\alpha\beta} = -\mathbf{p}_{\beta\alpha} \tag{4.93}$$

and

$$\Xi_{\sigma,\alpha\beta} = \Xi_{\sigma,\beta\alpha}. \tag{4.94}$$

The results hold because the momentum operator is purely imaginary, whereas the deformation potential is the matrix element of a real operator. The scattering tensors accordingly satisfy

$$R_\sigma^{ij}(\omega_S, -\omega_I, \omega) = R_\sigma^{ij}(-\omega_S, \omega_I, -\omega) = R_\sigma^{ji}(\omega_I, -\omega_S, -\omega) \tag{4.95}$$

and similarly for P^{ijh}. The scattering tensors for a Stokes process and its time-reverse anti-Stokes process are therefore equal, corresponding to the classical result (1.45), and the cross sections derived by microscopic theory conform to the general symmetry property discussed in Section 1.4.5. The tensors are not in general invariant under interchange of the polarizations of the incident and scattered photons, but (4.95) shows that this property does hold when ω can be neglected relative to ω_I and ω_S, which then become the same. This is the susceptibility-derivative limit, and the approximate symmetry (2.37) is only seriously violated in resonance scattering conditions.

4.2.3. Resonance Scattering

It is clear from (4.88) that the scattering cross sections become large in conditions of resonance between the incident and scattered frequencies and the electronic-transition frequencies. The large resonance scattering cross sections are experimentally useful for excitations whose light scattering would otherwise be unobservably weak. The resonant enhancement of the cross section is sometimes so large that new kinds of phenomena occur, for example multiple scattering by an excitation, or scattering by excitations normally forbidden by selection rules. The detailed frequency variation of the cross section depends on the nature of the electronic transition with which the light is in resonance, and a number of cases have been studied. The discussion here is limited to consideration of two simple types of electronic transition and descriptions of some typical experimental results. A comprehensive review of resonance scattering is given by Martin and Falicov (1975).

Consider the virtual transition scheme of Figure 4.11 where ω_I and ω_S are now assumed to be close to the lowest electronic transition frequency ω_G across the forbidden energy gap. Several terms in the scattering tensor (4.88) diverge as ω_I and ω_S approach ω_G, but the divergence is strongest in the third term in the bracket when α and β both refer to pair states with a hole in the top valence band and an electron in the bottom conduction band. If both these bands are parabolic with a reduced effective mass m^*, the dominant term in the scattering tensor has

$$\omega_\alpha = \omega_\beta = \omega_G + \frac{\hbar k^2}{2m^*}. \qquad (4.96)$$

The contribution of the third term in (4.88) can now be determined by integration over the wavevector $\mathbf{k}$. The integral is evaluated without difficulty if the wavevector dependence of the matrix elements is neglected (Loudon 1963, 1965), and the result for an incident frequency below the band gap is

$$R_\sigma^{ij}(\omega_S, -\omega_I, \omega) = \frac{1}{4\pi\omega}\left(\frac{2m^*}{\hbar}\right)^{3/2} p_{0\alpha}^i \Xi_{\sigma,\alpha\alpha} p_{\alpha 0}^j$$

$$\times \left[(\omega_G + \omega - \omega_I)^{1/2} - (\omega_G - \omega_I)^{1/2}\right] \quad (\omega_I \leqslant \omega_G). \quad (4.97)$$

The scattering tensor thus remains finite at the band gap but its slope diverges. The other kind of scattering tensor (4.89) does not have an analogous contribution since

$$p_{\alpha\alpha} = 0 \qquad (4.98)$$

as follows from (4.93).

A more complete theory should of course also take account of the weaker resonances in the scattering tensors and of the nonresonant contributions from other terms and other pairs of valence and conduction bands. These contributions can be very roughly allowed for by addition of a constant to the most resonant part (4.97) of the scattering tensor. The kind of predicted frequency variation of the cross section that results is shown by the curves of Figure 4.12. The zeros in the cross sections result from destructive interference between the resonant and nonresonant contributions. The experimental points in the figure come from measurements of scattering by CdS (Ralston et al. 1970, Callender et al. 1973). It is seen that the theory can produce a reasonable fit to the experimental points for the two transverse phonons by suitable choice of the relative signs and sizes of the resonant and nonresonant parts. It is not, however, possible to account for the frequency dependence of the longitudinal cross section by the same kind of theoretical expression, and Bendow et al. (1970) have made more sophisticated calculations for this case. Another experimental example is GaP (Weinstein and Cardona 1973), where it is necessary (Cardona 1971) to include the resonant contributions of the pair of spin-orbit split valence bands in the theory to account for the measured cross sections.

The most important electronic excitations in many crystals are to exciton states rather than to free electron-hole pair states of the kind assumed above. The excitons give rise to discrete lines in the absorption spectrum in contrast to the continuous spectrum of the free electron-hole

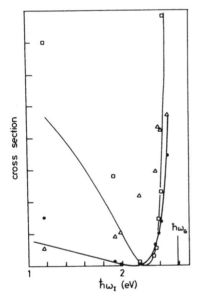

Figure 4.12 Resonant behavior of scattering cross sections for three phonons in CdS. The curves result from a theory described in text and the experimental points are Δ: $\omega_L^\perp$, ●: $\omega_T^\perp$, □: $\omega_T^\parallel$ in the notation of Table 4.4 (after Ralston et al. 1970).

pairs, and the absorption spectrum is dominated by a single exciton line in some cases. It is then more reasonable to take exciton states for the virtual intermediate states of the light-scattering process (Birman and Ganguly 1966, Ganguly and Birman 1967).

Consider the simplest case in which the scattering tensor is approximated by retaining only the transitions to a single exciton of frequency ω_{Ex}. The most resonant contribution again comes from the third term of (4.88), and with the introduction of an exciton damping parameter γ and a symbolic notation for the matrix elements, it gives

$$R_\sigma^{ij}(\omega_S, -\omega_I, \omega) = \frac{p_{Ex}^i \Xi_{\sigma, Ex} p_{Ex}^j / V}{(\omega_I - \omega - \omega_{Ex} + i\gamma)(\omega_I - \omega_{Ex} + i\gamma)}. \quad (4.99)$$

Figure 4.13 shows the variation with incident frequency of the cross section obtained from this form of scattering tensor. The double-peaked

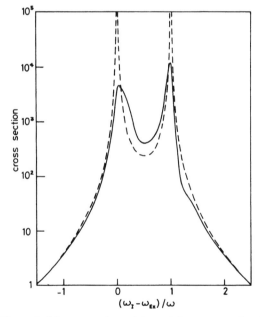

Figure 4.13 Theoretical frequency dependence of the cross section for an incident frequency in resonance with an exciton transition. The values of parameters assumed are typical of a semiconductor. The dashed curve is the zero-linewidth limit (after Barker and Loudon 1972).

structure results from the resonances of the incident and scattered frequencies with the exciton transition, and a wide variety of calculations produce qualitatively similar frequency variations (for a review, see Barker and Loudon 1972). The layer compound GaSe and the mixed crystals GaS_xSe_{1-x} are examples that show the double-peaked structure experimentally (Reydellet and Besson 1975, Chiang et al. 1976).

An unavoidable feature of resonance-scattering experiments is the significant variation of the refractive indices η_I and η_S with the incident frequency, and they also acquire imaginary parts in frequency regions where absorption of the light occurs. A significant part of the variation with frequency of the measured cross section may then be caused by such geometrical factors as the transmission of light through the crystal surfaces and its rate of attenuation within the crystal. It is not difficult in principle to take account of these factors and include their frequency variations in expressions for cross sections as measured outside the scattering crystal, but the results are complicated in the general case (Mills et al. 1968, 1970, Nkoma and Loudon 1975, Lax and Nelson 1976).

A simple result occurs for backscattering with incident and scattered beams normal to the flat surface of a crystal when any internal reflections from other surfaces can be neglected. As discussed in Section 1.4.3, the ubiquitous η_S/η_I factors in the scattering cross sections inside the crystal arise from their definition in terms of internal beam intensities. The cross section measured outside the crystal for the backscattering arrangement is obtained in the absence of any significant absorption by making the replacement

$$\frac{\eta_S}{\eta_I} \to \frac{16}{|1+\eta_I|^2|1+\eta_S|^2}. \qquad (4.100)$$

The same expression is also good for forward scattering through a parallel-sided slab. For stronger absorption where none of the incident light reaches crystal surfaces other than that illuminated from outside, the scattering volume must also be replaced

$$\mathfrak{v} \to \frac{Ac}{2(\eta_I''\omega_I + \eta_S''\omega_S)}. \qquad (4.101)$$

Here A is the illuminated area of crystal surface and the double primes denote imaginary parts. The scattering volume is thus the illuminated area divided by the sum of the absorption coefficients at the incident and scattered frequencies (defined as $\alpha_I = 2\eta_I''\omega_I/c$ and $\alpha_S = 2\eta_S''\omega_S/c$). These substitutions clearly introduce additional frequency dependence into the resonance cross sections measured experimentally (see also Section 8.3.2).

Let us turn now to the new lines that are sometimes detected in the resonance-scattering regime. These fall into two main categories, the first of which includes multiple scattering peaks. The first reported observation of resonance scattering in crystals by Leite and Porto (1966) on CdS recorded lines at shifts of two, three, and four times the frequency of a longitudinal optic phonon. Subsequent work (Leite et al. 1969, Klein and Porto 1969) extends the range of observed multiples of the longitudinal phonon frequency up to the ninth harmonic. The same phenomenon is seen in crystals other than CdS, and Scott et al. (1969) tabulate the observed frequencies of multiple longitudinal-optic-phonon lines. Harmonics of the transverse-optic-phonon frequency are not observed.

The second category of "new" line includes peaks at the frequencies of phonons that should normally be forbidden by selection rules for the scattering geometry employed, but appear nevertheless in resonance conditions. The effect again occurs for longitudinal phonons and the two purely longitudinal polar phonons of the wurtzite structure (see Table 4.4) provide an example. Consider an experiment in which the incident and scattered polarizations are both parallel to the same principal axis. Then it is seen from Table 4.2 that for the wurtzite point group $6mm$ or C_{6v}, the Γ_1 or A_1 longitudinal phonon of frequency $\omega_L^{\|}$ is allowed, but the Γ_5 or E_1 longitudinal phonon of frequency $\omega_L^{\perp}$ is forbidden. Contrary to these predictions, the two longitudinal phonons in CdS can produce scattering of comparable magnitude in resonance conditions (Klein and Porto 1969, Martin and Damen 1971). It should be emphasized for clarity that the scattering represented in Figure 4.12 is for allowed polarizations.

Various explanations are possible for the new lines seen in resonance scattering. The involvement only of longitudinal phonons suggests that the long-range or Fröhlich part of the electron-lattice interaction plays an important role. Its zero-order resonant contribution vanishes because of (4.98) but terms linear in q arise when account is taken of the small but finite change in electron or hole wavevector caused by phonon emission. The resulting first- and higher-order contributions can be quite significant if the other factors in the long-range term are large, and this mechanism seems able to account for at least some of the observed effects (Hamilton 1969, Mulazzi 1970, Malm and Haering 1971). The occurrence of multiple phonon lines is also accounted for by cascade processes that come into effect for ω_I greater than ω_G, when the excited electron-hole pair can make successive phonon-emitting transitions between real states before radiative recombination (Martin and Varma 1971).

The breakdown of the expected symmetry properties for first-order longitudinal phonon scattering is also accounted for by the terms of higher order in q in the long-range interaction. The symmetries listed in Tables 1.2 and 4.2 do not apply to such terms. Martin (1971) and Martin and

Damen (1971) show that the observed anomalous scattering in CdS is consistent with resonance of the incident light with a $1S$ exciton state whose isotropy produces an apparent Γ_1 or A_1 symmetry of scattering, despite the Γ_5 or E_1 symmetry of the phonon involved. The large dimensions of the exciton wavefunction enhance the importance of the terms in the scattering tensor that are linear in q. These terms cause a dependence of the cross section on the phonon wavevector, and for example the intensity of backward scattering is greater than that of forward scattering (Permogorov and Reznitsky 1976).

We have mainly used CdS as an illustration of resonance-scattering effects, but similar phenomena occur in other crystals where the detailed results and their interpretations may be different. Thus for example electric-quadrupole transitions lead to breakdown of the usual electric-dipole selection rules in experiments on Cu_2O with incident light in resonance with the well-known exciton transitions of this substance. There has been a great deal of work on resonance scattering in Cu_2O (see for example Yu and Shen 1975, Washington et al. 1977, and the references therein).

4.3 LIGHT SCATTERING BY POLARITONS

All the discussion since Section 4.1.1 refers to polar vibrational excitations whose wavevectors are large enough to satisfy (1.29), when the approximate form (4.14) of Fresnel's equation determines the mode frequencies. We complete the discussion by considering in the present section the small-wavevector regime where (1.29) is not satisfied and the full form (4.13) of Fresnel's equation must be used. The small-wavevector excitations are observed experimentally by scattering close to the forward direction in accordance with (1.21) and Figure 1.5; in practice, the full form of Fresnel's equation is needed for scattering angles ϕ of about $5°$ or smaller.

The mode frequencies are most easily determined for cubic symmetry crystals where (4.13) reduces to

$$\kappa\left(c^2q^2 - \kappa\omega^2\right)^2 = 0. \tag{4.102}$$

Thus one type of root of Fresnel's equation, corresponding to longitudinal modes, is obtained from the same equation (4.43) as in the large-wavevector limit; the longitudinal frequencies are independent of wavevector. The other, double, roots are obtained with the help of (4.52) from

$$\frac{c^2q^2}{\omega^2} = \kappa = \kappa_\infty \frac{\prod_L \left(\omega_L^2 - \omega^2\right)}{\prod_T \left(\omega_T^2 - \omega^2\right)}. \tag{4.103}$$

Light Scattering by Polaritons 193

These are the transverse excitation frequencies; they tend at large wavevectors to the frequencies ω_T discussed in Section 4.1.4.

Figure 4.14 shows the complete dispersion relations in the small-wavevector region for the simple case of zinc-blende structure GaAs. The mode frequencies at larger wavevectors off the right-hand side of the figure are just ω_L and ω_T, with the values listed in Table 4.3. The dispersion relation off the top of the figure tends to a large-frequency limit of (4.103),

$$\frac{cq}{\omega} = \kappa_\infty^{1/2} \qquad \omega \gg \omega_L, \omega_T. \tag{4.104}$$

This is just the usual relation for an electromagnetic wave in a medium of

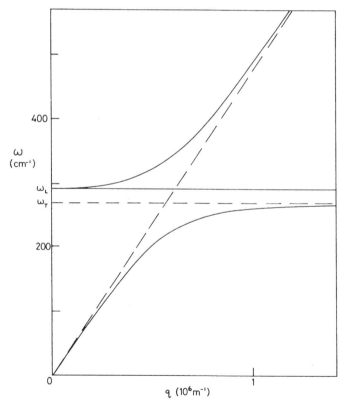

Figure 4.14 Calculated dispersion relations of the small-wavevector excitations in GaAs. The horizontal dashed line is at frequency ω_T; the oblique dashed line is a plot of (4.104); the horizontal full line is at the longitudinal excitation frequency ω_L; the full curves are the doubly degenerate polariton frequencies obtained from (4.103).

refractive index $\kappa_\infty^{1/2}$. The dashed lines in the figure show the extrapolated forms of these limiting dispersion relations of the transverse excitations, and these would represent the correct mode frequencies in the absence of any interactions between the vibrations and the electromagnetic waves. However, the interactions between transverse vibrations and transverse electromagnetic waves, included in the dispersion relation (4.103), produce a "repulsion" of the curves away from the point where they would otherwise cross. The correct excitations in this region are mixtures of vibrations and electromagnetic waves, or coupled phonons and photons, called *polaritons*. It is seen that the upper branch of the polariton-dispersion curve meets the longitudinal excitation at zero wavevector to produce a threefold degeneracy there. This agrees with the predictions of group theory, and the splitting of the longitudinal and transverse frequencies seen in scattering at larger angles is a consequence of the corresponding wavevectors being to the right of the crossover point in Figure 4.14.

The polariton dispersion relation for cubic crystals was first derived by Huang (1951, see also Born and Huang 1954). The extensive field of polariton studies is well reviewed by Mills and Burstein (1974), and the whole area of light-scattering work is clearly described by Claus et al. (1975). Figure 4.15 shows the results of the first experimental observation of light scattering by polaritons by Henry and Hopfield (1965) on GaP (see also Ushioda and McMullen 1972). The dashed lines are plots of (1.21) for several small values of the scattering angle; They are the same kind of

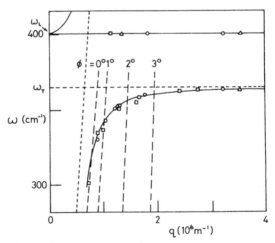

Figure 4.15 Measured points for polariton light scattering in GaP showing the theoretical dispersion curves (full lines) and the scans for various scattering angles (dashed lines) (after Henry and Hopfield 1965).

curve as shown in Figure 1.5, except that the refractive-index variation between the polariton-frequency region and that of the incident light is allowed for. It is seen that energy and momentum conservation exclude large parts of the polariton-dispersion relation from study by light scattering in this case.

Fresnel's equation (4.13) also factorizes for a uniaxial crystal (Loudon 1964), where there are two kinds of root, obtained from

$$c^2q^2 - \kappa^\perp \omega^2 = 0, \tag{4.105}$$

$$c^2q^2(\kappa^\perp \sin^2\theta + \kappa^\| \cos^2\theta) - \kappa^\perp \kappa^\| \omega^2 = 0, \tag{4.106}$$

with the same notation as in Section 4.1.5. The excitations corresponding to the solutions of (4.105) are transverse with their polarization perpendicular to the c axis, whereas the excitations associated with (4.106) do not have any simple polarization properties for general θ.

Figure 4.16 shows the resulting calculated dispersion relations for wurtzite-structure ZnO in the small-wavevector region for three directions θ of propagation relative to the c axis. The curves are qualitatively similar to those in the cubic case and many of the same general remarks apply. The limiting values of the polariton frequencies off the right-hand sides of the graphs match the larger-wavevector polar-mode frequencies shown in Figure 4.5. The remarks about the mixed polarization and symmetry characters of the angular-dependent polar modes made in Section 4.1.5 also apply to the corresponding polaritons. Figure 4.17 shows experimental results of Porto et al. (1966, see also Loose et al. 1976) for light scattering by polaritons in ZnO. This example illustrates the additional flexibility available in a uniaxial crystal where, as discussed in connection with (1.22), the accessible region of the dispersion curves is different for different choices of refractive index for the incident and scattered light. Although it is not shown in Figure 4.17, the experimental scan for $\theta=0$ and $\eta_I < \eta_S$ intersects the upper polariton branches and the corresponding peak in the cross section has been seen (Nicola et al. 1975).

There has been a great deal of work on light scattering by polaritons in crystals more complicated than those considered above. Thus, for example, in quartz, where there are many polariton branches (Loudon 1969, Merten 1969), the light scattering has been studied by Scott et al. (1967) and Scott and Ushioda (1969). Other uniaxial crystals investigated include ferroelectric $BaTiO_3$ (Pinczuk et al. 1969, Laughman et al. 1972, Heiman and Ushioda 1974), $LiIO_3$ (Otaguro et al. 1971, 1972), and BeO (Laughman and Davis 1974). The polariton frequencies in biaxial crystals, where Fresnel's equation in its most general form (4.13) must be used, were first treated by Merten (1968), and a representative experiment is that of Fukumoto et al.

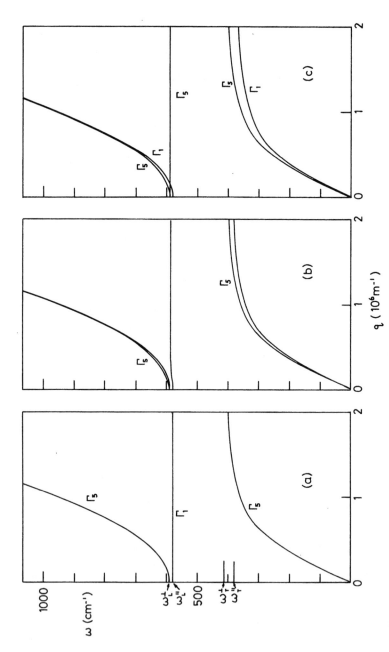

Figure 4.16 Calculated dispersion relations of the small-wavevector excitations in ZnO; (a) $\theta = 0$, (b) $\theta = 45°$, and (c) $\theta = 90°$.

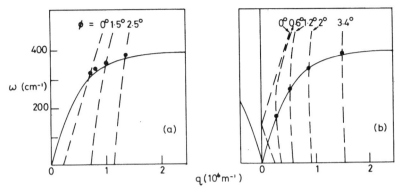

Figure 4.17 Measured points for transverse polariton light scattering in ZnO compared with the theoretical dispersion relation from (4.105); (a) $\eta_l > \eta_s$ and (b) $\eta_l < \eta_s$ (after Porto et al. 1966).

(1975) on biaxial $KNbO_3$. A much more detailed review of the field is given by Claus et al. (1975).

The theory of the cross section for light scattering by polaritons was first considered by Burstein et al. (1968, see Scott 1971 for a review of this and other earlier work). More comprehensive theories are given by Benson and Mills (1970a,b), Unger and Schaack (1971), and Barker and Loudon (1972). The method of the last reference is similar to that outlined for the calculation of the larger-wavevector polar-mode cross section in Section 4.1.2. The expression (4.22) is formally valid for the polaritons that occur at smaller wavevectors, but the calculation of the power spectra requires some generalization. An important feature of the cross-section expression is the prediction of peaks at frequencies determined by Fresnel's equation *with damping neglected*, as assumed in the expression (4.103) for the polariton-dispersion relation of a cubic crystal.

The polaritons considered so far are excitations that occur in the interior of a crystal sample. In addition to these "bulk" polaritons, there exist surface polaritons that propagate along the sample boundaries and have amplitudes that fall off exponentially toward the interior region. The surface-polariton frequencies and other properties depend upon the sample shape. Light scattering by surface polaritons has been observed (Evans et al. 1973, Prieur and Ushioda 1975) and treated theoretically (Nkoma 1975, Mills et al. 1976) for samples in the shape of a parallel-sided film.

REFERENCES

Anastassakis E., Iwasa S., and Burstein E. (1966), *Phys. Rev. Lett.* **17**, 1051.
Anda E. (1971), *Solid State Commun.* **9**, 1545.
Angress J. F., Cooke C., and Maiden A. J. (1968), *J. Phys.* **C1**, 1769.
Arguello C. A., Rousseau D. L., and Porto S. P. S. (1969), *Phys. Rev.* **181**, 1351.
Asawa C. K. and Barnoski M. K. (1970), *Phys. Rev.* **B2**, 205.
Bairamov B. K., Kitaev Y. E., Negodyiko V. K., and Khashkhozhev Z. M. (1974), *Sov. Phys. —Solid State* **16**, 725.
Barker A. S. (1964), *Phys. Rev.* **135**, A742.
Barker A. S. and Loudon R. (1972), *Rev. Mod. Phys.* **44**, 18.
Bendow B., Birman J. L., Ganguly A. K., Damen T. C., Leite R. C. C., and Scott J. F. (1970), *Opt. Commun.* **1**, 267.
Benson H. J. and Mills D. L. (1970a), *Phys. Rev.* **B1**, 4835.
Benson H. J. and Mills D. L. (1970b), *Solid State Commun.* **8**, 1387.
Bir G. L. and Pikus G. E. (1961), *Sov. Phys.—Solid State* **2**, 2039.
Birman J. L. and Ganguly A. K. (1966), *Phys. Rev. Lett.* **17**, 647.
Bloembergen N. (1965), *Nonlinear Optics* (New York: Benjamin).
Born M. and Huang K. (1954), *Dynamical Theory of Crystal Lattices* (Oxford: Clarendon Press).
Bottger G. L. and Damsgard C. V. (1972), *J. Chem. Phys.* **57**, 1215.
Brya W. J. (1971), *Phys. Rev. Lett.* **26**, 1114.
Burns G. and Scott B. A. (1970), *Phys. Rev. Lett.* **25**, 1191.
Burstein E. (1967), in R. Kubo and H. Kamimura, Eds., *Dynamical Processes in Solid State Optics* (New York: Benjamin), p. 34.
Burstein E., Ushioda S., and Pinczuk A. (1968), *Solid State Commun.* **6**, 407.
Callender R. H., Sussman S. S., Selders M., and Chang R. K. (1973), *Phys. Rev.* **B7**, 3788.
Canterford R. P. and Ninio F. (1975), *J. Phys.* **C8**, 1750.
Cardona M. (1971), *Solid State Commun.* **9**, 819.
Chiang T. C., Camassel J., Shen Y. R., and Voitchovsky J. P. (1976), *Solid State Commun.* **19**, 157.
Claus R., Merten L., and Brandmüller J. (1975), *Light Scattering by Phonon-Polaritons* (Berlin: Springer).
Cochran W. and Cowley R. A. (1962), *J. Phys. Chem. Solids* **23**, 447.
Couture L., Krauzman M., and Mathieu J. P. (1970), *Comptes Rendus* **270B**, 1246.
Dawson P. (1972), *J. Opt. Soc. Am.* **62**, 1049.
DiDomenico M., Wemple S. H., Porto S. P. S., and Bauman R. P. (1968), *Phys. Rev.* **174**, 522.
Evans D. J., Ushioda S., and McMullen J. D. (1973), *Phys. Rev. Lett.* **31**, 369.
Faust W. L. and Henry C. H. (1966), *Phys. Rev. Lett.* **17**, 1265.
Fröhlich H. (1954), *Adv. Phys.* **3**, 325.
Fukumoto T., Okamoto A., Hattori T., Mitsuishi A., and Fukuda T. (1975), *Solid State Commun.* **17**, 427.
Ganguly A. K. and Birman J. L. (1967), *Phys. Rev.* **162**, 806.
Gualberto G. M. and Arguello C. A. (1974), *Solid State Commun.* **14**, 911.

References

Hamilton D. C. (1969), *Phys. Rev.* **188**, 1221.
Hartwig C. M., Wiener-Avnear E., and Porto S. P. S. (1972), *Phys. Rev.* **B5**, 79.
Hartwig C. M., Wiener-Avnear E., Smit J., and Porto S. P. S. (1971), *Phys. Rev.* **B3**, 2078.
Heiman D. and Ushioda S. (1974), *Phys. Rev.* **B9**, 2122.
Henry C. H. and Hopfield J. J. (1965), *Phys. Rev. Lett.* **15**, 964.
Hobden M. V. and Russell J. P. (1964), *Phys. Lett.* **13**, 39.
Huang K. (1951), *Proc. R. Soc.* **A208**, 352.
Irwin J. C. and LaCombe J. (1970), *J. Appl. Phys.* **41**, 1444.
Johnston W. D. (1970), *Phys. Rev.* **B1**, 3494.
Johnston W. D. and Kaminow I. P. (1969), *Phys. Rev.* **188**, 1209.
Kaminow I. P. and Johnston W. D. (1967), *Phys. Rev.* **160**, 519.
Klein M. V. and Porto S. P. S. (1969), *Phys. Rev. Lett.* **22**, 782.
Kleinman D. A. (1962), *Phys. Rev.* **126**, 1977.
Laughman L. and Davis L. W. (1974), *Phys. Rev.* **B10**, 2590.
Laughman L., Davis L. W., and Nakamura T. (1972), *Phys. Rev.* **B6**, 3322.
Lax M. and Nelson D. F. (1976), in B. Bendow, J. L. Birman, and V. M. Agranovich, Eds., *Theory of Light Scattering in Condensed Matter* (New York: Plenum Press), p. 371.
Le Corre Y. (1963), *Comptes Rendus* **257**, 3352.
Leite R. C. C. and Porto S. P. S. (1966), *Phys. Rev. Lett.* **17**, 10.
Leite R. C. C., Scott J. F., and Damen T. C. (1969), *Phys. Rev. Lett.* **22**, 780.
Loose P., Wöhlecke M., Rölker B., and Rosenzweig M. (1976), *Solid State Commun.* **19**, 231.
Loudon R. (1963), *Proc. R. Soc.* **A275**, 218.
Loudon R. (1964), *Adv. Phys.* **13**, 423, erratum ibid. **14**, 621, 1965.
Loudon R. (1965), *J. Phys.* **26**, 677.
Loudon R. (1969), in G. B. Wright, Ed., *Light Scattering Spectra of Solids* (New York: Springer), p. 25.
Loudon R. (1973), *The Quantum Theory of Light* (Oxford: Clarendon Press).
Malm H. and Haering R. R. (1971), *Can. J. Phys.* **49**, 1823.
Martin R. M. (1971), *Phys. Rev.* **B4**, 3676.
Martin R. M. and Damen T. C. (1971), *Phys. Rev. Lett.* **26**, 86.
Martin R. M. and Falicov L. M. (1975), in M. Cardona, Ed., *Light Scattering in Solids* (Berlin: Springer), p. 79.
Martin R. M. and Varma C. M. (1971), *Phys. Rev. Lett.* **26**, 1241.
Martin T. P. and Genzel L. (1973), *Phys. Rev.* **B8**, 1630.
Mathieu J. P. and Krauzman M. (1971), *Ind. J. Pure. Appl. Phys.* **9**, 919.
Merten L. (1960), *Z. Naturforsch.* **15a**, 47.
Merten L. (1968), *Phys. Stat. Sol.* **30**, 449.
Merten L. (1969), *Z. Naturforsch.* **24a**, 1878.
Mills D. L. and Burstein E. (1974), *Rep. Prog. Phys.* **37**, 817.
Mills D. L., Chen Y. J., and Burstein E. (1976), *Phys. Rev.* **B13**, 4419.
Mills D. L., Maradudin A. A. and Burstein E. (1968), *Phys. Rev. Lett.* **21**, 1178.
Mills D. L., Maradudin A. A., and Burstein E. (1970), *Ann. Phys. N.Y.* **56**, 504.
Mooradian A. and Wright G. B. (1966), *Solid State Commun.* **4**, 431.

Mulazzi E. (1970), *Phys. Rev. Lett.* **25**, 228.
Nicola J. H., Freitas J. A., and Leite R. C. C. (1975), *Solid State Commun.* **17**, 1379.
Nilsen W. G. (1969), *Phys. Rev.* **182**, 838.
Nkoma J. S. (1975), *J. Phys.* **C8**, 3919.
Nkoma J. S. and Loudon R. (1975), *J. Phys.* **C8**, 1950.
Nye J. F. (1957), *Physical Properties of Crystals* (Oxford: Clarendon Press).
Otaguro W. S., Wiener-Avnear E., Arguello C. A., and Porto S. P. S. (1971), *Phys. Rev.* **B4**, 4542.
Otaguro W. S., Wiener-Avnear E., Porto S. P. S. and Smit J. (1972), *Phys. Rev.* **B6**, 3100.
Peierls R. E. (1955), *Quantum Theory of Solids* (Oxford: Clarendon Press).
Penna A. F., Chaves A., Andrade P. da R., and Porto S. P. S. (1976), *Phys. Rev.* **B13**, 4907.
Permogorov S. and Reznitsky A. (1976), *Solid State Commun.* **18**, 781.
Pinczuk A., Burstein E., and Ushioda S. (1969), *Solid State Commun.* **7**, 139.
Pine A. S. and Dresselhaus G. (1969), *Phys. Rev.* **188**, 1489.
Pine A. S. and Tannenwald P. E. (1969), *Phys. Rev.* **178**, 1424.
Porto S. P. S., Tell B., and Damen T. C. (1966), *Phys. Rev. Lett.* **16**, 450.
Poulet H. (1952), *Comptes Rendus* **234**, 2185.
Poulet H. (1955), *Ann. Phys. (Paris)* **10**, 908.
Prieur J. Y. and Ushioda S. (1975), *Phys. Rev. Lett.* **34**, 1012.
Quilichini M. (1975), *Phys. Stat. Sol. (b)* **68**, K155.
Quilichini M. and Poulet H. (1974), *Phys. Stat. Sol. (b)* **62**, 501.
Ralston J. M., Wadsack R. L., and Chang R. K. (1970), *Phys. Rev. Lett.* **25**, 814.
Reissland J. A. (1973), *The Physics of Phonons* (London: Wiley).
Reydellet J. and Besson J. M. (1975), *Solid State Commun.* **17**, 23.
Ruppin R. (1975), *J. Phys.* **C8**, 1969.
Ruppin R. and Englman R. (1970), *Rep. Prog. Phys.* **33**, 149.
Schlecht R. G. and Böckelmann H. K. (1973), *Phys. Rev. Lett.* **31**, 930.
Scott J. F. (1971), *Am. J. Phys.* **39**, 1360.
Scott J. F., Cheesman L. E., and Porto S. P. S. (1967), *Phys. Rev.* **162**, 834.
Scott J. F. and Damen T. C. (1972), *Opt. Commun.* **5**, 410.
Scott J. F., Damen T. C., and Shah J. (1971), *Opt. Commun.* **3**, 384.
Scott J. F., Leite R. C. C., and Damen T. C. (1969) *Phys. Rev.* **188**, 1285.
Scott J. F. and Porto S. P. S. (1967), *Phys. Rev.* **161**, 903.
Scott J. F. and Ushioda S. (1969), in G. B. Wright, Ed., *Light Scattering Spectra of Solids* (New York: Springer), p. 57.
Shapiro S. M. and Axe J. D. (1972), *Phys. Rev.* **B6**, 2420.
She C. Y., Masso J. D., and Edwards D. F. (1971), *J. Phys. Chem. Solids* **32**, 1887.
Tsuboi M., Terada M., and Kajiura T. (1969), *Bull. Chem. Soc. Jap.* **42**, 1871.
Unger B. and Schaack G. (1971), *Phys. Stat. Sol.* **48**, 285.
Ushioda S. and McMullen J. D. (1972), *Solid State Commun.* **11**, 299.
Washington M. A., Genack A. Z., Cummins H. Z., Bruce R. H., Compaan A., and Forman R. A. (1977), *Phys. Rev.* **B15**, 2145.
Weinstein B. A. and Cardona M. (1973), *Phys. Rev.* **B8**, 2795.
Yu P. Y. and Shen Y. R. (1975), *Phys. Rev.* **B12**, 1377.

CHAPTER FIVE
Structural Phase Changes

5.1 Soft Modes ———————————————————————— 206

5.2 Experimental Examples ———————————————— 213
    5.2.1 Quartz
    5.2.2 Perovskites
    5.2.3 Hydrogen-Bonded Ferroelectrics and Order-Disorder Transitions
    5.2.4 Cooperative Jahn-Teller Effects
    5.2.5 Acoustic Anomalies
    5.2.6 Light Scattering Near Zero Frequency

Light-scattering studies have made significant contributions to our understanding of the dynamics of structural phase changes and of cooperative magnetism in solids. In this chapter we are concerned purely with structural phase changes; magnetic phase changes are discussed in Chapter 6. The phase changes we are generally familiar with are of first order, e.g., melting, and are characterized by discontinuous changes in thermodynamic quantities such as the energy and the entropy. Second-order phase changes are by implication less abrupt and are characterized by continuous changes in the thermodynamic quantities just mentioned. Here we shall meet transitions of both types. The literature arising from the study of structural phase changes using light scattering and other complementary techniques, such as neutron scattering, has become so extensive in recent years that we cannot attempt to review it here (for reviews, see Scott 1974, Shirane 1974, Blinc and Žekš 1974, Fleury 1975). Our aim rather is to illustrate major areas of activity.

Since investigation of symmetry features prominently in light-scattering studies, the theory of symmetry restrictions on second-order phase transitions developed originally by Landau (1937) is of interest (see Landau and Lifshitz 1969). When a crystal changes structure it always has either one symmetry or another. Such a phase change may occur discontinuously through a sudden rearrangement of the atoms in the crystal giving a first-order phase change. However, the symmetry may also be changed by an arbitrarily small displacement of the atoms from their lattice points resulting in a phase transition of second order. At a first-order phase transition two different states are in equilibrium, but there is no predictable symmetry relationship between them. By contrast, in second-order phase transitions the states of the two phases are the same at the transition point, and it follows therefore that the symmetry of the body at the transition point must contain all the symmetry elements of both phases. If one assumes that the symmetry at the transition point is the same as the symmetry everywhere on one side of that point, we conclude that the symmetry of one phase must be higher than that of the other. In the great majority of phase transitions the more symmetrical phase corresponds to higher temperatures and the less symmetrical phase to lower temperatures.*

---

*The ferroelectric, Rochelle Salt, is an exception, changing from orthorhombic to monoclinic with increasing temperature on going through the lower Curie point ($-18°$C). A more straightforward example is provided by the mixed crystal $Tb_p Gd_{1-p} VO_4$, which for $p=0.365$ changes from tetragonal to orthorhombic symmetry at 8.5 K but reverts to tetragonal symmetry at 2.3 K; these changes are associated with the cooperative Jahn-Teller effect (Harley et al. 1974, see also Section 5.2.4).

Structural Phase Changes

In developing his theory of second-order structural phase changes, Landau (1937) considered any property of a crystal such as the ion density or electron density $\rho(\mathbf{r})$ that is invariant under the operations of the space group of the crystal (i.e., all translations, rotations, and reflections that leave $\rho(\mathbf{r})$ invariant). The first postulate in his theory is as follows:

1. If the crystal structure changes continuously from a high-symmetry to a lower-symmetry phase giving a new distribution $\rho(\mathbf{r}) = \rho_0(\mathbf{r}) + \Delta\rho(\mathbf{r})$, the symmetry group of $\rho(\mathbf{r})$ must be a subgroup of the symmetry group of $\rho_0(\mathbf{r})$, where $\rho_0(\mathbf{r})$ is invariant under operations of the high-symmetry space group. If we know the higher symmetry, it is then possible to predict the lower symmetries in second-order phase changes (for recent discussions see Cracknell et al. 1976, Petzelt and Dvorak 1976). It is of course possible to have structural phase changes that are, in fact, first order but that could be second order from a symmetry point of view, for example, $BaTiO_3$ (see Section 5.2.2).

Since any arbitrary function can be expanded as a sum of functions transforming as the irreducible representations of a symmetry group, it is possible to write

$$\Delta\rho(\mathbf{r}) = \sum_{n,i} C_{n,i}\phi_{n,i}(\mathbf{r}), \tag{5.1}$$

where the $\phi_{n,i}(\mathbf{r})$ form a basis for the nth irreducible representation of the symmetry group of $\rho_0(\mathbf{r})$. The dimension of the representation is the number of different i's for a particular n. Landau's second postulate is as follows:

2. Phase changes corresponding to different irreducible representations set in at the same temperature only by accident so that a second-order phase transition will generally correspond to a single irreducible representation. Hence (5.1) becomes

$$\Delta\rho(\mathbf{r}) = \sum_i C_i\phi_i(\mathbf{r}). \tag{5.2}$$

In describing phase transitions of second order, it is customary to define a quantity n, generally referred to as the order parameter, to represent the extent to which the configuration of atoms in the less symmetric phase differs from that in the more symmetric phase. In the latter case n is taken to be zero and has positive or negative values in the less symmetric phase. In transitions in which there is a displacement of atoms, n may be taken as the amount of displacement or any physical quantity proportional to it. In ferroelectrics, for example, n is taken to be P, the polarization that develops spontaneously at the Curie point. It should be emphasized, however, that recognition of the order parameter associated with a structural phase transition is not always straightforward.*

It follows from the continuous nature of phase transitions of second order that n must take arbitrarily small values near the transition point. In the neighborhood of this point it is assumed that the free energy F of a crystal can be expanded in powers of $n^\dagger$:

$$F = F_0 + \alpha n + An^2 + Bn^3 + Cn^4 + \cdots, \qquad (5.3)$$

where the coefficients α, A, B, C,... are functions of pressure and temperature; the value of n that occurs for a given pressure and temperature is determined from the condition that the free energy be a minimum. It can be shown (Landau and Lifshitz 1969, p. 433) that $\alpha = 0$ if the states with $n = 0$ and $n \neq 0$ have different symmetry. For a given value of temperature the equilibrium value n_0 of n is determined by the conditions

$$\frac{\partial F}{\partial n} = 0 \quad \text{and} \quad \frac{\partial^2 F}{\partial n^2} > 0. \qquad (5.4)$$

It follows from (5.3) and (5.4) that the state with $n = 0$ is stable for positive A, whereas for negative A values the stable state corresponds to $n \neq 0$. Hence $A = 0$ at the phase transition. If the crystal is stable for $A = 0$, $n = 0$, there must be an increase in F for small positive and negative changes of n. Hence we arrive at Landau's third postulate:

3. The coefficient of the cubic term in (5.3) must be zero if a second-order phase transition is to occur (B may, in fact, be identically zero for symmetry reasons and this represents a situation more likely to occur than the chance disappearance of B). There is a still further restriction that unless $C > 0$ the phase change will be of first order.

---

*In the so-called improper ferroelectrics, for example, the order parameter is not the polarization. It has been shown by neutron-scattering techniques (Dorner et al. 1972) that in the case of the improper ferroelectric $Tb_2(MO_4)_3$, which undergoes a phase transition at 159°C, the spontaneous polarization appears below T_C as a secondary effect because of the piezoelectric nature of the high-temperature phase. However, even though the macroscopic order parameter may be difficult to identify, the microscopic order parameter in a second-order phase transition is always the eigenvector amplitude of the soft mode (see Section 5.1)

$\dagger$Landau (1937) has discussed limitations on the use of this expansion, especially the use of high-order terms.

Structural Phase Changes

Each of the basis functions ϕ_i in (5.2) can be written in the form

$$\phi_i(\mathbf{r}) = u_{i,\mathbf{q}}(\mathbf{r})\exp(i\mathbf{q}\cdot\mathbf{r}) \qquad (5.5)$$

where $u_{i,\mathbf{q}}(\mathbf{r})$ has the periodicity of the lattice so that the irreducible representations are characterized by vectors $\mathbf{q}$ in the reciprocal lattice. Lifshitz (1941) added the fourth postulate:

4. Only values of q are allowed that are simple fractions of a reciprocal lattice vector so that the unit cell of the less symmetric lattice is a simple multiple of the more symmetric lattice cell* (for further discussion, see Section 5.2).

The Landau theory may be applied, with care, to transitions that are not completely second order but have some weak first-order character. The theory is of the mean-field type, that is, it is assumed that each particle moves in an average potential of all other particles. Such theories neglect critical phenomena associated with fluctuations in the order parameter (Stanley 1971); these fluctuations are large near critical points (for our purposes we take the critical temperature T_C to be the temperature at which n becomes zero, and, for convenience, we shall generally use T_C to refer to the temperature at which a phase transition occurs). The temperature region near T_C is referred to as the critical region and is described by the parameter $\varepsilon = (T - T_C)/T_C$, which is generally taken to be a small fraction. The dependence of physical properties on ε is related to power-law exponents generally referred to as critical exponents.

Although critical phenomena are not included in mean-field theory, singular behavior near T_C is described in this theory by the so-called classical critical exponents. Within the Landau framework the classical critical exponent β for the order parameter can be obtained by minimizing F with respect to n (classical theories of critical phenomena have in common the assumption that the free energy can be expanded as a power series in the order parameter near a critical point). In the mean-field approximation one can write (Stanley 1971) $A(T) = a(T - T_C)$ where $a = (\delta A/\delta T)_{T = T_C}$ is a constant. The coefficient $C(T)$ is assumed to be a

---

*This statement needs modification for some situations, e.g., helical periodicities may develop in magnetic spins that are incommensurate with the lattice (Bacon 1975). In these cases the three-dimensional translational periodicity of the crystal is lost so that the crystal symmetry cannot be described by a three-dimensional space group. Structural transitions associated with the development of charge-density waves also give rise to incommensurate effects (Wilson et al. 1975). Raman scattering connected with the charge-density wave in $TaSe_2$ has been described by Holy et al. (1976) (see also Steigmeier et al. 1975).

constant $C(T_C)$. The condition $\partial F/\partial n = 0$ then gives $(A + 2Cn^2) = 0$ and hence $n^2 = -A/2C = a(T_C - T)/2C$ or

$$n = \text{const.} \, (T_C - T)^{1/2}, \tag{5.6}$$

giving $\beta = \frac{1}{2}$ for the less symmetric phase. The classical critical exponent γ for static zero-field electric and magnetic susceptibilities

$$\chi(0) = \text{const.} \, (T_C - T)^{-\gamma} \tag{5.7}$$

has the value of 1.

Theories that depart from the mean-field approach give critical exponents that are referred to as nonclassical. There are numerous model calculations (Stanley 1971), for example, a three-dimensional Ising model gives the nonclassical values $\beta = 0.31$ and $\gamma = 1.25$. We shall see examples in Section 5.2, where both classical and nonclassical behavior are observed.

In Section 5.1 we outline the importance of specific phonon and exciton instabilities (the so-called soft modes) for the understanding of second-order structural phase changes and briefly discuss the so-called central modes. The latter are relaxation-type modes, centered on zero frequency, that have been observed near T_C in a number of crystal systems. Finally, in section 5.2, we present experimental examples of typical structural phase changes that have been studied in some depth using light-scattering techniques.

5.1 SOFT MODES

Structural phase changes in solids are quite often accompanied by observable acoustic anomalies (Section 5.2.5); in some cases a particular elastic constant or combination of elastic constants may fall to zero at the phase transition. In many materials such anomalies are due to interaction between acoustic modes and other lattice vibrational or electronic modes that fall in energy as the phase transition is approached from either side. Such modes are referred to as soft modes. Although more general approaches are possible, a connection between the order parameter referred to earlier and soft phonon modes may be made by regarding (5.3) as the potential of a one-dimensional oscillator expanded in powers of the displacement n. In the harmonic approximation this gives

$$\frac{\partial^2 F}{\partial n^2} = m\omega^2 = A,$$

so that

$$\omega = \text{const.} |T_C - T|^{1/2} \quad (5.8)$$

in the mean-field approximation (for a more extensive discussion, see Cochran 1971). Instability in a phonon frequency as $T \to T_C$ was first noted in the $\alpha - \beta$ phase transition in quartz (Section 5.2.1) by Raman and Nedungadi (1940). Later much interest centered on soft phonon modes associated with the change from the paraelectric to the ferroelectric phase (Sections 5.2.2 and 5.2.3). An early suggestion of a possible phonon instability associated with this change was made by Fröhlich (1949), who pointed out that the Lyddane-Sachs-Teller relationship, which has the form [see (4.45)]

$$\frac{\omega_L^2}{\omega_T^2} = \frac{\kappa_0}{\kappa_\infty}$$

for simple crystals, might imply an instability in the lattice vibrational spectrum of a ferroelectric; here ω_L and ω_T are zone-center transverse- and longitudinal-optic modes and κ_0 and κ_∞ are static and high-frequency relative permittivities. We shall see in Section 5.2 that the large increase in κ_0 at the phase transition is in fact associated with the fall in energy of a zone-center transverse-optic phonon. This instability was discussed in detail by Cochran (1960) using a lattice-dynamical approach. He concluded that ω_T can fall to zero because the temperature dependence of opposing short-range and long-range forces may lead to cancellation of the effective force constant. The association of structural phase transitions with phonon frequencies was also made by Anderson (1960) and was formalized by Cowley (1963) using a microscopic approach.

We shall give two examples here of the relationship between the soft-mode eigenvector and the structure of the crystal above and below the phase transition. Figure 5.1a shows the unit cell of the perovskite structure, representing a perovskite ferroelectric in its paraelectric phase. The displacements of the ions involved in the lowest-frequency TO mode are indicated in Fig. 5.1b. As the crystal is cooled toward the phase transition the frequency of the displacement of the ions involved in this mode gradually becomes smaller. Eventually the displacements are frozen in, reducing the symmetry of the crystal from cubic to tetragonal. It is clear that there is an electric-dipole moment associated with the frozen-in configuration and this is responsible for the ferroelectricity. Thus in a second-order transition the paraelectric and ferroelectric phases are connected in a continuous way by the gradual slowing down of the vibrational frequency of an infrared-active zone-center mode [this is an approximation to the situation in BaTiO$_3$ (section 5.2.2) where the ferroelectric phase transition has some first-order character]. It should be emphasized that distortions associated with zone-center modes (sometimes referred to as ferrodistortive modes) do not change the size of the Brillouin zone so that the number of ions in the unit cell does not change in the transition. Here the problem of determining possible distortions reduces to finding the irreducible representations of the crystal point group [i.e., all sets of rotations and reflections leaving $\rho_0(\mathbf{r})$ invariant].

In addition to zone-center instabilities there also exist zone-boundary soft modes. The most thoroughly explored of these is responsible for the

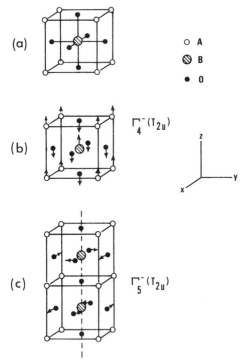

Figure 5.1 (a) Unit cell of the cubic perovskite structure (ABO$_3$); (b) eigenvector of the zone-center ($k=0$) soft ferroelectric mode which is associated with the cubic-to-tetragonal phase transition of BaTiO$_3$; (c) eigenvector of the zone-edge ($k=(\frac{1}{2},\frac{1}{2},\frac{1}{2})\pi/a$) soft mode which is associated with the cubic-to-tetragonal phase transition of SrTiO$_3$.

Soft Modes

cubic-to-tetragonal phase transition that occurs in $SrTiO_3$ at approximately 106 K (Section 5.2.2); its eigenvector is illustrated in Figure 5.1c. The oxygen octahedra rotate about a cube edge direction in opposite senses in adjacent cubic unit cells (Unoki and Sakudo 1967). We are dealing here, in fact, with a T_{2u} phonon that occurs at the R point ($\langle 111 \rangle$ corner) of the Brillouin zone (Cochran and Zia 1968). As the phase transition is approached from higher temperatures the frequency of the T_{2u} phonon falls and the eigenvector becomes frozen in below T_C. It is apparent from inspection of Figure 5.1c that the unit cell is now twice as large as the unit cell of the cubic perovskite structure (Figure 5.1a) and hence that the size of the Brillouin zone is halved.* In this process the zone-boundary point R of the cubic unit cell becomes the zone-center point Γ of the tetragonal unit cell. The order parameter may be taken to be the angle of rotation θ of the oxygen octahedra.

Structural phase changes may also be observed that are associated with soft tunneling modes (as in the case of the ferroelectric potassium dihydrogen phosphate (KDP) (Section 5.2.3) and with soft electronic modes (as in the case of the cooperative Jahn-Teller system $TbVO_4$ (Section 5.2.4).

To take our discussion of the dynamics of a second-order phase change a little further, we consider the response of noninteracting particles to an external force F (Thomas 1969; see Section 1.4 for a similar discussion). A static F produces a static output $X(F)$ that is generally a nonlinear function of F. Adding a small dynamic perturbation of amplitude f and frequency ω to the input, giving $F(t) = F + f\exp(-i\omega t)$, we get as a first approximation a linear response at the same frequency, centered on X:

$$X(t) = X + x\exp(-i\omega t). \tag{5.9}$$

The amplitude x is related to f by the generalized susceptibility $\chi_s(\omega)$:

$$x = \chi_s(\omega) f, \tag{5.10}$$

where the subscript s refers to a single particle. The frequency-dependence of the dynamic behavior shows in many cases a damped harmonic-oscillator response, similar to (1.88),

$$\chi_s(\omega) = \chi_s(0) \frac{\omega_s^2}{\omega_s^2 - \omega^2 - i\Gamma\omega}, \tag{5.11}$$

---

*Sb_5O_7I is also an interesting example of a material that undergoes a structural phase change as a result of condensation of a zone-boundary (M-point) phonon. This material develops mechanical strain at the transition point ($T_C = 481$ K) whose direction can be changed by external mechanical stress and is assigned to a class described as ferroelastic (Aizu 1969). For a description of light-scattering work on Sb_5O_7I, see Prettl and Rieder (1975).

where ω_s is the resonance frequency in the absence of damping and Γ is a damping constant (some contact with the rest of the system is in fact necessary for damping). Equation (5.11) may also be written in the form

$$\chi_s(\omega) = KT(\omega), \qquad (5.12)$$

where $K = \chi_s(0)\omega_s^2$ and $T(\omega) = (\omega_s^2 - \omega^2 - i\Gamma\omega)^{-1}$. If we now take account of the interaction $v_{nn'}$ of particles n' with particle n, we get an additional force on n

$$f_n^{\text{int}} = \sum_{n'} v_{nn'} x_{n'}, \qquad (5.13)$$

where we neglect the frequency dependence of $v_{nn'}$. In the mean-field approximation the response to the effective field $f_n + f_n^{\text{int}}$ is treated using linear response theory:

$$x_n = \chi_s(\omega)\left[f_n + \sum_{n'} v_{nn'} x_{n'}\right]. \qquad (5.14)$$

The N single-particle coordinates x_n can be replaced by N normal coordinates coupling to corresponding collective forces. In a crystal with translational symmetry the normal coordinates and forces are plane waves

$$x_n^q = x_q \exp(i\mathbf{q}\cdot\mathbf{R}_n)$$

$$f_n^q = f_q \exp(i\mathbf{q}\cdot\mathbf{R}_n), \qquad (5.15)$$

with wave vector $|q| = 2\pi/\lambda$. In the mean-field approximation each of these collective modes may be represented by a linear relationship between f and x

$$x_q = \chi(q,\omega) f_q. \qquad (5.16)$$

Taking the spatial Fourier transform of (5.14) gives

$$x_q = \chi_s(\omega)[f_q + \Delta_q x_q], \qquad (5.17)$$

where

$$\Delta_q = \sum_{n'} v_{nn'} \exp[-i\mathbf{q}\cdot(\mathbf{R}_n - \mathbf{R}_{n'})] \qquad (5.18)$$

Soft Modes

is referred to as the self-energy (Wehner 1966). This gives with (5.16)

$$\chi(q,\omega) = \frac{\chi_s(\omega)}{1 - \Delta_q \chi_s(\omega)}. \tag{5.19}$$

The collective response $\chi(q,\omega)$ of a system of interacting damped harmonic oscillators (for a stable system all normal modes must be damped; overdamped modes are referred to as relaxation modes) has the same form as that of a single oscillator (5.11) but with renormalized frequencies ω_q:

$$\chi(q,\omega) = \chi(q,0) \frac{\omega_q^2}{\omega_q^2 - \omega - i\Gamma\omega} \tag{5.20}$$

with

$$\frac{\omega_q^2}{\omega_s^2} = \frac{\chi_s(0)}{\chi(q,0)} = 1 - \Delta_q \chi_s(0) \tag{5.21}$$

(we have assumed that Γ is independent of q). The renormalization condition (5.21) gives a relationship between the static and dynamic properties of a system. It suggests that an instability $\chi(q,0) \to \infty$ is associated with a soft mode ($\omega_q \to 0$).

Contact now can be made between the discussion given above for external perturbations and thermally induced fluctuations of a system through the fluctuation-dissipation theorem. According to this theorem [see (1.82)] the inelastic scattering of light is proportional to

$$S(q,\omega) = \frac{\hbar}{\pi} [n(\omega) + 1] \operatorname{Im} \chi(q,\omega), \tag{5.22}$$

where $n(\omega)$ is given by (1.6). From (5.20)

$$S(q,\omega) = \text{const.} \, [n(\omega) + 1] \frac{\Gamma\omega}{(\omega_q^2 - \omega^2)^2 + \Gamma^2\omega^2}. \tag{5.23}$$

From (5.23) one finds that as $\omega_q \to 0$ the Stokes and anti-Stokes peaks move toward $\omega = 0$, merging into a single line; the height of this line gradually goes to infinity and the width to zero. Although our analysis cannot be used in the critical region (see p. 205), this conclusion suggests that the approach to instability is associated with low-frequency fluctuations of macroscopic size. These critical fluctuations drive the system into the new phase at the stability limit.

If the susceptibility diverges at the phase transition, the integrated spectral intensity will also diverge:

$$\frac{1}{2\pi}\int_{-\infty}^{+\infty} S(q,\omega)\,d\omega \propto k_B T \chi(q,0) \quad \text{for} \quad \hbar\omega \ll k_B T. \qquad (5.24)$$

In principle therefore critical phenomena can be studied either by measuring the temperature dependence of peak frequencies or the integrated intensity of the scattered light. However, (5.24) has very limited use, applying only to cases where the soft mode is Raman-active on both sides of T_C and is not coupled to another excitation having a finite scattering cross section of its own. It cannot therefore be applied to measure the critical exponent γ in systems generally studied, for example, $SrTiO_3$ or $BaTiO_3$ (Section 5.2.2) (see Lyons and Fleury 1978).

We now consider two excitations with uncoupled susceptibilities $\chi_1^0(q,\omega)$ and $\chi_2^0(q,\omega)$. If these excitations interact linearly the susceptibility $\chi_1(q,\omega)$ of the interacting system is, by analogy with (5.19),

$$\chi_1(q,\omega) = \frac{\chi_1^0(q,\omega)}{1 - \Delta_q \chi_1^0(q,\omega)}, \qquad (5.25)$$

where $\Delta_q = \Lambda \chi_2^0(q,\omega)$ and Λ is a constant. For example, in tetragonal $BaTiO_3$ mode 1 is a Raman-active soft optic phonon and mode 2 is a transverse acoustic phonon (Section 5.2.2). The spectral profile for the coupled soft-optic mode is given by (5.22), the susceptibility now being $\chi_1(q,\omega)$, and a similar expression may be written for the coupled acoustic mode. Such expressions account reasonably well for the complex Raman-Brillouin spectra observed for $BaTiO_3$ (Section 5.2.2). This complexity illustrates the difficulty of extracting details of critical phenomena from soft-mode spectra.

In the more complex situations arising from anharmonic coupling, one is concerned with the detailed form of the self-energy term Δ. It was pointed out by Cowley et al. (1971, see also Levanyuk and Sobyanin 1968, Coombs and Cowley 1973, Cowley and Coombs 1973) that in piezoelectric crystals anharmonic coupling of the soft optic mode to the phonon density of states produces a self-energy that leads to a response function $T(\omega)$ (see 5.12) of the form

$$T(\omega) = \left(\omega_0^2 - \omega^2 - i\Gamma\omega - \frac{\alpha\tau}{1 + i\omega\tau}\right)^{-1}, \qquad (5.26)$$

where α is a positive constant determined by the coupling strength and τ is a relaxation time describing the rate at which the soft mode comes into

Experimental Examples

equilibrium with the phonon density [when $\omega\tau \gg 1$ the response (5.26) becomes identical with that of a damped harmonic oscillator since the coupled systems do not have time to interact]. A response of the form (5.26), irrespective of its origin, will give rise to quasielastic scattering near T_C in addition to soft-mode sidebands.* Indeed for certain relationships between the parameters in (5.26), light scattering may be regarded as two separate responses, one a damped oscillator response (5.12) and the other a relaxation-type response centered on $\omega=0$ (see, e.g., Shapiro et al. 1972). The scattering at $\omega=0$ is generally described as a central mode. The central peak is always weak except close to T_C, where mixing with the soft mode is larger. Because of this interaction the frequency of the soft mode does not fall to zero at T_C [see (5.7)] and the central mode ultimately becomes the unstable mode associated with the transition.

It should be emphasized that the central mode described above is not the result of critical fluctuations (for possible mechanisms for central modes see Schwabl 1972, 1973; Silberglitt 1972, Feder 1973, and Section 5.2.6). However, fluctuations may give rise to quasielastic scattering above T_C by producing the low-temperature phase locally with some lifetime τ (Müller et al. 1976).

5.2 EXPERIMENTAL EXAMPLES

5.2.1 Quartz

We have already pointed out in Section 5.1 that the first example of a soft phonon mode in Raman spectroscopy was described by Raman and Nedungadi (1940). While investigating the first-order transition in quartz from the α (low temperature, D_3^4) to the β (high temperature D_6^5) phase at ~570 C they found that a phonon mode with A_1 symmetry that occurs at 220 cm^{-1} at low temperatures decreased in frequency as the transition temperature is approached, and they concluded that excitations of this mode had a direct bearing on the phase transition. After this show of insight work on quartz languished until Scott and Porto (1967) made a thorough investigation of the phonon spectrum, assigning a mode at 147 cm^{-1} to a two-phonon excitation (see also Section 4.1.5). Shapiro et al. (1967) found that the behavior of the 220 cm^{-1} band was quite complex and Scott (1968) subsequently explained the complexity by invoking

---

*An expression similar to (5.26) has been used by Mountain (1966) to describe quasielastic scattering from density fluctuations in liquids; in this case coupling occurs to internal degrees of freedom of the fluid (see Chapter 8).

anharmonic coupling between the 220 and 147 cm^{-1} excitations. Höchli and Scott (1971) showed that there is a slight hysteresis associated with the phase transition.

In general, it is not easy to verify experimentally the presence of effects in solids due to critical phenomena. In this connection it should be mentioned that quartz shows a strong increase in the intensity of elastic light scattering in the vicinity of the $\alpha - \beta$ phase transition and this has been assigned in the past to critical opalescence, a phenomenon readily observable in fluids (Chapter 8). It seems, however, that the effect in quartz may be due to the growth of small domains, probably associated with the coexistence of the α and β structures in the hysteresis region (Shapiro and Cummins 1968, Höchli and Scott 1971).

In the high-temperature β phase of quartz the soft mode is neither infrared nor Raman active. Its temperature dependence, however, has been investigated using neutron-scattering techniques by Axe and Shirane (1970) who found it overdamped. The neutron-scattering technique is a powerful tool for the study of structural phase changes, and most of the systems discussed in this chapter have also been investigated by this method (Shirane 1974).

Quartz is of historical interest in the study of soft-mode systems, and it shows many of the phenomena encountered in such systems. Other materials with the same structure, for example, $AlPO_4$ and GeO_2 have also received attention (Scott 1974). Interesting line-shape effects arising from anharmonic interaction between phonons have been investigated in $AlPO_4$ by Scott (1970) and by Zawadowski and Ruvalds (1970) and Ruvalds and Zawadowski (1970).

5.2.2 Perovskites

Although most ferroelectric materials, such as Rochelle salt, have very complex structures, the discovery about 30 years ago that $BaTiO_3$ (which has the relatively simple perovskite structure, Figure 5.1) was ferroelectric stimulated interest in the fundamental nature of ferroelectricity and in structural instabilities in perovskites [for a discussion of ferroelectricity in $BaTiO_3$ and also in KH_2PO_4 and $NaNO_2$ (Section 5.2.3) see Cochran 1969]. The first studies on low-frequency phonons in perovskites were carried out by Barker and Tinkham (1962) on $SrTiO_3$ and by Spitzer et al. (1962) on $SrTiO_3$ and $BaTiO_3$ using infrared measurements. At about the same time inelastic neutron-scattering experiments were carried out on $SrTiO_3$ by Cowley (1962, 1964), who found that the temperature dependence of the lowest-frequency TO phonon agreed with the temperature dependence of the dielectric constant. Subsequently detailed light-scattering studies of $KTaO_3$ and $SrTiO_3$ were carried out by Fleury and Worlock

(1968) and of $BaTiO_3$ by Fleury and Lazay (1971) and Lazay and Fleury (1971).

$BaTiO_3$ exhibits a first-order phase transition from the high-temperature cubic (O_h) structure to a structure with tetragonal (C_{4v}) symmetry at 129.5°C; the tetragonal structure is ferroelectric. There is, however, considerable second-order character to the transition (this mixture of first-order and second-order character is a feature of phase changes in perovskites in general) and an overdamped soft optic mode plays a major role. Since the first-order Raman effect is forbidden in the cubic perovskite structure (every ion is at a center of symmetry), the light-scattering experiments were carried out only in the ferroelectric phase (see Table 3.1). However neutron-scattering measurements have been carried out in the cubic phase by Harada et al. (1971) showing soft-mode behavior. Some striking features were observed in the light-scattering studies including (a) a Brillouin-scattering cross section (see Section 8.2) enhanced by a factor of $\sim 10^3$ on going from the cubic phase to the ferroelectric phase cooled to 7°C, (b) a dip in the spectral shape of the overlapping Raman and Brillouin spectra suggesting interference between the underdamped acoustic mode and the overdamped soft-optic mode, which has E symmetry (Figure 5.2), (c) an increase in the Brillouin frequency of $\sim 20\%$ and a reduction of the Brillouin linewidth by a factor of four on going from room temperature to the phase transition.

As the temperature increases in the ferroelectric phase, the peak height of the soft-mode spectrum decreases and its width increases. At the transition to the cubic phase, light scattering by the E-symmetry soft mode

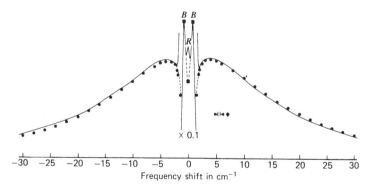

Figure 5.2 Raman-Brillouin spectrum of $BaTiO_3$ in the tetragonal phase at 22.7° C in z(xz)y geometry. The solid curve gives the experimental line shape. Brillouin peaks are labeled B and the elastic peak is labeled R. The theoretical lineshape is indicated by round dots, square dots, and dashes. Square dots correspond to 10 times less gain than round dots (after Lazay and Fleury 1971).

vanishes for symmetry reasons. These various effects have been explained (Fleury and Lazay, 1971, Lazay and Fleury, 1971) on the basis of a piezoelectric interaction between the strain (long wavelength acoustic phonon) and the polarization fluctuations (the soft optic mode) using an expression for coupled modes of the form (5.25).

Although $BaTiO_3$ presents an interesting example of a displacive phase transition, it is not an ideal one from a light-scattering point of view since the soft mode is overdamped. By contrast, well-behaved phonon modes are found in the ferroelectric perovskite $PbTiO_3$ in both neutron (Shirane et al. 1970) and light-scattering (Burns and Scott 1973) experiments. This material undergoes a phase transition at $T_C = 493°C$ similar to that of $BaTiO_3$ but with more pronounced first-order character. In $PbTiO_3$, however, the transition is characterized by an underdamped soft mode.

In an interesting study of $KTaO_3$ and $SrTiO_3$ carried out by Fleury and Worlock (1968), first-order Raman scattering was induced in the cubic paraelectric phase by application of an electric field. An external electric field applied in a $\langle 100 \rangle$ direction distorts the unit cell, removing all centers of symmetry, and making all zone-center optic phonons Raman active. It was then found that the lowest-frequency TO phonon in $KTaO_3$ showed soft-mode behavior as the temperature was reduced (Figure 5.3) indicating a tendency to instability; the frequency decreases from 85 cm^{-1} at 300 K to 18 cm^{-1} at 10 K for low applied fields (fields down to 50 V/cm were used). The soft-mode frequency does not vanish in $KTaO_3$ at the lowest temperatures, and in fact the crystal does not show a phase transition to a ferroelectric state. However, the temperature dependence of the soft mode

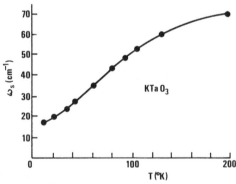

Figure 5.3 Temperature dependence of the soft ferroelectric mode in the cubic phase of $KTaO_3$. The solid curve was calculated using the Lyddane-Sachs-Teller relationship [see (4.45)] (after Fleury, 1972).

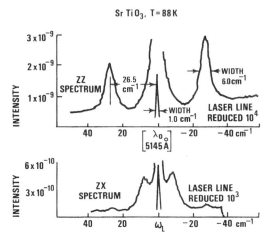

Figure 5.4 Raman spectrum of SrTiO$_3$ in the tetragonal phase showing a singly degenerate (A_{1g}) phonon (upper trace) and a doubly degenerate (E_g) phonon (lower trace) (after Worlock and Olsen 1971)

(Figure 5.3) is consistent with predictions of (4.45) based on measured values of the dielectric constant and the crystal can be induced to become ferroelectric by application of uniaxial stress (Uwe and Sakudo 1975).

SrTiO$_3$ undergoes a transition from the cubic (O_h) to a tetragonal (D_{4h}) phase at ~106 K, which is closely second order in nature. We have already pointed out (Section 5.1; Figure 5.1c) that this transition is due to condensation of a zone-edge phonon resulting in doubling of the unit cell below T_C. The tetragonal symmetry below T_C reduces the triple degeneracy of the T_{2u} soft mode to a singly degenerate (A_{1g}) and doubly-degenerate (E_g) mode whose frequencies move apart with decreasing temperature.* Both these modes have been observed in Raman scattering by Fleury et al. (1968) and by Worlock and Olsen (1971) (Figure 5.4), and their frequencies are plotted as a function of temperature in Figure 5.5. Above T_C the wavevector of the triply degenerate T_{2u} soft mode is too large to permit study by optical methods. However, the soft mode has been studied in this temperature range by neutron scattering (Shirane and Yamada 1969, Cowley et al. 1969) and its temperature dependence is also illustrated in Figure 5.5.

*Below T_C the symmetry-restoring mode and the soft mode are identical only if the soft mode above T_C is nondegenerate. If the soft mode is degenerate above T_C, as in the case of SrTiO$_3$, it splits below T_C and the symmetry-restoring mode is a linear combination of the component soft modes.

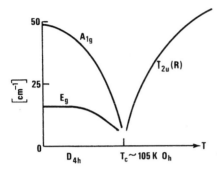

Figure 5.5 Temperature dependence of the frequency of the A_{1g} and E_g modes in the tetragonal phase of $SrTiO_3$ and of the soft T_{2u} mode, measured by neutron scattering, in the cubic phase.

It has been pointed out by Worlock (1969) that since in second-order phase transitions the new structure is a subgroup of the old structure there will always be a zone-center A_{1g} component of the soft mode below T_C regardless of the symmetry of the soft mode in the more symmetric phase (this point was subsequently discussed in a formal way by Birman 1973 and by Shigenari 1973). Hence Raman scattering is generally applicable to the study of second-order transitions below T_C. Neutron scattering has of course a more general application.

In $SrTiO_3$ the average value of the angle of rotation of the oxygen octahedra is the order parameter for the transition at ~ 105 K (Section 5.1) and the fluctuations in θ are the soft mode. The temperature dependence of $\langle \theta \rangle$ in the vicinity of the phase transition has received considerable attention. Müller and Berlinger (1971), using epr methods, found $\langle \theta \rangle^3 \alpha \propto |T - T_C|$ in the temperature range 98 to 105.5 K, suggesting a value of the critical exponent β of $\frac{1}{3}$. The value of $\beta = \frac{1}{3}$ has been found for a wide variety of physical systems near T_C and is often taken as evidence for the presence of critical fluctuations [for measurements on magnetic systems near T_C see, for example, the nmr studies by Heller and Benedek (1962, 1965) on the antiferromagnet MnF_2 and the ferromagnet EuS].

In concluding this section we outline briefly a simple model of phase transitions in perovskites due to Thomas and Müller (1968) in terms of the order parameter θ describing the motion of the oxygen octahedron in the T_{2u} mode (Figure 5.1c). The energy associated with this motion, expanded to fourth order in θ, is (see also Devonshire 1949, Feder and Pytte 1970)

$$V(T) = \tfrac{1}{2}a(T)\left[\theta_x^2 + \theta_y^2 + \theta_z^2\right] + \tfrac{1}{4}b(T)\left[\theta_x^4 + \theta_y^4 + \theta_z^4\right] \\ + \tfrac{1}{2}c(T)\left[\theta_x^2\theta_y^2 + \theta_y^2\theta_z^2 + \theta_z^2\theta_x^2\right], \quad (5.27)$$

where $x = [100]$, $y = [010]$, and $z = [001]$ and a, b, c are parameters dependent on temperature. Minimizing $V(T)$ with respect to θ shows that a tetragonal distortion (as observed in SrTiO$_3$) is stable below T_C if $c > b > 0$. However, in the region enclosed by the inequalities $b > c$ and $b > -2c$, a trigonal distortion sets in below T_C, as observed in LaAlO$_3$ and a number of rare-earth isomorphs (Scott 1974). The tetragonal and trigonal distortions correspond to rotations of the oxygen octahedron about $\langle 100 \rangle$ and $\langle 111 \rangle$ axes, respectively. Of the aluminate isomorphs PrAlO$_3$ is unusual. It is trigonally distorted at room temperature but exhibits a number of additional structural phase changes below room temperature (Harley et al. 1973, Fleury et al. 1974, a related paper by Harley 1977, Birgeneau et al. 1974). This additional complexity is due to interaction of the low-lying Pr^{3+} electronic energy levels with rotations of the (AlO$_6$)$^{3-}$ octahedron and is a manifestation of the cooperative Jahn-Teller effect (Section 5.2.4).

In SrTiO$_3$ $c \cong b$, so that this material under normal conditions has only a marginal preference for a tetragonal rather than a trigonal distortion. In fact by applying uniaxial stress in various directions, it is possible to go from both the cubic and tetragonal phase to the trigonal phase and also to produce the ferroelectric phase (Uwe and Sakudo 1976).

5.2.3 Hydrogen-Bonded Ferroelectrics and Order-Disorder Transitions

There are many hydrogen-bonded ferroelectrics, but of these potassium dihydrogen phosphate (KH$_2$PO$_4$, generally abbreviated to KDP) and its isomorphs have attracted most attention from spectroscopists. At room temperature KDP has tetragonal symmetry and is piezoelectric. It undergoes a largely second-order transition at 122 K to a ferroelectric phase with orthorhombic symmetry; the polar axis lies along the c axis of the tetragonal phase. Deuteration changes T_C to 213 K but does not appreciably affect the saturation polarization. In the ferroelectric phase the K$^+$, P^{5+}, and O^{2-} ions are displaced along the c axis relative to their positions in the paraelectric phase, and these displacements largely account for the saturation polarization. The eigenvector proposed by Cochran (1961) for the soft mode associated with the phase transitions is shown in Figure 5.6; it is apparent that the protons move in directions closely at right angles to the K-PO$_4$ ions.

The protons in KDP are in a double-well potential between the two oxygen ions in the O-H-O bond (Figure 5.6). There are two characteristic frequencies associated with such a double-well, one the quasiharmonic frequency of motion within the individual wells and a second, lower frequency of tunneling between the two wells. Kobayashi (1968) suggested that the low-frequency tunneling mode is mixed with the optic-type motion of the K-PO$_4$ ions via Coulomb interaction, forming two coupled modes.

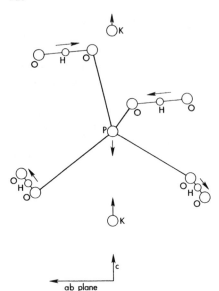

Figure 5.6 Eigenvector proposed by Cochran (1961) for the soft mode associated with the ferroelectric phase transition in KDP.

As the temperature is reduced the energy of the proton system is reduced by taking up a symmetric rather than a disordered arrangement relative to the PO_4 octahedra so that the protons approach a disorder-to-order phase transition. At the same time the low-frequency component of the coupled mode becomes unstable causing a spontaneous polarization along the c axis. Kobayashi, in effect, suggests that in this very complex system the transition to ferroelectricity is accompanied by both displacive and order-disorder behavior. His theory does not include refinements such as damping, necessary for experimental interpretation.

The first detailed investigation of KDP by Raman scattering was made by Kaminow and Damen (1968) who found a strong over-damped response in the low-frequency spectrum (Figure 5.7a). This is the soft ferroelectric mode, consistent with results of earlier infrared measurements by Barker and Tinkham (1963). Kaminow and Damen discussed their results using a single damped-harmonic-oscillator model, but it was shown later by She et al. (1972) that coupling between the soft mode and a zone-center transverse optic mode of the same (B_2) symmetry at about 180 cm^{-1} (Figure 5.7b) is important. The work of She et al. followed the approach of Katiyar et al. (1971) who had earlier reached similar conclusions in a light-scattering study of the isomorphs CsH_2AsO_4 and KH_2AsO_4. These authors found strong anharmonic coupling between the soft mode in CsH_2AsO_4 ($T_C = 143$ K) and an optic mode of B_2 symmetry at ~ 100

cm^{-1}, giving interference effects similar to those found in AlPO$_4$. These interference effects were somewhat similar to interference effects seen earlier in the infrared absorption spectra of perovskites by Barker and Hopfield (1964, this paper should be consulted for analysis of coupled-mode spectra). From their coupled-mode analysis, She et al. (1972) found that the frequency of the uncoupled soft mode varies as $\omega^2 = 67.7\,(T-30)$ cm^{-2}, going to zero at 30 K. However, extrapolation of their measurements suggests that the lower of the coupled modes is driven to zero

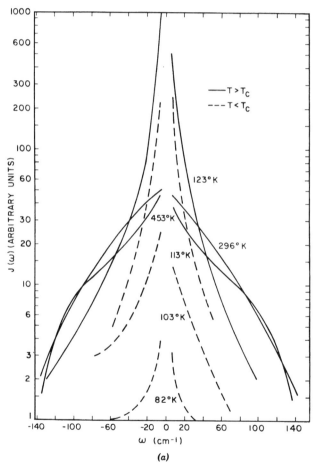

Figure 5.7 (a) Low-frequency Raman spectrum of KDP (after Kaminow and Damen 1968). $J(\omega)$ is the observed spectral intensity divided by the Bose factor.

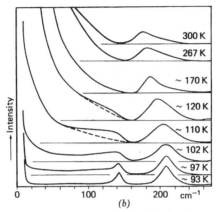

Figure 5.7 (b) Measured Raman intensity of KDP (after Shigenari and Takagi 1971). The dashed curves are extrapolations of the central peak (see Figure 5.7a). Corrected temperatures are taken from Butsuri, *30*, 733, 1975.

frequency at about 116 K, close to the true transition temperature of 122 K; such differences of temperature are generally discussed in terms of coupling of the soft mode to acoustic modes, which may result in the acoustic mode becoming unstable before the soft-mode frequency goes to zero (for example, see Cowley et al. 1971, Žeks and Barreto 1975, also Section 5.2.5).

The question always arises in interpreting overdamped spectra of whether a description in terms of damped harmonic oscillation (5.11) or Debye relaxation

$$\chi(\omega) = \chi(0) \frac{1}{1 + i\omega\tau} \qquad (5.28)$$

is more appropriate. In the limit of very strong damping ($\Gamma \gg \omega_s$), the two are equivalent, with the Debye relaxation time τ related to the damped harmonic oscillator parameters Γ and ω_s by $\tau = \Gamma/\omega_s^2$. In the case of KDP Peercy (1973) found that when hydrostatic pressure is applied at room temperature the soft-mode Raman response became underdamped for pressures greater than 6 kbar. At 9.3 kbar a peak developed in the response at 50 cm^{-1} and from a coupled-mode analysis Peercy found $\omega = 136$ cm^{-1} and $\Gamma = 160$ cm^{-1} for the soft mode. This result suggests that the soft mode should be regarded as a propagating wave rather than a diffusive excitation so that a harmonic-oscillator description is preferable to a Debye model and is a good example of the usefulness of hydrostatic-pressure techniques for the study of structural phase changes (Peercy 1975).

In the Raman spectrum of KDP below T_C an additional underdamped mode with A_1 symmetry is observed. This occurs at about 140 cm^{-1} at ~100 K (Shigenari and Takagi 1971) and moves to higher energy with decreasing T (Fig. 5.7b). Calculations by Lavrencic et al. (1972) suggest that this excitation may be derived from the soft-mode in the paraelectric phase (also, see Blinc & Žekš 1974, p. 148).

A variety of materials undergo order-disorder phase transitions associated with reorientation of complex ions. Of these the best known is NaNO$_2$, which is ferroelectric in its room-temperature, orthorhombic (C_{2v}) noncentrosymmetric phase (also, see Section 4.1.6). On warming to about 164°C it undergoes a first-order transition to the paraelectric orthorhombic (D_{2h}) centrosymmetric phase with an intermediate structure forming about 1° lower in temperature. The transition to the ferroelectric phase involves rearrangement of the NO$_2^-$ ions between two orientations and may be described by a double-well potential. Each of the NO$_2^-$ ions has an electric dipole moment, which, in the ordered phase, largely determines the observed polarization. Dielectric studies by Hatta and Sawada (1966) showed that the tunneling motion of the NO$_2^-$ ions (the soft mode of the system) is both slow and overdamped and can be characterized by a single Debye relaxation time $\tau(T) \sim 10^{-10}$ s at 22°C. This result is consistent with the quasielastic neutron-scattering studies carried out by Sakurai et al. (1970). Such low-frequency responses in solids can be studied by light-scattering techniques (see Section 5.2.6), but no measurements have been reported for NaNO$_2$. A similar comment may be made about order-disorder transitions in materials such as NH$_4$Cl and NH$_4$Br, associated with rotational disorder of the NH$_4^+$ ions (see Lazay et al. 1969, Wang and Fleury 1969, Geisel et al. 1975).

In concluding this section we mention a growing interest in the use of light scattering for studying dynamic disorder in fast-ion conductors. These materials show unusually large ionic conductivity associated with extensive disorder in a component sublattice, and they have practical applications (Mahan and Roth 1976). The motion of the ions contributing to the conductivity has been studied by light scattering in a number of materials including AgI (Ag$^+$ ions are mobile; Delaney and Ushioda 1976), Na$_2$O.11Al$_2$O$_3$ where large departures from stoichiometry occur in the sodium sublattice (Na$^+$ ions are mobile; Chase et al. 1976) and crystals with the fluorite structure, for example PbF$_2$ (F$^-$ ions are mobile; Harley et al. 1975). A review of progress in this area has been given by Chase (1976).

5.2.4 Cooperative Jahn-Teller Effects

The cooperative Jahn-Teller effect (CJTE) is caused by interaction between low-lying electronic states of one type of ion in a solid and the lattice vibrations (for a review, see Gehring and Gehring 1975). It may be of first or second order, and in both cases it involves a symmetry-lowering distortion of the crystal lattice and a splitting of electronic levels. Spinels containing some transition metal ions show this effect but these materials are not transparent and not readily studied by optical means. However, vanadates and arsenates of some rare-earth ions (for example $DyVO_4$) also show the CJTE (Cooke et al. 1970). These materials are optically transparent and crystallize with the tetragonal zircon structure. In this symmetry an orbital (non-Kramers) doublet can exist and such a doublet couples only to singly-degenerate vibrational modes. This is in contrast with the cubic-symmetry spinels in which orbitally-degenerate electronic states can couple only to degenerate lattice modes, resulting in more complex situations.

The theorem of Jahn and Teller (1937) states that the energy of an ion with non-Kramers degeneracy will be reduced by at least one distortion of its environment, which lowers the symmetry and raises the degeneracy in first order (linear molecules are an exception). The elastic energy of the system increases quadratically with the amplitude of the distortion so that eventually an equilibrium value for the distortion amplitude is reached. The Jahn-Teller effect associated with isolated ions in solids has been extensively studied by spectroscopic methods (for reviews see Sturge 1967, Abragam and Bleaney 1970, Ham 1972, Englman 1972). The coupling between isolated ions and the crystal lattice occurs through both lattice strain and phonons. If the active ions are present in a crystal in sufficiently high concentration the distortions give rise to an elastic interaction between neighboring ions and the entire crystal can become unstable with respect to a distortion arising from the cooperative effect of these interactions. As in the case of perovskites (Section 5.2.2), the phase transitions resulting from these interactions can lead to parallel alignment of the distortions (ferrodistortive) or to more complicated geometric arrangements. In the ferrodistortive case domains may be formed leading to difficulties in the study of polarization effects in light scattering.

$DyVO_4$ and $TbVO_4$ have been extensively studied by a variety of methods (Gehring and Gehring, 1975). These crystals undergo a second-order phase transition due to the CJTE at 14 and 34 K, respectively. The crystal space group above T_C is tetragonal D_{4h}; there are two equivalent rare-earth ions in the unit cell each with site symmetry D_{2d}. The crystal space group below T_C is D_{2h}, and this change to orthorhombic symmetry is a consequence of coupling to a zone-center B_{1g} vibrational mode in the case of $DyVO_4$ and a zone-center B_{2g} vibrational mode in the case of $TbVO_4$. A detailed study of the Raman spectra of these crystals shows (Elliott et al. 1972) that the phonon spectra are not appreciably changed by the transition except for a splitting of the doubly degenerate E_g modes below T_C. In particular, there is no evidence for the existence of a soft

Experimental Examples

phonon mode above or below T_C. However, in each crystal a soft electronic mode, associated with the Jahn-Teller splitting of lower electronic states, falls to low frequency as the temperature is increased to T_C.

The low-lying electronic levels of the rare-earth ions in $DyVO_4$ and $TbVO_4$ are shown in Figure 5.8 at temperatures well below and well above T_C. Dy^{3+} ($4f^9, {}^6H_{15/2}$) has an odd number of electrons and the CJTE occurs because the lowest states are two Kramers doublets whose separation is sufficiently small in comparison with the strength of the cooperative interaction. Although Tb^{3+} ($4f^8, {}^7F_6$) has an even number of electrons there is an actual degeneracy above T_C, which is raised at lower temperatures. The measured temperature dependence of one of the low-lying electronic transitions of $TbVO_4$ ($A_1 - A_1$ in Figure 5.8b) is shown in Figure 5.9.

Although detailed theoretical treatments of the CJTE now exist in the literature (Kanamori 1960, Allen 1968, Pytte 1971, Elliott et al. 1972), we look here only at some mean-field aspects. The ion-ion coupling arises

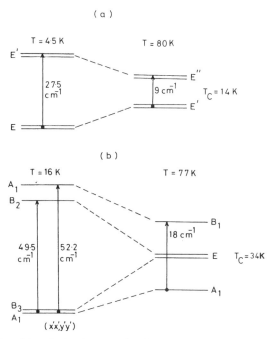

Figure 5.8 Low-lying electronic energy levels of (a) $DyVO_4$ and (b) $TbVO_4$ well below and well above T_C.

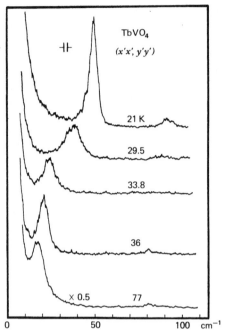

Figure 5.9 Temperature dependence of the electronic Raman transition $A_1 - A_1$ of TbVO$_4$ (see Figure 5.8 (b)) (after Elliott et al. 1972). $T_C = 34$ K.

through interactions via optic and acoustic phonons and also through macroscopic strain. The effect of the latter, which gives rise to long-range interaction, is predominant in most of the vanadates investigated (DyVO$_4$ is an exception) and accounts for the remarkable agreement between predictions of mean-field theory and experiment. For a degenerate orbital doublet the mean-field Hamiltonian may be written

$$\hat{H} = -\tfrac{1}{2} \sum_{l \neq l'} J(l,l') \sigma^z(l) \sigma^z(l'), \tag{5.29}$$

when the states $|\sigma^z(l) = \pm 1\rangle$ are associated with alignment of the distortion at site l along the x axis $(+1)$ or the y axis (-1)*: the tetragonal axis of the crystal is the z axis. This coupling between local distortions gives rise to an order-disorder transition.[†] If $\langle \sigma^z(l) \rangle$ takes a nonzero value, we

*The σ's are referred to as pseudo-spin operators. For discussions of their use see, e.g., Cochran (1969) and Elliott (1971).
[†]The vanadates were not included in the previous section because of the special nature of the CJTE and because of the detailed understanding achieved.

have an effective molecular field as in the magnetic case. This effective field $\hat{H}_{\mathrm{eff}} = \lambda \langle \sigma^z \rangle \sigma^z$ splits the degenerate states by $\pm \lambda \langle \sigma^z \rangle$, where $\lambda = \sum_{l \neq l'} J(l,l')$, and we have explicitly assumed that the ordering takes place in a zone-center mode. $\langle \sigma^z \rangle$ is the order parameter for the transition, and its actual value for a particular temperature is determined by the usual self-consistency condition of mean-field theory,

$$\langle \sigma^z \rangle = \tanh \frac{\lambda \langle \sigma^z \rangle}{k_B T} \tag{5.30}$$

Equation (5.29) leads to a typical Brillouin variation of $\langle \sigma^z \rangle$, falling from unity at $T=0$ to zero at a transition temperature given by $k_B T_C = \lambda$. This situation applies to TmVO$_4$ where the ground state is a degenerate doublet, well separated from higher levels (see Gehring and Gehring; also Section 5.2.5). If, as in the case of DyVO$_4$, the accidental degeneracy is not quite exact above T_C a term $\varepsilon \sigma^x$ (representing a splitting of the levels by 2ε above T_C) must be added to (5.29) and (5.30) now becomes, for $T < T_C$,

$$\left[\varepsilon^2 + \lambda^2 \langle \sigma^z \rangle^2 \right]^{1/2} = \lambda \tanh \frac{(\varepsilon^2 + \lambda^2 \langle \sigma^z \rangle^2)^{1/2}}{k_B T}. \tag{5.31}$$

The quantity $2[\varepsilon^2 + \lambda^2 \langle \sigma^z \rangle^2]^{1/2}$ is the splitting between the two electronic levels and this does not become zero at $T = T_C$. The absence of a predicted soft mode is a consequence of the use of the mean-field approximation. The full dynamic solution (Elliott et al. 1972) shows in fact that in the case of DyVO$_4$ the soft mode is a zone-center acoustic phonon driven flat by coupling to the electronic mode (also, see Section 5.2.5). On the other hand, in TbVO$_4$, the soft mode is in fact the transition associated with the splitting of the E doublet (Figure 5.8) rather than the coupled acoustic phonon.

The temperature dependence of the splitting of E-type phonons and of the change in electronic excitation energies (Figure 5.9) on going through the phase transition was extensively studied by Elliott et al. (1972) using Raman scattering. They showed that the splitting of the E-type phonons was determined by the order parameter $\langle \sigma^z \rangle$. They concluded that the behavior of TbVO$_4$ was closely mean field in character but that DyVO$_4$ showed departures from mean-field behavior. Results of Raman-scattering measurements on TmAsO$_4$ and TmVO$_4$ also indicated mean-field behavior (Harley et al. 1972). The critical exponent β was measured by optical methods in TbVO$_4$ and DyVO$_4$ by Harley and MacFarlane (1975) to within 30 mdeg of T_C; they found that TbVO$_4$ shows classical mean-field behavior with $\beta = 0.50$, whereas $\beta = 0.34$ in DyVO$_4$.

Differences in the low-lying Raman and infrared electronic excitation spectrum of $TbVO_4$ observed in the distorted phase has been assigned by Ergun et al. (1976) to Davydov splitting. This arises because there are two equivalent rare-earth ions in the unit cell, related by inversion, and interaction between them splits excitations into levels of even and odd parity (for further discussion, see Section 3.1.3 and Section 7.1). The former are observed in Raman scattering and the latter in infrared absorption.

5.2.5 Acoustic Anomalies

Many crystals show pronounced acoustic anomalies in the region of structural phase transitions and in some cases the anomalies are due to interaction between soft optic modes and acoustic modes. The acoustic anomalies may be observed by measuring sound velocities (and hence elastic constants) using ultrasonic or Brillouin-scattering techniques (see Chapter 8). The manner in which such an interaction affects sound velocities is illustrated schematically in Figure 5.10 where the solid lines represent TO and TA modes at temperatures well removed from a phase transition. In a situation characteristic of ferroelectrics, the TO frequencies at $q \cong 0$ will fall as T_C is approached. If symmetry conditions allow coupling between the two branches, the optic mode repels the acoustic branch leading to a decrease in slope at $q \cong 0$ and hence to a decrease in sound velocity. The slope of the acoustic branch at $q \cong 0$ may in fact fall to zero before the frequency of the TO mode falls to zero so that the structural instability is ultimately associated with the mode that is mainly acoustic rather than the mode that is mainly optic (see below). Only a single elastic constant (or a unique combination of elastic constants), determined by the symmetry properties of the coupled-mode system, shows an anomaly as T_C is approached (see p. 203).

The coupling between a soft mode and an acoustic mode may be linear or nonlinear in nature (Fleury 1971). Linear coupling takes the form COA where O represents the amplitude of the soft optic mode, A the amplitude of the acoustic mode, and C is a coupling tensor. The coefficient C is

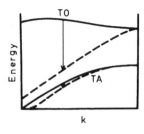

Figure 5.10 Schematic representation of coupling between a transverse-optic soft mode (TO) and a transverse acoustic mode (TA).

Experimental Examples

wavevector dependent and may be expanded in powers of q:

$$C = C_1 q + C_2 q^2 + C_3 q^3 + \cdots . \tag{5.32}$$

A term independent of q is not included because it involves a uniform translation of the crystal as a whole. For crystals containing lattice sites without inversion symmetry, the coefficient C_1 is dominant. If we retain only the first term in (5.32) the frequencies of the coupled modes are obtained from

$$(\omega^2 - \omega_A^2)(\omega^2 - \omega_O^2 - i\Gamma_O \omega) - C_1^2 q^2 / M_O M_A = 0, \tag{5.33}$$

where ω_O and ω_A are the uncoupled frequencies, Γ_O is the optic-mode damping, and M_O and M_A are the effective masses. If, for purposes of illustration, we ignore Γ_O we obtain for the coupled acoustic frequency $\omega_C (\omega_C \cong \omega_A \ll \omega_O)$

$$\omega_C^2 - \omega_A^2 \cong -\frac{1}{\omega_O^2} \left(\frac{C_1^2 q^2}{M_O M_A} \right), \tag{5.34}$$

equivalent to a coupled acoustic-mode velocity

$$v_C^2 = v_A^2 - \frac{1}{\omega_O^2} \left(\frac{C_1^2}{M_O M_A} \right), \tag{5.35}$$

so that in a second-order transition it is possible for the second term on the right of (5.35) to equal the first term before ω_O falls to zero. In this case the coupled acoustic mode becomes unstable before the optic-mode instability is complete. This conclusion is not appreciably affected by retaining damping. This type of behavior occurs in KDP (Brody and Cummins 1968), where the zone-center soft-optic mode (Section 5.2.3) is linearly coupled to transverse acoustic modes; the linear coupling is a consequence of the piezoelectric nature of KDP (see Dvořák 1968). The softening of the appropriate elastic constant C_{66} is shown in Figure 5.11a. The rise in the velocity of sound below T_C is a consequence of the development of a new structure stable against motions of the original C_{66} type.

Miller and Axe (1967) reach the interesting conclusion that only unstable modes that are Raman active will interact linearly with acoustic modes. They also conclude that with linear coupling dominating, optic-mode instability will always drive the acoustic mode unstable first. The soft optic mode in KDP is Raman active on both sides of T_C, accounting for the dip in C_{66} at T_C [Figure 5.11a; Elliott et al. (1971) discuss the detailed shape of

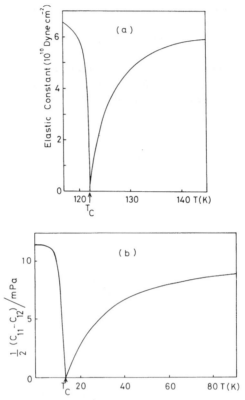

Figure 5.11 Temperature dependence of elastic constants measured by Brillouin scattering (a) C_{66} in KDP (after Brody and Cummins 1968) (b) $\frac{1}{2}(C_{11}-C_{12})$ in DyVO$_4$ (after Sandercock et al. 1972).

this curve]. Axe et al. (1970) show that in the case of centrosymmetric crystals it is necessary to carry the expansion (5.32) to fourth order in q to obtain linear coupling between a soft optic mode and acoustic modes. They found, using neutron scattering, that in the case of the centrosymmetric KTaO$_3$ there is a strong interaction between the soft TO mode (Section 5.2.2) and a TA branch at about one tenth the distance to the edge of the Brillouin zone.

For centrosymmetric crystals it is necessary to include nonlinear coupling between optic and acoustic modes to generate acoustic anomalies at $q=0$. The lowest-order interaction is of the form DO^2A. The effects of such nonlinear terms have been discussed by Fleury (1971) and will not be enlarged on here. It should be pointed out, however, that in centrosymmetric crystals such as SrTiO$_3$, in which the soft mode is not Raman active above T_C (Section 5.2.2), the acoustic anomaly at $q=0$ occurs as a step rather than a dip (Bell and Rupprecht 1963, Pytte 1970).

It also happens in crystals undergoing phase changes due to the cooperative Jahn-Teller effect (Section 5.2.4) that coupling between a zone-center soft electronic mode and zone-center acoustic phonons gives rise to an acoustic anomaly as T_C is approached. Again this instability is associated with softening of an elastic constant of well-defined symmetry and measurement of its temperature dependence by ultrasonic and Brillouin scattering techniques provides detailed information about the electron-lattice interaction (for a review, see Melcher 1976).

The temperature dependence of elastic constants of DyVO$_4$ has been measured using ultrasonic techniques ($\nu \cong 3 \times 10^7$ Hz) by Melcher and Scott (1972) and by Sandercock et al. (1972) using Brillouin scattering ($\nu \cong 10^{10}$ Hz). The lattice distortion associated with the phase change has B_{1g} symmetry (Section 5.2.4). The combination of elastic constants with this symmetry is $C_{11} - C_{12}$ and Sandercock et al. found that this combination fell to zero at T_C (Figure 5.11b) in agreement with the ultrasonic measurements of Melcher and Scott. Both sets of authors observed that the temperature dependence of $C_{11} - C_{12}$ was not consistent with mean-field behavior and Sandercock et al. showed that agreement with experiment above T_C could be improved by allowing for the existence of short-range order.

The phase transition in TbVO$_4$ corresponds to condensation of a zone-center lattice distortion of B_{2g} symmetry (Section 5.2.4). The temperature dependence of the associated elastic constant C_{66} has been investigated by both ultrasonic and Brillouin methods by Sandercock et al. (1972). They found that the results were consistent with mean-field behavior, indicating the dominance of zone-center acoustic phonons in the Jahn-Teller interaction. Similar conclusions were reached in the case of TmVO$_4$ ($T_C = 2.15$ K) by Melcher et al. (1973).

A thermodynamic variable such as an elastic constant has different values depending on whether the frequency of measurement is high or low compared to the inverse of some internal relaxation time. Relaxation times of interest in the Jahn-Teller case relate to equilibrium between Jahn-Teller distortions on different sites (referred to as spin-spin interaction) and between single-ion Jahn-Teller distortions and the lattice (referred to as spin-lattice relaxation). The frequency dependence of C_{66} in TbVO$_4$ has been measured by Sandercock et al. (1972) and discussed in terms of the above relaxation times.

Finally we mention a very restricted range of second-order transitions in which strain is the only order parameter. Application of Landau's general theory of second-order phase transitions by Anderson and Blount (1965) showed that such transitions can only occur if (a) the strain results in a change of symmetry and (b) all cubic, trigonal, and most hexagonal symmetries are not involved. A pressure-induced phase change in TeO_2 has been assigned to this class by Peercy and Fritz (1974), and Fritz and Peercy (1975).

5.2.6 Light Scattering Near Zero Frequency

Neutron-scattering experiments on many materials undergoing structural phase transitions, for example, $SrTiO_3$, Nb_3Sn, $KMnF_3$, and $LaAlO_3$, have shown that an extremely narrow quasielastic component grows rapidly in strength as the phase transition is approached and reaches maximum intensity at or very close to T_C (Axe et al. 1974, Cowley 1974). This quasielastic scattering is generally referred to as a central peak (Section 5.1). The neutron-scattering experiments suggest that the central peak and the soft mode are closely related, the dynamics being characterized by a phonon-like oscillating response and also some other slow response. However, the energy resolution currently available with neutron techniques is not adequate to place a limitation on the width of observed central modes and the possibility of a nondynamic origin arises. Axe et al. (1974) discussed scattering caused by strains associated with point defects (these are responsible for Huang diffuse scattering of X rays) but concluded that this type of static scattering was not consistent with observations. Halperin and Varma (1976) investigated in detail the possibility that moving defects may give rise to a very narrow dynamic central peak.

A central mode was observed in KDP by Lagakos and Cummins (1974) near $T_C = 122$ K using Brillouin scattering. They fitted their spectra (Figure 5.12) using an expression of the form (5.26) but their resolution ($\Delta \nu \cong 1$ GHz) was not adequate to measure the linewidth of the central peak. By studying the structure of the light-scattering column at different temperatures near T_C in the paraelectric phase, Durvasula and Gammon (1977) concluded that the central peak in KDP was due to scattering by static defects. It is also of interest to note that Lagakos and Cummins (1975) failed to observe a central peak in CsH_2AsO_4 and that Durvasula and Gammon (1975) found quasielastic scattering in KH_2AsO_4 due to domain formation only.

Lyons and Fleury (1976) using Brillouin scattering have observed a dynamic central peak of width 2.3 GHz in $KTaO_3$ at 300 K superimposed on a broader peak of width 108 GHz. They suggest that the former is due to entropy fluctuations (see Chapter 8) and that the latter is due to

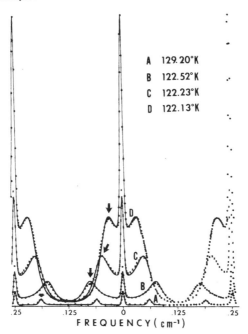

Figure 5.12 Brillouin spectrum of KDP at different temperatures near $T_C = 122$ K (after Lagakos and Cummins 1974).

two-phonon difference processes associated with large-wavevector phonons; neither peak is associated with a phase transition. Lyons and Fleury (1978) have identified both a static and a dynamic central peak in lead germanate, the former being assigned to defects and the latter to phonon density fluctuations.

It has already been pointed out (Section 5.2.4) that in the TbVO$_4$ lattice, phonons of C_{66} symmetry have the strongest coupling to the lowest four electronic levels of the Tb^{3+} ion; for $T > T_C$ these comprise two singlets separated by 18 cm^{-1} and a doublet midway between them (Figure 5.8). The phonon coupling is almost equally strong to an excitation between the singlets and to an excitation between the levels of the doublet (i.e., a zero-frequency excitation). In neutron-scattering experiments Hutchings et al. (1975) found a quasielastic peak near T_C, which they assigned to excitations of the doublet; excitation between the singlets gave rise to a coupled mode with the C_{66} acoustic phonon. It would appear that further experiments are required to establish with certainty the detailed origin of the quasielastic peak. The possibility of observing a central mode in such systems by light scattering has been discussed by Pytte (1973) and by Smith (1975), but no experiments have been reported.

REFERENCES

Abragam A. and Bleaney B. (1970), *Electron Paramagnetic Resonance of Transition Ions* (Oxford: Clarendon Press).

Aizu K. (1969), *J. Phys. Soc. Jap.* **27**, 387.

Allen S. J. (1968), *Phys. Rev.* **166**, 530; also **167**, 492.

Anderson P. W. (1960), *Fiz. Dielectr. Acad. Nauk. SSSR* 290.

Anderson P. W. and Blount E. I. (1965), *Phys. Rev. Lett.* **14**, 217.

Axe J. D., Harada J., and Shirane G. (1970), *Phys. Rev.* **B1**, 1227.

Axe J. D., Shapiro S. M., Shirane G., and Riste T. (1974), in T. Riste, Ed., *Anharmonic Lattices, Structural Transitions and Melting* (Leiden Noordhoff), p.23.

Axe J. D. and Shirane G. (1970), *Phys. Rev.* **B1**, 342.

Bacon G. E. (1975), *Neutron Diffraction*, 3rd ed. (Oxford: Clarendon Press).

Barker Jr., A. S. and Hopfield J. J. (1964), *Phys. Rev.* **135**, A1732.

Barker Jr., A. S. and Tinkham M. (1962), *Phys. Rev.* **125**, 1527.

Barker Jr., A. S. and Tinkham M. (1963), *J. Chem. Phys.* **38**, 2257.

Bell R. O. and Rupprecht G. (1963), *Phys. Rev.* **129**, 90.

Birgeneau R. J., Kjems J. K., Shirane G., and Van Uitert L. G. (1974), *Phys. Rev.* **B10**, 2512.

Birman J. L. (1973), *Phys. Lett.* **45A**, 196.

Blinc R. and Žekš B. (1974), in E. P. Wohlfarth, Ed., *Soft Modes in Ferroelectrics and Antiferroelectrics, Vol. XIII, Selected Topics in Solid State Physics* (Amsterdam: North Holland).

Brody E. M. and Cummins H. Z. (1968), *Phys. Rev. Lett.* **21**, 1263.

Burns G. and Scott B. A. (1973), *Phys. Rev.* **B7**, 3088.

Chase L. L. (1976), in G. D. Mahan and W. L. Roth, Eds., *Superionic Conductors* (New York: Plenum Press), p.299.

Chase L. L., Hao C. H., and Mahan G. D. (1976), *Solid State Commun.* **18**, 401.

Cochran W. (1960), *Adv. Phys.* **9**, 387.

Cochran W. (1961), *Adv. Phys.* **10**, 401.

Cochran W. (1969), *Adv. Phys.* **18**, 157.

Cochran W. (1971), in E. J. Samuelsen, E. Andersen, and J. Feder, Eds., *Structural Phase Transitions and Soft Modes* (Oslo: Universitetsforlaget), p.1.

Cochran W. and Zia A. (1968), *Phys. Stat. Sol.* **25**, 273.

Cooke A. H., Ellis C. J., Gehring K. A., Leask M. J. M., Martin D. M., Wanklyn B. M., Wells M. R., and White R. L. (1970) *Solid State Commun.* **8**, 689.

Coombs G. J. and Cowley R. A. (1973), *J. Phys. C, Solid State Phys.* **6**, 121.

Cowley R. A. (1962), *Phys. Rev. Lett.* **9**, 159.

Cowley R. A. (1963), *Adv. Phys.* **12**, 421.

References

Cowley R. A. (1964), *Phys. Rev.* **A134**, 981.
Cowley R. A., Buyers W. J. L., and Dolling G. (1969), *Solid State Commun.* **7**, 181.
Cowley R. A. (1974), *Ferroelectrics* **6**, 163.
Cowley R. A. and Coombs G. J. (1973) *J. Phys. C., Solid State Phys.* **6**, 143.
Cowley R. A., Coombs G. J., Katiyar R. S., Ryan J. F., and Scott J. F. (1971), *J. Phys. C., Solid State Phys.* **4**, L203.
Cracknell A. P., Lorenc J., and Przystawa J. A. (1976) *J. Phys. C, Solid State Phys.* **9**, 1731.
Delaney M. J. and Ushioda S. (1976), *Solid State Commun.* **19**, 297.
Devonshire A. F. (1949), *Philos. Mag.* **40**, 1040.
Dorner B., Axe J. D., and Shirane G. (1972), *Phys. Rev.* **B6**, 1950.
Durvasula L. N. and Gammon R. W. (1975), in M. Balkanski, R. C. C. Leite, and S. P. S. Porto, Eds., *Light Scattering in Solids* (Paris: Flammarion), p. 775.
Durvasula L. N. and Gammon R. W. (1977) *Phys. Rev. Lett.* **38**, 1081.
Dvořák V. (1968), *Phys. Rev.* **167**, 525.
Elliott R. J. (1971), in E. J. Samuelsen, E. Andersen and J. Feder, Eds., *Structural Phase Transitions and Soft Modes* (Oslo: Universitetsforlaget), p. 235.
Elliott R. J., Harley R. T., Hayes W., and Smith S. R. P. (1972), *Proc. R. Soc.* **A328**, 217.
Elliott R. J., Young A. P., and Smith S. R. P. (1971) *J. Phys. C, Solid State Phys.* **4**, L317.
Englman R. (1972), *The Jahn-Teller Effect in Molecules and Crystals* (London: Wiley).
Ergun H. B., Gehring K. A., and Gehring G. A. (1976) *J. Phys. C, Solid State Phys.* **9**, 1101.
Feder J. (1973), *Solid State Commun.* **13**, 1039.
Feder J. and Pytte E. (1970), *Phys. Rev.* **B1**, 4803.
Fleury P. A. (1971), *J. Acoust. Soc. Am.* **49**, 1041.
Fleury P. A. (1975), in M. Balkanski, R. C. C. Leite, and S. P. S. Porto, Eds., *Light Scattering in Solids* (Paris: Flammarion) p. 747.
Fleury P. A. and Lazay P. D. (1971), *Phys. Rev. Lett.* **26**, 1331.
Fleury P. A., Lazay P. D., and Van Uitert L. G. (1974), *Phys. Rev. Lett.* **33**, 492.
Fleury P. A., Scott J. F., and Worlock J. M. (1968), *Phys. Rev. Lett.* **21**, 16.
Fleury P. A. and Worlock J. M. (1968), *Phys. Rev.* **174**, 613.
Fritz I. J. and Peercy P. S. (1975), *Solid State Commun.* **16**, 1197.
Fröhlich H. (1949) *Theory of Dielectrics* (Oxford: Clarendon Press).
Gehring G. A. and Gehring K. A. (1975), *Rep. Prog. Phys.* **38**, 1.
Geisel T., Hochheimer H. D., and Keller J. (1975), in M. Balkanski, R. C. C. Leite, and S. P. S. Porto, Eds., *Light Scattering in Solids* (Paris: Flammarion), p. 801.
Halperin B. I. and Varma C. M. (1976), *Phys. Rev.* **B14**, 4030.
Ham F. S. (1972), in S. Geschwind, Ed., *Electron Paramagnetic Resonance* (New York: Plenum Press), p. 1.
Harada J., Axe J. D., and Shirane G. (1971), *Phys. Rev.* **B4**, 155.
Harley R. T. (1977), *J. Phys. C, Solid State Phys.* **10**, L205.
Harley R. T., Hayes W., Perry A. M., and Smith S. R. P. (1973), *J. Phys. C, Solid State Phys.* **6**, 2382.
Harley R. T., Hayes W., Perry A. M., Smith S. R. P., Elliott R. J., and Saville I. D. (1974), *J. Phys. C, Solid State Phys.* **7**, 3145.

Harley R. T., Hayes W., Rushworth A. J., and Ryan J. F. (1975), *J. Phys. C, Solid State Phys.* **8**, L530.
Harley R. T., Hayes W., and Smith S. R. P. (1972), *J. Phys. C, Solid State Phys.* **5**, 1501.
Harley R. T. and MacFarlane R. M. (1975), *J. Phys. C, Solid State Phys.* **8**, L451.
Hatta I. and Sawada S. (1966), *Jap. J. Appl. Phys.* **4**, 389.
Heller P. and Benedek G. B. (1962), *Phys. Rev. Lett.* **8**, 428.
Heller P. and Benedek G. B. (1965), *Phys. Rev. Lett.* **14**, 71.
Höchli U. T. and Scott J. F. (1971), *Phys. Rev. Lett.* **26**, 1627.
Holy J. A., Klein M. V., McMillan W. L., and Meyer S. F. (1976), *Phys. Rev. Lett.* **37**, 1145.
Hutchings M. T., Scherm R., Smith S. H., and Smith S. R. P. (1975). *J. Phys. C, Solid State Phys.* **8**, L393.
Jahn H. A. and Teller E. (1937), *Proc. R. Soc.* **A161**, 220.
Kaminow I. P. and Damen T. C. (1968), *Phys. Rev. Lett.* **20**, 1105.
Kanamori K. (1960), *J. Appl. Phys.* **31**, 14.
Katiyar R. S., Ryan J. F. and Scott J. F. (1971), *Phys. Rev.* **B4**, 2635.
Kobayashi K. (1968), *J. Phys. Soc. Jap.* **24**, 497.
Lagakos N. and Cummins H. Z. (1974), *Phys. Rev.* **B10**, 1063.
Lagakos N. and Cummins H. Z. (1975), *Phys. Rev. Lett.* **34**, 883.
Landau L. D. (1937), *Phys. Z. Sowjetunion* **11**, 26.
Landau L. D. and Lifshitz E. M. (1969), *Statistical Physics* (New York: Pergamon Press), Chapter 14.
Lazay P. D. and Fleury P. A. (1971), in M. Balkanski, Ed., *Light Scattering in Solids* (Paris: Flammarion), p. 406.
Lazay P. D., Lunacek J. H., Clark N. A., and Benedek G. B. (1969), in G. B. Wright, Ed., *Light Scattering Spectra in Solids* (New York: Springer-Verlag), p. 593.
Lavrencic B., Levstek I., Žeks B., and Blinc R. (1972), *Adv. Raman Spectr.* **1**, 157.
Levanyuk A. P. and Sobyanin A. A. (1968), *JETP* **26**, 612.
Lifshitz E. M. (1941), *Z. Eksp. Teor. Fiz.* **11**, 255, 269.
Lyons K. B. and Fleury P. A. (1976), *Phys. Rev. Lett.* **37**, 161.
Lyons K. B. and Fleury P. A. (1978), *Phys. Rev.* **B17**, 2403
Mahan G. D. and Roth W. L., Eds. (1976) *Superionic Conductors* (New York: Plenum Press).
Melcher R. L. (1976), *Phys. Acoust.* **12**, 1.
Melcher R. L., Pytte E., and Scott B. A. (1973), *Phys. Rev. Lett.* **31**, 307.
Melcher R. L. and Scott B. A. (1972), *Phys. Rev. Lett.* **28**, 607.
Miller P. B. and Axe J. D. (1967), *Phys. Rev.* **163**, 924.
Mountain R. D. (1966), *J. Res. Natl. Bur. Standards (U.S.)* **A70**, 207.
Müller K. A., Dalal N. S., and Berlinger W. (1976), *Phys. Rev. Lett.* **36**, 1504.
Müller K. A. and Berlinger W. (1971), *Phys. Rev. Lett.* **26**, 13.
Peercy P. (1973), *Phys. Rev. Lett.* **31**, 379.
Peercy P. (1975), in M. Balkanski, R. C. C. Leite, and S. P. S. Porto, Eds., *Light Scattering in Solids* (Paris: Flammarion), p. 782.
Peercy P. and Fritz I. J. (1974), *Phys. Rev. Lett.* **32**, 466.
Petzelt J. and Dvořák V. (1976), *J. Phys. C, Solid State Phys.* **9**, 1571, 1587.

References

Prettl K. and Rieder K. H. (1975) in M. Balkanski, R. C. C. Leite and S. P. S. Porto, Eds., *Light Scattering in Solids* (Paris: Flammarion), p. 939.
Pytte E. (1970), *Phys. Rev.* **B1**, 924.
Pytte E. (1971), *Phys. Rev.* **B3**, 3503.
Pytte E. (1973), *Phys. Rev.* **B8**, 3954.
Raman C. V. and Nedungadi T. M. K. (1940), *Nature* **145**, 147.
Ruvalds J. and Zawadowski A. (1970), *Phys. Rev.* **B2**, 1172.
Sakurai J., Cowley R. A., and Dolling G. (1970), *J. Phys. Soc. Jap.* **28**, 1426.
Sandercock J. R., Palmer S. B., Elliott R. J., Hayes W., Smith S. R. P., and Young A. P. (1972), *J. Phys. C, Solid State Phys.* **5**, 3126.
Schwabl F. (1972), *Phys. Rev. Lett.* **28**, 500.
Schwabl F. (1973), *Solid State Commun.* **13**, 181.
Scott J. F. (1968), *Phys. Rev. Lett.* **21**, 907.
Scott J. F. (1970), *Phys. Rev. Lett.* **24**, 1107.
Scott J. F. (1974), *Rev. Mod. Phys.* **46**, 83.
Scott J. F. and Porto S. P. S. (1967), *Phys. Rev.* **161**, 903.
Shapiro S. M., O'Shea D. C., and Cummins H. Z. (1967), *Phys. Rev. Lett.* **19**, 361.
Shapiro S. M., Axe J. D., Shirane G., and Riste T. (1972), *Phys. Rev.* **B6**, 4332.
Shapiro S. M. and Cummins H. Z. (1968), *Phys. Rev. Lett.* **21**, 1578.
She C. Y., Broberg T. W., Wall L. S., and Edwards D. F. (1972), *Phys. Rev.* **B6**, 1847.
Shigenari T. (1973), *Phys. Lett.* **46A**, 243.
Shigenari T. and Takagi Y. (1971), *J. Phys. Soc. Jap.* **31**, 312.
Shirane G. (1974), *Rev. Mod. Phys.* **46**, 437.
Shirane G., Axe J. D., and Harada J., and Remeika J. P. (1970), *Phys. Rev.* **B2**, 155.
Shirane G. and Yamada Y. (1969), *Phys. Rev.* **177**, 858.
Silberglitt R. (1972), *Solid State Commun.* **11**, 247.
Smith S. R. P. (1975) in M. Balkanski, R. C. C. Leite and S. P. S. Porto, Eds., *Light Scattering in Solids* (Paris: Flammarion), p. 329.
Spitzer W. G., Miller R. C., Kleinman D. Z., and Howarth L. E. (1962) *Phys. Rev* **126**, 1710.
Stanley H. E. (1971), *Introduction to Phase Transitions and Critical Phenomena* (Oxford: Clarendon Press).
Steigmeier E. F., Loudon R., Harbeke G., and Auderset H. (1975), *Solid State Commun.* **17**, 1447.
Sturge M. D. (1967), *Solid State Phys.* **20**, 91.
Thomas H. (1969), *IEEE Trans. Magn.* **5**, 874.
Thomas H. and Müller K. A. (1968), *Phys. Rev. Lett.* **21**, 1256.
Unoki H. and Sakudo T. (1967), *J. Phys. Soc. Jap.* **23**, 546.
Uwe H. and Sakudo T. (1975), *J. Phys. Soc. Jap.* **38**, 183.
Uwe H. and Sakudo T (1976), *Phys. Rev.* **B13**, 271.
Wang C. H. and Fleury P. A. (1969), in G. B. Wright, Ed., *Light Scattering Spectra in Solids* (New York: Springer-Verlag), p. 651.
Wehner R. K. (1966), *Phys. Stat. Sol.* **15**, 725.
Wilson J. H., DiSalvo F. J., and Mahajan S. (1975) *Adv. Phys.* **24**, 117.

Worlock J. M. (1969), *Bull. Am. Phys. Soc.* **14**, 368.

Worlock J. M. and Olsen D. H. (1971), in M. Balkanski, Ed., *Light Scattering in Solids* (Paris: Flammarion), p. 410.

Zawadowski A. and Ruvalds J. (1970), *Phys. Rev. Lett.* **24**, 1111.

Žeks B. and de Sa Barreto F. C. (1975), in M. Balkanski, R. C. C. Leite, and S. P. S. Porto, Eds., *Light Scattering in Solids* (Paris: Flammarion), p. 822.

CHAPTER SIX
Magnetic Scattering

6.1 Scattering by Simple Paramagnets and Ferromagnets ———————— 240
    6.1.1 Faraday Rotation
    6.1.2 The Scattering Cross Section
    6.1.3 Microscopic Theory

6.2 First-Order Light Scattering by Antiferromagnets ———————— 256
    6.2.1 Antiferromagnetic Magnons
    6.2.2 Antiferromagnetic Cross Section

6.3 Second-Order Light Scattering by Antiferromagnets ———————— 266
    6.3.1 Scattering Mechanism
    6.3.2 Magnon Interaction Effects

6.4 Magnetic Defect Scattering ———————— 276
    6.4.1 Scattering by Point Defects
    6.4.2 Scattering by Mixed Crystals

Magnetic excitations in crystals generally have smaller light-scattering cross sections than do vibrational excitations, and the observation of magnetic scattering had to wait for the advent of laser sources in Raman spectroscopy. The first successful experiment was that of Fleury et al. (1966) on antiferromagnetic FeF_2; light scattering has since become one of the standard methods for determining the frequencies and other properties of magnetic excitations.

Magnetic scattering occurs in paramagnetic materials and in the various types of magnetically ordered structure at low temperatures. The magnetic excitations have the form of spin waves or magnons at temperatures small compared to the transition temperature of an ordered magnetic material. Magnon frequencies are typically smaller than phonon frequencies, and ferromagnets and ferrimagnets always have a very-low-frequency excitation which is best observed by Brillouin scattering techniques. The magnon frequencies in antiferromagnets are usually rather higher and observable by Raman-scattering spectroscopy; most measurements are made on antiferromagnetic crystals (for a review of light scattering by antiferromagnets, see Balucani and Tognetti 1976). Ferromagnets and ferrimagnets that contain two or more magnetic ions in the magnetic primitive cell also have higher frequency excitations that sometimes contribute to the light-scattering spectrum (Khater 1978a,b).

The theory of the magnetic-scattering cross section is similar in principle to the vibrational case. It is possible to formulate macroscopic and microscopic theories that provide complementary accounts of the physical origins of the scattering. Magnetic scattering is, however, rather simpler to treat than vibrational scattering. This is a consequence of the basically simpler structure of the magnetic excitations, which can be adequately described in terms of a small number of macroscopic or microscopic parameters.

Second-order magnetic scattering is also observed in some crystals. The spectra can be interpreted in a quantitative manner that contrasts with the somewhat qualitative information obtained from second-order vibrational spectra.

Additional or modified scattering occurs in magnetic crystals doped with suitable magnetic or nonmagnetic impurities. The magnetic defect-induced scattering is analogous to the vibrational scattering discussed in Section 3.3.

6.1 SCATTERING BY SIMPLE PARAMAGNETS AND FERROMAGNETS

6.1.1 Faraday Rotation

We consider first the scattering by a cubic or isotropic material that undergoes a transition from a paramagnetic to a ferromagnetic state as the temperature is lowered through T_C. The macroscopic theory of magnetic scattering follows essentially the path described in Sections 3.1.4 and 4.1.2 for vibrational scattering, but it is necessary to take account of important differences between the susceptibilities of magnetic and nonmagnetic media. It is convenient to discuss the properties of the susceptibility in terms of the Faraday rotation, an effect closely related to the magnetic scattering. The macroscopic theory of the magnetic cross section follows in Section 6.1.2 and the corresponding microscopic theory is given in Section 6.1.3. References for the original work on the macroscopic theory include Pershan (1967), L'vov (1968a), LeGall (1971), and LeGall and Jamet (1971).

The electric susceptibility of a magnetic material has the symmetry property (Landau and Lifshitz 1960, Section 82)

$$\chi^{ij}(\mathbf{M}) = \chi^{ji}(-\mathbf{M}), \qquad (6.1)$$

where $\mathbf{M}$ is the magnetization defined so that its contribution to the magnetic induction is $\mu_0 \mathbf{M}$. We consider only experiments in which the light is in spectral regions where the material has no linear absorption, and the susceptibility then satisfies the additional relation

$$\chi^{ij}(\mathbf{M}) = \chi^{ji*}(\mathbf{M}). \qquad (6.2)$$

The susceptibility can be expanded in a power series in the components of the magnetization, and the first two terms are often sufficient,

$$\chi^{ij}(\mathbf{M}) = \chi^{ij} + \frac{\partial \chi^{ij}(\mathbf{M})}{\partial M^h} M^h. \qquad (6.3)$$

Application of the symmetry relations (6.1) and (6.2) then shows that the first term on the right is real and symmetric under interchange of i and j, similar to the susceptibility considered in (3.22). By contrast, the second, magnetic contribution in (6.3) is imaginary and antisymmetric, changing sign when i and j are interchanged.

The susceptibility components are further restricted by the spatial symmetry, and for a cubic or isotropic material they can be written in the matrix form

$$\chi^{ij}(\mathbf{M}) = \begin{bmatrix} \chi & -iGM^z & iGM^y \\ iGM^z & \chi & -iGM^x \\ -iGM^y & iGM^x & \chi \end{bmatrix}. \qquad (6.4)$$

Here χ is the real symmetric magnetization-independent susceptibility, whose symmetry properties are given more generally in Table 4.1, and

$$iG = \partial \chi^{ij}(\mathbf{M})/\partial M^h \qquad (6.5)$$

(for *ijh* any cyclic permutation of *xyz*). The relation between field and polarization that follows from the susceptibility (6.4) is

$$\mathbf{P} = \epsilon_0 \chi \mathbf{E} + i\epsilon_0 G \mathbf{M} \times \mathbf{E}. \qquad (6.6)$$

The components of the magnetization have thermal fluctuations whose power spectra determine the frequency distribution of the scattered light. However, one component of the magnetization has a nonzero average value for a paramagnet in an applied magnetic field or for a ferromagnet below the transition temperature. We take this to be the z component, and denote the average value $\langle M^z \rangle$. Its static contribution to the second term on the right of (6.6) is the cause of the Faraday rotation.

It follows from (6.4) or (6.6) that the refractive indices for light waves of right and left circular polarization propagating parallel to z are, respectively,

$$\eta^+ = [1 + \chi + G\langle M^z \rangle]^{1/2}, \qquad (6.7)$$

$$\eta^- = [1 + \chi - G\langle M^z \rangle]^{1/2}. \qquad (6.8)$$

The different velocities of the two circular components of a plane-polarized wave produce a Faraday rotation of the plane of polarization. The rotation angle per unit pathlength is

$$\theta_E = \frac{\omega(\eta^+ - \eta^-)}{2c} \approx \frac{\omega G \langle M^z \rangle}{2c\eta}, \qquad (6.9)$$

where

$$\eta = (1 + \chi)^{1/2} \qquad (6.10)$$

is the refractive index in the absence of a static magnetization, and the magnetic terms in (6.7) and (6.8) are assumed small enough for expansion of the square roots.

The subscript on the rotation angle symbolizes its origin in the effect of the magnetization on the *electric* susceptibility. A further contribution to the rotation arises from the *magnetic* susceptibility. The equation of motion of the magnetization in the presence of an applied magnetic field $\mathbf{B}_O$ is (see for example Chapter 16 of Kittel 1966 or Chapter 3 of Abragam 1961)

$$\frac{d\mathbf{M}}{dt} = \gamma \mathbf{M} \times \mathbf{B}_O. \tag{6.11}$$

Here the gyromagnetic ratio is given by

$$\gamma = \frac{g\beta}{\hbar}, \tag{6.12}$$

where β is the Bohr magneton and g is the ratio of Bohr magnetons to units $\hbar$ of angular momentum carried by the magnetic ions. It follows from (6.11) that the magnetization induced by an applied field of frequency ω is

$$\frac{i\gamma \mathbf{M} \times \mathbf{B}_O}{\omega}, \tag{6.13}$$

equivalent to a magnetic susceptibility ($\mu_0 M / B_O$)

$$\frac{i\gamma\mu_0}{\omega} \begin{bmatrix} 0 & -M^z & M^y \\ M^z & 0 & -M^x \\ -M^y & M^x & 0 \end{bmatrix}. \tag{6.14}$$

The antisymmetric off-diagonal terms in the magnetic susceptibility have a similar effect to the corresponding terms in the electric susceptibility (6.4). The resulting Faraday rotation angle analogous to (6.9) is

$$\theta_M \approx \frac{\eta\gamma\mu_0 \langle M^z \rangle}{2c}. \tag{6.15}$$

The total Faraday rotation is a sum of the electric and magnetic contributions. The magnetic part is independent of the frequency of the electromagnetic wave. The electric part is proportional to the frequency, and the susceptibility derivative G also tends to increase with increasing frequency [see (6.53)]. The electric Faraday rotation typically exceeds the magnetic Faraday rotation for frequencies larger than approximately 10^{14} Hz or 3000 cm^{-1}.

6.1.2 The Scattering Cross Section

The Faraday rotations are caused by the contributions of the *average* magnetization component $\langle M^z \rangle$ to the polarization (6.6) and the magnetization (6.13). Light scattering is caused by the contributions of the *fluctuating* magnetization components to the same polarization and magnetization.

Consider first the electric contribution. The relation (1.43) between the incident electric field and Stokes polarization takes a form determined by (6.6),

$$\mathbf{P}_S = i\epsilon_O G \mathbf{M}^* \times \mathbf{E}_I, \qquad (6.16)$$

where the fluctuating magnetization now plays the role of system variable. The corresponding cross section from (1.70) is

$$\frac{d^2\sigma}{d\Omega d\omega_S} = \frac{\omega_I \omega_S^3 \mathfrak{v} V \eta_S \epsilon_O^2 G^2 (\boldsymbol{\varepsilon}_S \times \boldsymbol{\varepsilon}_I)^i (\boldsymbol{\varepsilon}_S \times \boldsymbol{\varepsilon}_I)^j}{(4\pi\epsilon_O)^2 c^4 \eta_I} \langle M^i M^{j*} \rangle_\omega. \qquad (6.17)$$

Only the fluctuating part of the magnetization causes light scattering and the quantity

$$M^z - \langle M^z \rangle \qquad (6.18)$$

must be used for z components in the power spectra.

The magnetic contribution arises similarly from a component $\mathbf{M}_S$ of the magnetization that oscillates at the Stokes frequency and is given by (6.13) as

$$\mathbf{M}_S = i\gamma \mathbf{M}^* \times \mathbf{B}_I / \omega_I, \qquad (6.19)$$

where $\mathbf{B}_I$ is the magnetic vector of the incident light. The Stokes magnetization radiates scattered light in much the same way as the Stokes polarization. A calculation similar to that of Section 1.4.2 leads to a general result

$$\frac{d^2\sigma}{d\Omega d\omega_S} = \left(\frac{\mu_0}{4\pi}\right)^2 \frac{\omega_I \omega_S^3 \mathfrak{v} V \eta_I \eta_S^3 \langle \mathbf{b}_S \cdot \mathbf{M}_S^* \, \mathbf{b}_S \cdot \mathbf{M}_S \rangle_{\omega_S}}{c^4 |\mathbf{B}_I|^2} \qquad (6.20)$$

analogous to (1.68), where $\mathbf{b}_S$ is a unit vector parallel to the magnetic vector $\mathbf{B}_S$ of the scattered light. Insertion of the Stokes magnetization from (6.19) gives

$$\frac{d^2\sigma}{d\Omega d\omega_S} = \left(\frac{\mu_0}{4\pi}\right)^2 \frac{\omega_S^3 \mathfrak{v} V \eta_I \eta_S^3 \gamma^2 (\mathbf{b}_S \times \mathbf{b}_I)^i (\mathbf{b}_S \times \mathbf{b}_I)^j}{c^4 \omega_I} \langle M^i M^{j*} \rangle_\omega. \qquad (6.21)$$

The electric and magnetic contributions to the magnetic cross section were identified theoretically by Elliott and Loudon (1963) and Bass and Kaganov (1960), respectively. The total cross section is a sum of (6.17) and

(6.21) together with some cross terms between the electric and magnetic parts. However, the ratio of the electric and magnetic contributions to the cross section is approximately equal to the square of the ratio of the corresponding contributions to the Faraday rotation. Thus the magnetic part of the cross section is usually very small for the frequencies ordinarily used in light-scattering experiments. In what follows we restrict attention to the electric part (6.17). Alternative accounts of the theory are given by Moriya (1967, 1968) and Loudon (1970).

Expressions for the magnetization power spectra are derived by means of the fluctuation-dissipation theory of Section 1.4.4. Suppose that a fictitious magnetic field **B** is applied to a magnetic crystal. The energy of interaction is

$$H = -VM^i B^i, \qquad (6.22)$$

and the generalized force on the ith magnetization component is accordingly

$$F^i = VB^i. \qquad (6.23)$$

The linear response functions T^{ij} for the magnetization are defined by

$$M^i - \langle M^i \rangle = T^{ij} VB^j. \qquad (6.24)$$

Expressions for them are obtained by solution of the equation of motion for the magnetization.

Suppose that the crystal is subjected to a real static field $\mathbf{B}_O$ parallel to the z axis in addition to the fictitious field **B**. It is assumed that $\mathbf{B}_O$ is much larger than the fictitious field **B**. The basic equation of motion (6.11) can be improved by insertion of terms to represent relaxation of the magnetization components toward their average values ($\langle M^z \rangle B^x / B_O$, $\langle M^z \rangle B^y / B_O, \langle M^z \rangle$) in the presence of the applied fields. The components of the equation of motion then become the Bloch equations (see for example Chapter 3 of Abragam 1961), and their forms in the present case are

$$\frac{dM^x}{dt} = \gamma M^y B_O + \gamma (M^y B^z - M^z B^y) - \frac{M^x - (\langle M^z \rangle B^x / B_O)}{\tau_2} \qquad (6.25)$$

$$\frac{dM^y}{dt} = -\gamma M^x B_O + \gamma (M^z B^x - M^x B^z) - \frac{M^y - (\langle M^z \rangle B^y / B_O)}{\tau_2} \qquad (6.26)$$

$$\frac{dM^z}{dt} = \gamma (M^x B^y - M^y B^x) - \frac{M^z - \langle M^z \rangle}{\tau_1}, \qquad (6.27)$$

where the longitudinal relaxation time τ_1 is usually larger than the transverse relaxation time τ_2.

The first two Bloch equations combine to give

$$-i\omega M^+ = -i\omega_0 M^+ + i\gamma(M^z B^+ - M^+ B^z) - \frac{M^+}{\tau_2} + \frac{\langle M^z \rangle B^+}{B_0 \tau_2}, \quad (6.28)$$

where

$$M^+ = M^x + iM^y, \quad (6.29)$$

B^+ is similarly defined, ω is the frequency of the applied field **B**, and

$$\omega_O = \gamma B_O. \quad (6.30)$$

The Bloch equation in this form can easily be solved approximately. Since the fictitious field **B** is much smaller than $\mathbf{B}_O$, the x and y components of the magnetization are then also small and the average value of the z component is close to

$$\langle M^z \rangle = M_O + \frac{\chi_M B_O}{\mu_0}. \quad (6.31)$$

Here M_O is the spontaneous magnetization, which is nonzero only in the ferromagnetic state below temperature T_C, and

$$\chi_M = \mu_0 \frac{\partial \langle M^z \rangle}{\partial B_O} \quad (6.32)$$

is the differential magnetic susceptibility.

The Bloch equation (6.28) is therefore solved approximately by replacement of M^z by its average value (6.31) and neglect of the term $M^+ B^z$, which is of second order in the components of the small field **B**. The generalized force associated with M^+ is found by rearrangement of the interaction energy (6.22) to be

$$F^+ = \tfrac{1}{2} V B^+. \quad (6.33)$$

Thus the response function obtained by comparison of the approximate form of (6.28) with the defining equation (6.24) is

$$T^{++} = \frac{(2\gamma \langle M^z \rangle / V \omega_O)[\omega_O - (i/\tau_2)]}{\omega_O - \omega - (i/\tau_2)}. \quad (6.34)$$

Scattering by Simple Paramagnets and Ferromagnets

A similar treatment of (6.27) leads to

$$T^{zz} = \frac{\chi_M/V\mu_0\tau_1}{-i\omega + (1/\tau_1)}, \qquad (6.35)$$

and the remaining response functions are all zero.

The magnetization power spectra required for the cross section (6.17) are determined by the fluctuation-dissipation theorem (1.82). The contributions to the cross section from the two types of response (6.34) and (6.35) are

$$\frac{d^2\sigma}{d\Omega\,d\omega_S} = \frac{\omega_I\omega_S^3\mathfrak{v}\eta_S\epsilon_0^2 G^2\hbar\omega\gamma\langle M^z\rangle|(\boldsymbol{\epsilon}_S\times\boldsymbol{\epsilon}_I)^+|^2[n(\omega)+1]}{(4\pi\epsilon_0)^2 2c^4\eta_I\omega_O}$$

$$\times \frac{1/\pi\tau_2}{(\omega_O-\omega)^2 + (1/\tau_2)^2} \qquad (6.36)$$

and

$$\frac{d^2\sigma}{d\Omega\,d\omega_S} = \frac{\omega_I\omega_S^3\mathfrak{v}\eta_S\epsilon_0^2 G^2\hbar\omega\chi_M[(\boldsymbol{\epsilon}_S\times\boldsymbol{\epsilon}_I)^z]^2[n(\omega)+1]}{(4\pi\epsilon_0)^2 c^4\eta_I\mu_0}\cdot\frac{1/\pi\tau_1}{\omega^2+(1/\tau_1)^2}. \qquad (6.37)$$

These are known as the transverse and longitudinal scattering cross sections; they can be separated experimentally by appropriate choices of polarizations.

The longitudinal cross section (6.37) has a Lorentzian peak centered on zero frequency shift. The temperature dependence of the scattering is determined by the Bose-Einstein factor and the magnetic susceptibility. It is permissible to put

$$n(\omega) + 1 \approx \frac{k_B T}{\hbar\omega} \qquad (6.38)$$

for the very small frequency shifts involved. The mean field or molecular field expression (5.7) for the susceptibility has the form

$$\chi_M = \frac{N\beta^2\mu_0}{Vk_B(T-T_C)} \qquad (6.39)$$

for temperatures higher than T_C in a crystal with spins of $\frac{1}{2}$ and g factor equal to 2. At lower temperatures, the susceptibility has a maximum in the region of T_C and falls to zero at zero temperature. Figure 6.1 shows the calculated temperature dependence of the cross section. The occurrence of "critical" scattering close to T_C in the longitudinal case was predicted by L'vov (1968b) and Moriya (1968) but has not been observed experimentally.

The transverse cross section (6.36) also has a Lorentzian shape, now centred on the frequency ω_O defined in (6.30) and with a width $2/\tau_2$. The frequency ω_O is close to 1 cm^{-1} for a typical applied field of 1 tesla, and Brillouin spectroscopic techniques are used for observation of the scattering. Figure 6.2 shows both the Stokes and anti-Stokes spectra of CrBr$_3$. The lines L are caused by longitudinal phonons and the lines M_1 and M_2 are the result of simultaneous detection of magnetic scattering at two different scattering angles ϕ. These lines have different frequency shifts as a result of anisotropy effects not included in the above derivation (for more general response functions, see Anda 1973, 1976). The strong magnetic scattering by CrBr$_3$ is associated with its extremely high Faraday rotation.

The calculated temperature dependence of the transverse scattering is shown in Figure 6.1. The thermal factor can again be expanded as in (6.38)

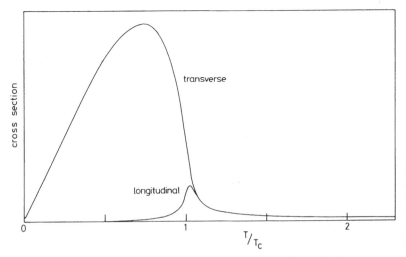

Figure 6.1 Temperature dependences of the transverse and longitudinal cross sections according to molecular field theory for a crystal that has $S=\frac{1}{2}$ and $g=2$. The magnitude of the applied field is such that $100\hbar\omega_O = k_B T_C$.

Scattering by Simple Paramagnets and Ferromagnets

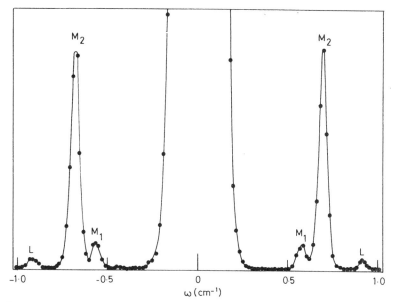

Figure 6.2 Brillouin spectrum of CrBr$_3$ in an applied field of 0.79 tesla at temperature ~8 K (from Sandercock 1974).

except at very low temperatures and with the help of (6.30), the scattered intensity from (6.36) is proportional to

$$\frac{\langle M^z \rangle T}{B_O}. \tag{6.40}$$

The strengths of the transverse and longitudinal scattering are equal for temperatures above T_C where the spontaneous magnetization M_O vanishes, but the transverse scattering far exceeds the longitudinal scattering below T_C as a result of the nonzero M_O. Figure 6.3 compares the calculated temperature dependence of the cross section with experimental points for CrBr$_3$. The experiment confirms the broad region of intense scattering in the ferromagnetic phase and the nonzero limiting value at high temperatures in the paramagnetic phase.

The polarization factor in (6.36) is

$$|(\varepsilon_S \times \varepsilon_I)^+|^2 = |\varepsilon_S^y \varepsilon_I^z - \varepsilon_S^z \varepsilon_I^y + i\varepsilon_S^z \varepsilon_I^x - i\varepsilon_S^x \varepsilon_I^z|^2$$
$$= (\varepsilon_S^y \varepsilon_I^z - \varepsilon_S^z \varepsilon_I^y)^2 + (\varepsilon_S^z \varepsilon_I^x - \varepsilon_S^x \varepsilon_I^z)^2. \tag{6.41}$$

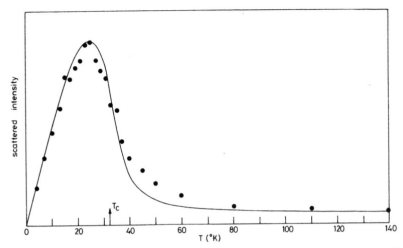

Figure 6.3 Comparison of theory with experiment for the temperature dependence of the intensity of magnetic scattering in CrBr$_3$ (from Sandercock 1974).

The magnetic scattering thus couples linear and circular polarizations of the incident and scattered light. For example, incident light linearly polarized along z produces scattered light circularly polarized in the xy-plane. The magnetic scattering is purely antisymmetric in the polarizations of incident and scattered light, a consequence of the antisymmetric form of the susceptibility derivative shown in (6.4) [see also (2.38)]. The longitudinal scattering (6.37) is also antisymmetric.

Magnetic scattering by yttrium iron garnet (YIG) has been studied by Sandercock and Wettling (1973). This is a ferrimagnet with many magnetic ions in the primitive cell and hence many modes of magnetic excitation. One of the modes is associated with oscillation of the macroscopic magnetization vector at a frequency close to ω_O, and the above theory is expected to apply in a first approximation. The experiments show, however, that there are two aspects in which the simple theory must be improved to provide an adequate description of light scattering by YIG.

The first of these is the termination of the susceptibility derivative expansion (6.3) at the linear term. The quadratic term is also important in YIG, particularly the contributions proportional to $M^z M^h$. These combine with the linear term to produce effective susceptibility derivatives of the forms

$$\frac{\partial \chi^{ij}(\mathbf{M})}{\partial M^h} + \frac{\partial^2 \chi^{ij}(\mathbf{M})}{\partial M^h \partial M^z} \langle M^z \rangle. \qquad (6.42)$$

Scattering by Simple Paramagnets and Ferromagnets

The quadratic term also gives rise to the Cotton-Mouton and Voigt effects (LeGall and Jamet 1971), and its possible contribution to light scattering was considered by LeGall et al. (1971) and Hu and Morgenthaler (1971). Its presence causes an interesting "anomaly" in the comparison of magnetic Stokes and anti-Stokes spectra, since the sign of the second term in (6.42) is reversed for the anti-Stokes cross section. This is a result of the reversal of the magnetization upon time reversal, as discussed after (1.94), and it removes the usual simple connection between Stokes and anti-Stokes intensities. Sandercock and Wettling (1973) observed the anomalous ratio of Stokes and anti-Stokes intensities in YIG.

The second aspect of the simple theory invalid for YIG is the assumption (6.2) of zero absorption. Detailed interpretation of a series of experiments (Wettling et al. 1975) shows that absorptive contributions to the susceptibility are important and indeed dominate the cross section for incident frequencies toward the upper end of the visible spectrum.

6.1.3 Microscopic Theory

The macroscopic theory provides no basis for calculating the susceptibility derivatives or for understanding their variation with the incident frequency, and such information must be sought in a microscopic theory of the scattering process. We consider here the scattering at temperatures well below the transition temperature where the magnetic excitations are quantized spin waves, or magnons, and only the transverse cross section has a significant magnitude. The coupling of incident and scattered light to the magnons occurs by virtual excitation of higher electronic states, as in vibrational scattering.

The theory of magnons in a ferromagnet is given, for example, by Kittel (1963), and the results are quoted here without proof. The magnetic crystals studied by light scattering are almost all insulators where the magnetic moments belong to transition metal ions. The spin angular momentum of each ion generates a magnetic moment $g\beta S$, and we assume that the ionic ground state has zero orbital angular momentum. The Hamiltonian of the crystal in an applied field as before is

$$\hat{H} = -2J \sum_{\langle \mu, \nu \rangle} \hat{\mathbf{S}}_\mu \cdot \hat{\mathbf{S}}_\nu - g\beta B_O \sum_\mu \hat{S}^z_\mu, \qquad (6.43)$$

where μ and ν label magnetic ions and J is the exchange coupling, assumed to be significant only for nearest-neighbor spins. The first summation counts each pair of neighboring spins only once.

The Hamiltonian is approximately diagonalized by the transformation

$$\hat{S}^-_\mu = \hat{S}^x_\mu - i\hat{S}^y_\mu = \left(\frac{2S}{N}\right)^{1/2} \sum_\mathbf{q} \exp(-i\mathbf{q} \cdot \mathbf{r}_\mu) \hat{b}^\dagger_\mathbf{q}, \qquad (6.44)$$

where the N primitive cells in the crystal each contain one magnetic ion, $\mathbf{r}_\mu$ is the position of spin $\mathbf{S}_\mu$, and $\hat{b}_\mathbf{q}^\dagger$ is the creation operator for a magnon of wavevector $\mathbf{q}$. The latter satisfies the usual commutation relation with the corresponding destruction operator $\hat{b}_\mathbf{q}$. The diagonalized Hamiltonian is

$$\hat{H} = \sum_\mathbf{q} \hbar \omega_\mathbf{q} \hat{b}_\mathbf{q}^\dagger \hat{b}_\mathbf{q}, \qquad (6.45)$$

where the magnon frequencies are

$$\omega_\mathbf{q} = \gamma B_E (1 - \gamma_\mathbf{q}) + \gamma B_O. \qquad (6.46)$$

The exchange field B_E is defined by

$$g\beta B_E = 2JzS \qquad (6.47)$$

and

$$\gamma_\mathbf{q} = \frac{1}{z} \sum_\delta \exp(i\mathbf{q} \cdot \boldsymbol{\delta}), \qquad (6.48)$$

where the $\boldsymbol{\delta}$ are the set of vectors connecting a spin to its z nearest neighbors; the functional form of $\gamma_\mathbf{q}$ depends only on the geometric arrangement of magnetic ions in the crystal.

The ground state of the Hamiltonian is one in which no magnons are excited and all the ionic magnetic moments are aligned parallel to the z axis. This is the state of the crystal at zero temperature. At higher temperatures, each thermally excited magnon reduces the total z component of the crystal spin by one unit, leading to the well-known Bloch law for the temperature dependence of the magnetization, the connection between magnetization and average ionic spin component being

$$\langle M^z \rangle = \frac{Ng\beta \langle S^z \rangle}{V}. \qquad (6.49)$$

For light-scattering experiments, the magnitude of the wavevector $\mathbf{q}$ is usually so small that it can be set equal to zero in the magnon dispersion relation. Then noting that γ_O is unity, the magnon frequency ω_O obtained from (6.46) is the same as the excitation frequency (6.30) derived in the macroscopic theory. More accurate magnon theory (Kittel 1963) takes account of dipole-dipole forces in addition to the exchange coupling included in (6.43), leading for cubic crystals to the small wavevector result

$$\omega_\mathbf{q} = \gamma \left[(B_O - \mu_O N^z \langle M^z \rangle)(B_O - \mu_O N^z \langle M^z \rangle + \mu_O \langle M^z \rangle \sin^2\theta) \right]^{1/2}. \qquad (6.50)$$

Here N^z is the sample demagnetization factor parallel to z and θ is the angle between **q** and the z axis. The more accurate frequency is orientation dependent, analogous to the polar vibration frequencies discussed in Section 4.1. Additional angular dependence of the magnon frequency occurs for lower symmetry crystals, such as the hexagonal $CrBr_3$ mentioned in Section 6.1.2.

The main mechanism for the scattering of light by magnons occurs by electric-dipole coupling of radiation to the crystal, and relies on the coupling of spin and orbital motions of the magnetic ions (Elliott and Loudon 1963). The mechanism is described here in terms of the ionic energy-level structure shown in Figure 6.4, and further details are provided by Shen and Bloembergen (1966) and Fleury and Loudon (1968).

As shown in Figure 6.4, the ionic ground state of spin S and zero orbital momentum L is split into $2S+1$ components of spacing $\hbar\omega_O$ by the applied field B_O. The ion is assumed to possess an excited orbital P state with $L=1$ and the same spin S as the ground state. Spin-orbit interaction of the usual form $\lambda \mathbf{L} \cdot \mathbf{S}$ splits the excited state into three components with typical spacings of order 10^2 cm^{-1}. The much smaller magnetic field splittings (~ 1 cm^{-1}) can be ignored in the excited state. The energy eigenfunctions $|J,J^z\rangle$ for the split excited states are expressed as linear combinations of the unperturbed $|L^z, S^z\rangle$ eigenfunctions in the usual way (e.g., Edmonds 1957).

We consider first the scattering of light by a spin transition of a single

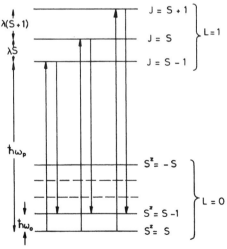

Figure 6.4 Electronic energy levels showing pairs of virtual transitions. Note that $\hbar\omega_P$ is normally much larger than λS and $\hbar\omega_O$ (after Fleury and Loudon 1968).

ion from the $S^z = S$ to the $S^z = S-1$ component of its orbital ground state. The transition can proceed by the pairs of allowed electric-dipole transitions shown in Figure 6.4, where each of the three spin-orbit split excited states can act as the virtual intermediate state. An incident photon is absorbed in one transition and a scattered photon is emitted in the other transition of the pair.

The differential cross section for a single-ion transition is given in general by (1.117), and it is not difficult to carry out the intermediate-state summation explicitly for the present simple example. The calculation is easiest when S has its smallest possible value of $\frac{1}{2}$. The $J = S-1$ intermediate state is then absent, and the cross section is

$$\frac{d\sigma}{d\Omega} = \frac{e^4 \omega_I \omega_S^3 \eta_S}{(4\pi\epsilon_O)^2 18\hbar^4 c^4 \eta_I} |(\boldsymbol{\varepsilon}_S \times \boldsymbol{\varepsilon}_I)^+|^2 |\langle 0,0|D^z|1,0\rangle\langle 1,-1|D^-|0,0\rangle|^2$$

$$\times \left\{ \frac{1}{\omega_P - \omega_I} - \frac{1}{\omega_P + \omega_S} - \frac{1}{\omega_P + \frac{3}{2}\lambda - \omega_I} + \frac{1}{\omega_P + \frac{3}{2}\lambda + \omega_S} \right\}^2, \quad (6.51)$$

where the bras and kets now show the $|L, L^z\rangle$ orbital states and **D** is defined in (1.118).

It is seen that the cross section depends critically on the size of the spin-orbit coupling parameter λ, and vanishes when λ is zero. The spin-orbit interaction is much smaller than the excitation energy $\hbar\omega_P$ of the virtual intermediate states in most magnetic crystals. The cross section can thus be expanded in powers of λ and only the lowest-order contribution retained. The result for general S, after a fair amount of algebra, can be written

$$\frac{d\sigma}{d\Omega} = \frac{\omega_I \omega_S^3 \eta_S \epsilon_O^2 g^2 \beta^2 G^2 S}{(4\pi\epsilon_O)^2 2c^4 \eta_I} |(\boldsymbol{\varepsilon}_S \times \boldsymbol{\varepsilon}_I)^+|^2, \quad (6.52)$$

where

$$G^2 = \frac{e^4 \lambda^2}{2\epsilon_O^2 \hbar^4 g^2 \beta^2} |\langle 0,0|D^z|1,0\rangle\langle 1,-1|D^-|0,0\rangle|^2$$

$$\times \left\{ \frac{1}{(\omega_P - \omega_I)^2} - \frac{1}{(\omega_P + \omega_S)^2} \right\}^2. \quad (6.53)$$

It is sometimes convenient to represent the total matrix element for the transition between the single-ion spin states by means of a spin Hamiltonian

$$\hat{H}_{\text{spin}} = \epsilon_0 g \beta G \varepsilon_S \times \varepsilon_I \cdot \hat{\mathbf{S}}, \tag{6.54}$$

defined so that the total matrix element is equal to

$$\langle S, S-1|\hat{H}_{\text{spin}}|S, S\rangle. \tag{6.55}$$

This is indeed the case with the above definition of the spin Hamiltonian since the only component of the spin operator that connects the spin states gives

$$\langle S, S-1|\hat{S}^-|S, S\rangle = (2S)^{1/2}. \tag{6.56}$$

The above discussion concerns the scattering by a spin transition on a single ion, but it is now a trivial matter to obtain the cross section for scattering by a magnon. From (6.44), the sum of spin transition operators $\hat{S}_\mu^-$ over all the magnetic ions produces the creation operator of the zero-wavevector magnon. Thus the magnon cross section is obtained from the single-ion cross section (6.52) by inclusion of a factor N obtained by comparison of (6.44) and (6.56), the magnon thermal factor $n(\omega_O)+1$, and a volume factor $\mathfrak{v}/V$, which takes account of the fraction of crystal used in the light scattering. The result is in exact agreement with the macroscopic expression (6.36) after removal of the final factor to obtain the differential cross section, replacement of $\langle M^z\rangle$ by its low temperature value of $Ng\beta S/V$, and of ω by ω_O. Thus the quantity G defined in (6.53) is the quantum-mechanical equivalent of the phenomenological G introduced in the discussion of Faraday rotation in Section 6.1.1.

The explicit form of G shows that it increases as ω_I is increased toward the excitation frequency ω_P, where resonance magnetic scattering effects occur, similar to the vibrational case discussed in Section 4.2.3. With typical values of the crystal parameters chosen as in the vibrational estimate of Section 4.2.2, the differential cross section estimated from (6.52) and (6.53) is comparable to or smaller than the vibrational cross section (4.90), but there is a wide range of uncertainty. More reliable estimates can of course be made for crystals of known Faraday rotation.

The above calculation has been particularly simplified by the assumption of a single group of spin-orbit split excited states. A more realistic calculation should allow the intermediate state summation to range over all excited states accessible by electric-dipole transitions, including for example charge transfer and vibronic transitions. This modifies the expression for G, but the overall form of the cross section, including its dependence on the spin-orbit coupling and the polarization vectors remains the same. Magnetic ions with ground states of zero orbital angular momentum include Mn^{++}, Fe^{+++}, Eu^{++}, and Gd^{+++}. The scattering symmetry shown in (6.41) or (6.52) is also valid for a ground state that has quenched orbital angular momentum (see Appendix H of Kittel 1966 for an explanation of quenching).

The magnetic excitations themselves are much more difficult to treat for crystals whose magnetic ions do not have $L=0$ ground states. The orbital degeneracy of the ground state may be lifted by the crystal fields and each low-lying crystal-field level gives rise to magnetic excitations that generally involve changes in both orbital and spin components. The magnons described above are a special case of these magnetic "excitons." Spin-orbit coupling is not essential for light scattering by excitons in which the ionic ground and excited levels have spin states in common. The purely antisymmetric polarization factor (6.41) does not necessarily hold in the more general case, and the only restrictions are those imposed by the symmetry of the excitation and of the crystal structure. Cracknell (1969) has extended the matrices of Table 1.2 to include the symmetry-restricted forms of the second-order susceptibility appropriate to excitations in ordered magnetic crystals.

Although the quantum-mechanical theory of light scattering outlined here is concerned with ferromagnets well below their transition temperatures, the calculation can be extended to all temperatures, including the paramagnetic region (Loudon 1970). The results agree with the macroscopic theory of Section 6.1.2. In particular, the frequency shift (6.30) remains independent of temperature.

It should be emphasized that light-scattering experiments examine true running-wave excitations of the magnetic system since their wavevector $\mathbf{q}$, although small, is usually much larger than the inverse of the sample dimensions. This contrasts with ferromagnetic resonance experiments in the microwave region, where the wavelengths used are often large compared with sample dimensions, and the relevant spin excitations are magnetostatic modes having frequencies characteristic of the sample shape employed.

6.2 FIRST-ORDER LIGHT SCATTERING BY ANTIFERROMAGNETS

6.2.1 Antiferromagnetic Magnons

The antiferromagnetic analogue of the Hamiltonian (6.43) is (Kittel 1963)

$$\hat{H} = 2J \sum_{\langle \mu, \nu \rangle} \hat{\mathbf{S}}_\mu \cdot \hat{\mathbf{S}}_\nu - g\beta B_A \left[\sum_\mu \hat{S}_\mu^z - \sum_\nu \hat{S}_\nu^z \right] - g\beta B_O \left[\sum_\mu \hat{S}_\mu^z + \sum_\nu \hat{S}_\nu^z \right]. \quad (6.57)$$

First-Order Light Scattering by Antiferromagnets

The first term is the exchange interaction between nearest neighbor spins, and its sign is such that the spins order on two sublattices. The spins on one sublattice (label μ) point parallel to z, whereas the spins on the other sublattice (label ν) point antiparallel to z. The second term represents the effect of crystal-field and dipole-dipole forces by an anisotropy field B_A that acts in opposite directions on the two sublattices; the form used here is only valid at temperatures below the antiferromagnetic transition. The final term in the Hamiltonian is the energy of the spins in an applied field B_O parallel to z.

The first part of Figure 6.5 is a schematic representation of the antiferromagnetic ground state. The z components of the total spins on the two sublattices are equal and opposite, so that the entire crystal has a spin component of zero. Each magnetic primitive cell contains two magnetic ions, one on each sublattice, and there are correspondingly two types of magnetic excitation for each wavevector $\mathbf{q}$. The second and third parts of Figure 6.5 represent these two types of magnon. One magnon changes the spin z component of the entire crystal to -1 and is denoted by labels $\downarrow \mathbf{q}$, whereas the other magnon corresponds to spin component $+1$ and is labelled $\uparrow \mathbf{q}$. As shown in Figure 6.5, both sublattices participate in both kinds of magnon. The degree of participation varies with wavevector and is

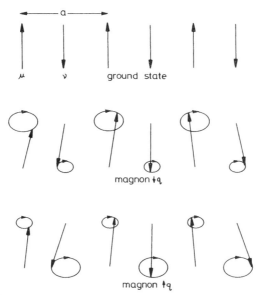

Figure 6.5 Schematic representations of the ground state and the two kinds of magnon in a two-sublattice antiferromagnet. The z axis points up the page.

described by two functions $u_\mathbf{q}$ and $v_\mathbf{q}$ (Kittel 1963). Each magnon changes the spin z component of one sublattice by $\pm u_\mathbf{q}^2$ and of the other sublattice by $\mp v_\mathbf{q}^2$, where the requirement

$$u_\mathbf{q}^2 - v_\mathbf{q}^2 = 1 \tag{6.58}$$

ensures that the magnon changes the spin component of the entire crystal by ± 1. Note that the sublattice with change in spin component $\pm u_\mathbf{q}^2$ has the larger circle of precession in Figure 6.5. Figure 6.6 shows a typical wavevector dependence of $u_\mathbf{q}^2$ and $v_\mathbf{q}^2$, with increasing confinement of the magnons to one or the other sublattice at large wavevectors.

The Hamiltonian (6.57) is diagonalized by a transformation from spin to magnon operators, similar to but more complicated than the ferromagnetic

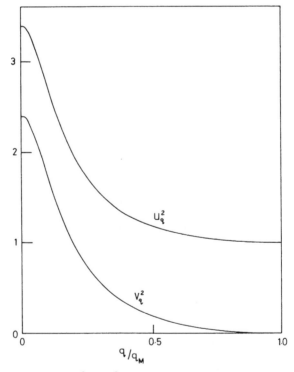

Figure 6.6 Variations of u_q^2 and v_q^2 with $\mathbf{q}$ in the (100) direction in MnF$_2$. The values at the zone boundary wavevector q_M are the same for any direction (from Fleury and Loudon 1968).

transformation (6.44). The result is (Kittel 1963)

$$\hat{H} = \sum_{\mathbf{q}} \{\hbar\omega_{\downarrow\mathbf{q}}\hat{b}^\dagger_{\downarrow\mathbf{q}}\hat{b}_{\downarrow\mathbf{q}} + \hbar\omega_{\uparrow\mathbf{q}}\hat{b}^\dagger_{\uparrow\mathbf{q}}\hat{b}_{\uparrow\mathbf{q}}\}, \tag{6.59}$$

where

$$\omega_{\downarrow\mathbf{q}} = \omega_{\mathbf{q}} + \gamma B_O, \tag{6.60}$$

$$\omega_{\uparrow\mathbf{q}} = \omega_{\mathbf{q}} - \gamma B_O, \tag{6.61}$$

and

$$\omega_{\mathbf{q}} = \gamma\{(B_E + B_A)^2 - (B_E\gamma_{\mathbf{q}})^2\}^{1/2}, \tag{6.62}$$

with B_E and $\gamma_{\mathbf{q}}$ given as before by (6.47) and (6.48). The two magnon frequencies are degenerate in the absence of an applied field. The frequencies of the zero-wavevector magnons observed in light scattering are determined by

$$\omega_O = \gamma\{(2B_E + B_A)B_A\}^{1/2}, \tag{6.63}$$

since γ_O is equal to unity, and an applied field splits the zero-field line into two symmetric components.

Much of the experimental work has been concerned with the rutile-structure antiferromagnets whose magnetic ordering is shown in Figure 6.7. The vibrational scattering by rutile is discussed in Section 3.1.6. The crystal symmetry operations in the antiferromagnetic state must leave invariant not only the positions of the ions but also their magnetic moments, and magnetic ordering reduces the space group from that of rutile to P_{nnm} or D_{2h}^{12}. The strong antiferromagnetic exchange interaction in these crystals occurs between a spin and its eight *second-nearest* neighbors, the interaction with the two nearest neighbors being ferromagnetic and much smaller. The above theory remains valid if this latter interaction is ignored, and it is not difficult to show with the help of Figure 6.7 that (6.48) gives

$$\gamma_{\mathbf{q}} = \cos\frac{aq_x}{2}\cos\frac{aq_y}{2}\cos\frac{cq_z}{2}. \tag{6.64}$$

A feature of the rutile structure is the vanishing of $\gamma_{\mathbf{q}}$ on the Brillouin zone boundary, where the maximum magnon frequency obtained from (6.62) is

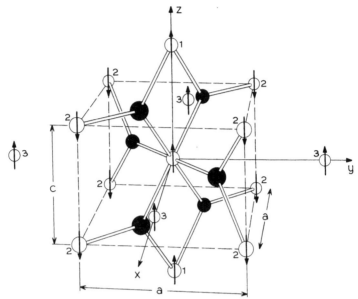

Figure 6.7 Primitive cell of the rutile structure indicated by the dashed lines. The first, second, and third magnetic neighbors of the central ion are shown. The arrows represent the magnetic ordering of MnF$_2$, FeF$_2$, and CoF$_2$ (from Fleury and Loudon 1968).

accordingly

$$\omega_M = \gamma(B_E + B_A). \tag{6.65}$$

Figure 6.8 shows the measured wavevector dependence of the magnon frequencies in the rutile-structure antiferromagnet MnF$_2$ in zero applied field. The separation between the two curves, not predicted by simple magnon theory, is a consequence of the weaker exchange coupling to the nearest and perhaps the third-nearest neighbors. The theory is easily generalized to include additional exchange couplings and the resulting formula fits the measured curves very well. The zero wavevector magnon frequency of 8.7 cm^{-1} in MnF$_2$ is not affected by the additional coupling.

The simple magnon theory also applies to FeF$_2$ since the crystal field splitting of the Fe^{++} ground state produces a lowest level that has quenched orbital angular momentum and can be treated as an $L=0$ state. The zero wavevector magnon frequency has the large value of 52 cm^{-1}, convenient for light-scattering spectroscopy. The higher frequency compo-

First-Order Light Scattering by Antiferromagnets

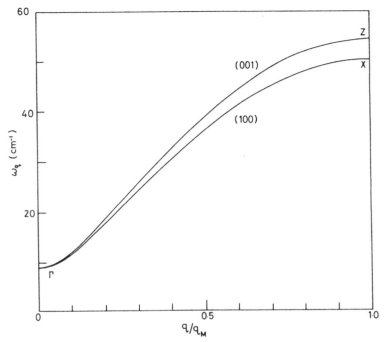

Figure 6.8 Magnon dispersion in MnF_2 at 4·2 K in the (100) and (001) directions as determined by neutron scattering (after Okazaki et al. 1964).

nents of the crystal-field split Fe^{++} ground state combine in the crystal to produce magnetic excitons with frequencies of 1000 cm^{-1} or more. The success of simple magnon theory in the crystal field ground state depends upon its comparatively large separation from the higher states.

By contrast, the simple magnon theory does not apply to CoF_2, where the Co^{++} ion has two low-lying Kramers doublets in the crystal field that are split by exchange interactions. The resulting magnetic excitons are combinations of transitions from the ground state of each magnetic ion to its three upper levels. The way in which excitations of the two Co^{++} ions in the primitive cell combine to produce magnetic excitations of the entire crystal can be determined by group theory (Loudon 1968). The calculation is similar to that for internal lattice vibrations described in Section 3.1.3 (see also Section 7.1). Figure 6.9 shows the measured wavevector dependence of the exciton frequencies in CoF_2. The excitons that result from the single-ion transitions to levels 1 and 3 remain twofold degenerate, whereas

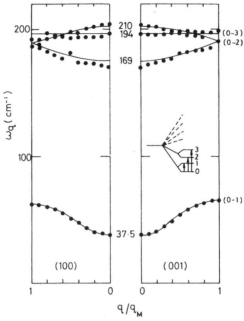

Figure 6.9 Wavevector dependence of the magnetic exciton frequencies in CoF_2 as determined by neutron scattering. The insert shows the parent single-ion transitions (after Cowley et al. 1973).

those associated with the transition to level 2 are Davydov split except on the Brillouin zone boundary.

The magnon theory outlined above applies well below the Néel temperature T_N of the antiferromagnetic phase transition. The magnon frequencies decrease with increasing temperature. In the paramagnetic phase above T_N, the excitation frequency is given by the same expression (6.30) as applies to a ferromagnet.

6.2.2 Antiferromagnetic Cross Section

The theory of light scattering by an antiferromagnet is very similar to that for a ferromagnet outlined in Section 6.1. The macroscopic approach derives a cross section expressed in terms of the magnetooptical properties of the crystal, whereas the microscopic calculation parallels that of Section 6.1.3. We consider here only the latter, and summarize the theory developed by Elliott and Loudon (1963), Shen and Bloembergen (1966), Moriya (1967, 1968), and Fleury and Loudon (1968).

First-Order Light Scattering by Antiferromagnets

The microscopic calculation of the single-ion cross section (6.52) applies equally to a magnetic ion in a ferromagnet and an antiferromagnet. However, the subsequent conversion to a magnon cross section is more complicated in the antiferromagnet because of the more complex relation between spin and magnon operators (Kittel 1963). Nevertheless, the principles of the conversion remain the same, and the cross section obtained for scattering by the $\downarrow$ magnon is

$$\frac{d\sigma}{d\Omega} = \frac{\omega_I \omega_S^3 \mathfrak{v} \eta_S \epsilon_O^2 G^2 \hbar \gamma M_O (u_O + v_O)^2 |(\epsilon_S \times \epsilon_I)^+|^2 [n(\omega_{\downarrow O}) + 1]}{(4\pi\epsilon_O)^2 2c^4 \eta_I}, \quad (6.66)$$

where M_O is the magnetization $Ng\beta S/V$ of a fully aligned sublattice. The cross section for the $\uparrow$ magnon is the same, except that $\omega_{\uparrow O}$ appears in the thermal factor and the polarization factor is changed to

$$|(\epsilon_S \times \epsilon_I)^-|^2. \quad (6.67)$$

Thus for incident light linearly polarized along z, the two kinds of magnon produce scattered light of opposite circular polarizations in the xy-plane. The derivation of these cross sections assumes that magnetic ions on the two sublattices occupy equivalent kinds of site.

The theoretical cross section for an antiferromagnet is therefore almost the same as that for a ferromagnet in (6.36) except for the appearance of an extra factor $(u_O + v_O)^2$. Typical values of u_O^2 and v_O^2 are shown in Figure 6.6. However, u_O and v_O have opposite signs and the extra factor causes a reduction in the cross section. Most antiferromagnets satisfy the inequality

$$B_E \gg B_A, \quad (6.68)$$

and Fleury and Loudon (1968) show that in this case

$$(u_O + v_O)^2 \approx \left(\frac{B_A}{2B_E}\right)^{1/2}. \quad (6.69)$$

It is seen from (6.63) that both the magnon frequency and the cross section are proportional to $B_A^{1/2}$ in the limit (6.68), and as a rough rule, antiferromagnets with smaller $q=0$ magnon frequencies tend to be poorer light scatterers. Representative numerical values are

$$\begin{array}{lll} \text{MnF}_2: & \omega_O = 8.7 \text{ cm}^{-1} & (u_O + v_O)^2 = 0.08 \\ \text{FeF}_2: & 52 \text{ cm}^{-1} & 0.4. \end{array} \quad (6.70)$$

The relatively large frequency shift and cross section of FeF_2 led to its role in the first successful observation of magnetic scattering (Fleury et al. 1966). Figure 6.10 shows the single peak in the spectrum for zero applied field together with the symmetric splitting into a pair of peaks in an applied field of 5.2 tesla. The antisymmetric dependence of the cross section on the incident and scattered polarization vectors predicted by (6.66) and (6.67) has been verified by a series of experiments with varying polarization, similar in principle to those shown in Figure 2.23. The same experiments show the cross section for the phonon at 257 cm^{-1} (see Table 3.5) to be symmetric in the polarization vectors.

The temperature dependence of the magnetic cross section in FeF_2 measured by Fleury (1971) shows a fall-off as the temperature is increased toward T_N. The temperature dependence can be calculated by the methods

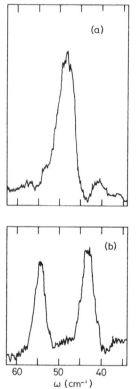

Figure 6.10 First-order magnetic scattering by FeF_2 at 20 K. (a) $B_O = 0$. (b) $B_O = 5 \cdot 2$ T parallel to the c axis (from Fleury and Loudon 1968).

described for the ferromagnetic case in Section 6.1. It is found (Cottam 1975) that agreement with the experiments on FeF_2 can only be achieved by inclusion of quadratic terms in the susceptibility-derivative expansion, as in the case of YIG described in Section 6.1.2. The measured low-temperature value of the cross section is of order 10^{-10} m^2, considerably smaller than a typical vibrational cross section.

Light scattering by the magnetic excitons derived from the higher frequency crystal-field levels in FeF_2 has also been observed. Chinn and Zeiger (1971) have detected such exciton scattering in the vicinity of 1000 cm^{-1}. More spectacularly, Macfarlane (1974) has measured excitons at shifts of 6381 and 6401 cm^{-1} in FeF_2; these are the highest-frequency excitations to be measured by light-scattering spectroscopy.

Detailed observations of the light scattering by CoF_2 have also been made (Macfarlane 1970, Moch et al. 1971). The zero-wavevector frequencies of all the exciton branches in Figure 6.9 have been determined, with the results entered on the figure. The order of magnitude of the observed cross sections is 10^{-7} m^2 and thus much larger than for FeF_2. The single-ion levels of Co^{++} are a case in which spin-orbit coupling is not essential for light scattering, as described at the end of Section 6.1.3. The antisymmetric property of the simple cross section (6.66) does not hold for the corresponding magnetic excitons, and the appropriate matrices from the tabulation of Cracknell (1969) must be used instead. Ishikawa and Moriya (1971) have made a detailed theory of the CoF_2 scattering cross section, which agrees well with the measurements. Further exciton branches in the frequency range 800 to 1400 cm^{-1} have also been measured by light-scattering spectroscopy (Macfarlane and Morawitz 1971, Moch et al. 1971).

Other antiferromagnetic crystals whose magnon or magnetic exciton spectra have been observed by light scattering include NiF_2 (Fleury 1969, 1971, Hutchings et al. 1970), FeF_3 (Shepherd 1973), $KCoF_3$, [$RbCoF_3$, $TlCoF_3$ (Nouet et al. 1973, Toms et al. 1973, Ryan et al. 1973), $FeCl_2 \cdot 2H_2O$ (Kinne et al. 1975), $FeBO_3$ (Jantz et al. 1976) and CoO (Chou and Fan 1976). Some of these crystals (NiF_2, FeF_3, and $FeBO_3$) are canted antiferromagnets and the total spins of their sublattices are not exactly antiparallel. Extensions of the theory are required for the canted structure. The spin arrangement of the ideal antiferromagnet described in Section 6.2.1 can be seriously perturbed by application of sufficiently high magnetic fields, leading eventually to the "spin-flop" transition and to a different magnetic structure. The cross section is expected to vary rapidly with applied field in the vicinity of the transition (Rezende 1973, Terakawa and Okiji 1975).

6.3 SECOND-ORDER LIGHT SCATTERING BY ANTIFERROMAGNETS

6.3.1 Scattering Mechanism

The shift and intensity of the first-order light scattering by magnons in MnF_2 indicated in (6.70) are too small for convenient experimental detection. However, MnF_2 does show additional scattering in its antiferromagnetic phase, and spectra for two polarizations are shown in Figure 6.11. It is seen by reference to Figure 6.8 that the peak scattering occurs at frequency shifts in the region of twice the zone-boundary magnon frequencies. This second-order or two-magnon scattering is similar in principle to the second-order scattering by phonons discussed in Section 3.2. Second-order magnon scattering was first seen in FeF_2 by Fleury et al. (1966) and it has since been observed in a wide range of antiferromagnetic crystals.

Calculation of the second-order magnon scattering is very much easier than in the phonon case. There is only one magnon branch in a simple antiferromagnet in the absence of an applied field, and the phonon result

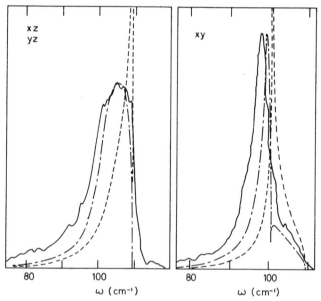

Figure 6.11 Theoretical and experimental spectra for two-magnon scattering by MnF_2 at 10 K. Continuous line: experiments of Fleury et al. (1967); dashed line: noninteracting magnon theory; dot-dash line: interacting magnon theory (from Thorpe 1970).

(3.53) is replaced by

$$2\omega_q = \omega_I - \omega_S. \tag{6.71}$$

The density of states, which largely determines the second-order magnon spectrum, is therefore

$$\rho_2(\omega) = \sum_q \delta(\omega - 2\omega_q), \tag{6.72}$$

which is a simpler special case of the phonon result (3.54). This two-magnon density of states is the same as the single-magnon density of states scaled in frequency by a factor of two. Figure 6.12 shows the single-magnon density of states for MnF_2 constructed from the theoretical magnon dispersion relation with exchange and anisotropy fields determined by fits to the neutron scattering results. The Van Hove singularities are labelled with their associated points of high symmetry in the MnF_2 Brillouin zone.

As in the phonon case, the intensity of second-order scattering reflects the magnitude of the two-magnon density of states, but also involves a **q**-dependent weighting function determined by the detailed scattering mechanism. It is possible to pursue a macroscopic approach to the second-order magnetic scattering, using second-order susceptibility derivatives similar to that in (6.42). However, we follow here the microscopic theory that has proved much more fruitful in the magnetic case.

The spin-orbit mechanism used in the treatment of first-order scattering in Section 6.1.3 can be extended to second order. It is necessary to consider a transition scheme like that of Figure 6.4, but with the down-pointing arrows ending on the $S^z = S - 2$ component of the orbital ground state. The resulting second-order cross section analogous to (6.66) is of

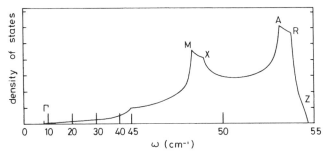

Figure 6.12 Magnon density of states in MnF_2. Note the change of scale at 45 cm^{-1} (from Nikotin et al. 1969).

fourth order in the spin-orbit coupling λ, is proportional to the x and y components of ε_S and ε_I, and is normally several orders of magnitude smaller than the first-order cross section.

Experiments show, however, that the intensity of second-order magnetic scattering is comparable to or even larger than the intensity of first-order scattering. For example, Figure 6.13 shows the first- and second-order lines of FeF_2 in the same spectrum. These results suggest that the second-order scattering arises not from a first-order mechanism taken to higher order (as in the phonon case) but from a new mechanism that occurs only in second order. Such a mechanism, known as the exchange scattering mechanism, seems to account for all the observed features of the second-order scattering. The basic idea, which was originated by Tanabe et al. (1965) in a theory of second-order absorption by antiferromagnetic magnons, was applied to second-order scattering by Fleury et al. (1967) and has been developed by Fleury and Loudon (1968), Moriya (1967, 1968), and L'vov (1968c).

The microscopic origin of the exchange-scattering mechanism is illustrated in Figure 6.14. Consider the specific case of MnF_2 where the ground orbital state of Mn^{++} has positive parity and a spin of 5/2. Figure 6.14 shows some energy levels of two representative ions μ and ν, one on each magnetic sublattice. The series of transitions indicated accomplishes the absorption of an incident photon, the emission of a scattered photon, and the excitation of both ions from their ground states to their first excited spin states.

The quantum-mechanical transition matrix element arises in third-order perturbation theory. Suppose that the ground state of ion μ has an electron $\mathbf{r}_1$ with $s^z = \frac{1}{2}$ accommodated in an orbital $|\mu\rangle$, while ion ν has an electron $\mathbf{r}_2$ with $s^z = -\frac{1}{2}$ in an orbital $|\nu\rangle$. The Hamiltonian for the electric-dipole interactions of the incident and scattered fields with the electrons,

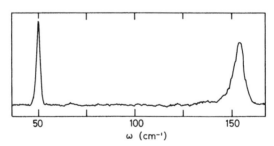

Figure 6.13 First and second-order magnetic scattering by FeF_2 at 15 K (from Fleury and Loudon 1968).

Second-Order Light Scattering by Antiferromagnets

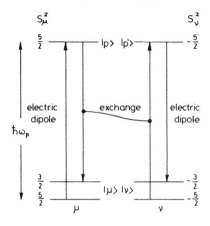

Figure 6.14 Successive transitions of the exchange scattering mechanism for the case of two Mn^{++} ions.

together with the Coulomb interaction of the electrons, is

$$\hat{H} = e(\hat{\mathbf{E}}_I + \hat{\mathbf{E}}_S) \cdot (\mathbf{r}_1 + \mathbf{r}_2) + \frac{e^2}{4\pi\epsilon_0|\mathbf{r}_1 - \mathbf{r}_2|}. \qquad (6.73)$$

The initial and final two-ion states for the transition of Figure 6.14 can be written $|\mu_+(\mathbf{r}_1)\nu_-(\mathbf{r}_2)\rangle$ and $|\mu_-(\mathbf{r}_2)\nu_+(\mathbf{r}_1)\rangle$, respectively, where the $+$ and $-$ subscripts indicate spin components of $+\frac{1}{2}$ and $-\frac{1}{2}$. There are many different third-order matrix elements of the Hamiltonian (6.73) that connect the required initial and final states. A representative contribution is

$$\langle \nu_+(\mathbf{r}_1)|e\hat{\mathbf{E}}_S\cdot\mathbf{r}_1|p'_+(\mathbf{r}_1)\rangle \left\langle p'_+(\mathbf{r}_1)\mu_-(\mathbf{r}_2)\left|\frac{e^2}{4\pi\epsilon_0|\mathbf{r}_1-\mathbf{r}_2|}\right|\nu_-(\mathbf{r}_2)p_+(\mathbf{r}_1)\right\rangle$$

$$\times \frac{\langle p_+(\mathbf{r}_1)|e\hat{\mathbf{E}}_I\cdot\mathbf{r}_1|\mu_+(\mathbf{r}_1)\rangle}{\hbar^2(\omega_I - \omega_P)^2}, \qquad (6.74)$$

where $|p\rangle$ and $|p'\rangle$ are one-electron orbitals of an odd-parity P-state of excitation energy $\hbar\omega_P$, and the relatively small splittings of states with different spin components have been neglected in the denominator.

The essential feature of the two-ion scattering process is the double spin-flip caused by the exchange of electrons in the central matrix element of (6.74). This matrix element of the Coulomb interaction is analogous to the exchange matrix elements between components of the ionic orbital ground states that give rise to the exchange couplings in the Hamiltonians (6.43) and (6.57). The total matrix element for the transition between the initial and final two-ion states can be represented by means of a spin Hamiltonian of the form

$$\hat{H}_{\text{spin}} = F^{ij}\hat{E}_S^i \hat{E}_I^j \hat{S}_\mu^- \hat{S}_\nu^+, \tag{6.75}$$

where the coefficients F^{ij} are chosen so that the matrix element

$$\left\langle S_\mu^z = \frac{3}{2}, S_\nu^z = -\frac{3}{2} \middle| \hat{H}_{\text{spin}} \middle| S_\mu^z = \frac{5}{2}, S_\nu^z = -\frac{5}{2} \right\rangle \tag{6.76}$$

is equal to the sum of all contributions (6.74).

The exchange scattering mechanism acts only for pairs of ions on opposite magnetic sublattices since the Hamiltonian (6.73) conserves the total z component of the electron spins. Such conservation is evident in the matrix element of (6.76) but does not occur for the corresponding excitation of a pair of ions on the same sublattice. In addition, because of the short range of exchange coupling, the strength of the scattering mechanism falls off rapidly with the ionic separation, and its effect is usually restricted to nearest neighbors on the opposite sublattices. The total two-ion spin Hamiltonian of the crystal is thus obtained by summation of (6.75) over all such pairs of ions, taking account of restrictions on the phases and spatial components i and j of the various contributions imposed by the symmetry of the lattice.

The case of MnF_2 is treated in great detail by Fleury and Loudon (1968) and the lengthy spin Hamiltonian is there derived with the help of group theory. An expression for the cross section is now obtained by the same steps as in first-order magnon scattering. It is necessary to convert the spin operators to magnon operators, and it is found that only the combination

$$\hat{b}^\dagger_{\downarrow \mathbf{q}} \hat{b}^\dagger_{\uparrow -\mathbf{q}} + \hat{b}^\dagger_{\uparrow \mathbf{q}} \hat{b}^\dagger_{\downarrow -\mathbf{q}} \tag{6.77}$$

occurs. The exchange scattering mechanism thus excites one magnon of each kind, and the sum of their frequencies from (6.60) and (6.61) is

$$\omega_{\downarrow \mathbf{q}} + \omega_{\uparrow -\mathbf{q}} = 2\omega_{\mathbf{q}}, \tag{6.78}$$

independent of the applied field B_O. Thus (6.71) holds even in the presence of an applied field, and experiments confirm that no changes in second-order spectra occur in applied fields of $5T$, unlike the splitting of the first-order spectrum shown in Figure 6.10.

Second-Order Light Scattering by Antiferromagnets

The cross section derived for second-order scattering with incident and scattered polarizations in the xy plane is

$$\frac{d^2\sigma}{d\Omega d\omega_S} = \frac{\omega_I \omega_S^3 \mathfrak{v} \eta_S 64 S^2 (F^{xy})^2 (\varepsilon_S^x \varepsilon_I^y + \varepsilon_S^y \varepsilon_I^x)^2}{(4\pi\epsilon_0)^2 c^4 \eta_I V}$$

$$\times \sum_{\mathbf{q}} [n(\omega_{\mathbf{q}}) + 1]^2 (u_{\mathbf{q}}^2 + v_{\mathbf{q}}^2)^2 \sin^2\frac{aq_x}{2} \sin^2\frac{aq_y}{2} \cos^2\frac{cq_z}{2} \delta(\omega - 2\omega_{\mathbf{q}}). \quad (6.79)$$

The trigonometric factor arises from the summation of the spin Hamiltonian over the second-neighbor pairs of Mn^{++} ions shown in Figure 6.7. Other polarizations of the light give similar results but with different trigonometric factors. Numerical estimates of the magnitude of the cross section based on the form (6.74) of the transition matrix element are consistent with the size of the measured cross section (Fleury and Loudon 1968).

The summation in the second line of (6.79) is a weighted two-magnon density of states, similar to the weighted density of states in the phonon cross section (3.63), but having a much more explicit form that can easily be calculated numerically. The first factor in the sum is close to unity at low temperatures, the second factor is also close to unity in the outer part of the Brillouin zone, where the bulk of the magnon density of states resides (see Figure 6.6), and the trigonometric factors provide the main weighting effect. They suppress contributions from the density of states in the vicinities of some critical points but not others, depending on the polarizations of the incident and scattered light. They determine the participation of pairs of critical point magnons in the scattering in accordance with selection rules that can be calculated group theoretically in the manner described for phonons in Section 3.2.1.

The theoretical spectra calculated from (6.79) and the analogous expression for polarizations in the zx or yz planes are shown by the dashed curves in Figure 6.11. First and second neighbor interactions are included in the magnon dispersion relation. The theory provides an approximate fit to the measurements, but magnon interaction effects must be included for a good quantitative agreement.

6.3.2 Magnon Interaction Effects

A really satisfactory theory of the second-order magnetic light scattering must treat the two-magnon state with a greater degree of sophistication than is called for in simple addition of the properties of two single-magnon states. The two magnons created in a scattering event interact, and their energy differs from a sum of the individual magnon energies.

A simplified picture of the source of the magnon interaction energy is shown in Figure 6.15. Consider just the exchange term in the Hamiltonian (6.57) and retain only the z components of the spins as in the Ising model of an ordered magnet. The first line of the figure represents the antiferromagnetic ground state and the second and third lines represent successive steps in changing the z components of two neighboring spins by one unit each. The energies required for the two steps are

$$2JzS = \hbar\gamma B_E \tag{6.80}$$

and

$$2J(zS-1). \tag{6.81}$$

The presence of the first spin-change leads to a reduction in the energy needed for the second spin-change and the total energy required is

$$2J(2zS-1), \tag{6.82}$$

being smaller than the energy of spin changes on two separated ions by an amount $2J$.

The spin Hamiltonian (6.75) for the second-order scattering involves

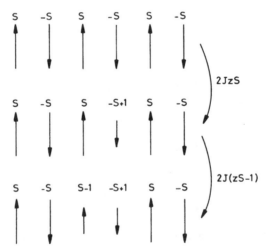

Figure 6.15 Successive steps in making a double spin flip on adjacent ions.

exactly the same kind of spin changes on neighboring ions. Although the Ising model is a rough approximation, the energy (6.80) correctly reproduces the exchange part of the magnon frequency (6.65) at the Brillouin zone boundary, where the peaks in the magnon density of states have their origins. The requirement of the exchange-scattering mechanism that the magnons are created on neighboring ions thus leads to a reduction in the energy of the two-magnon state by an amount of order $2J$ compared to twice the energy of a single magnon.

A more accurate calculation of the second-order spectrum must take account of the x and y components of the spins, leading to magnon frequencies that vary with wavevector. The requirements of crystal symmetry in forming suitable sums of the two-ion interaction (6.75) over the various pairs of neighbors must be satisfied in any realistic calculation of magnon interaction effects. A convenient and powerful method of calculation is provided by a fully quantum-mechanical version of the fluctuation-dissipation theory of Section 1.4, in which the classical response functions are replaced by the thermal Green functions mentioned in Section 1.5.1. The magnon interaction effects are taken into account in a proper calculation of the Green functions. Full details of the Green function calculation are given by Elliott and Thorpe (1969), and here we summarize the results.

A simpler crystal to consider first is the perovskite-structure antiferromagnet $RbMnF_3$, which has the calculational advantages over MnF_2 of a simple cubic arrangement of magnetic ions, an absence of any significant anisotropy fields, and only a single significant exchange interaction (of known magnitude). Group theory shows that the exchange-scattering mechanism produces a second-order spectrum with the single symmetry Γ_3^+ or E_g in the cubic group $m3m$ or O_h. The dashed curve in Figure 6.16 shows the spectrum calculated without inclusion of magnon interaction effects, analogous to the MnF_2 calculation described in Section 6.3.1. The smooth continuous curve includes the magnon interactions (Elliott et al. 1968, Elliott and Thorpe 1969) and shows the shift in intensity toward smaller frequencies expected from the simple arguments based on (6.82). The theoretical curves do not make use of any adjustable parameters other than their vertical scale. Finally, the experimental curve in Figure 6.16 is the work of Fleury (1968). The excellent match between theory and experiment provides a clear demonstration of the importance of magnon interactions in determining the detailed shapes of second-order spectra.

A similarly good agreement between theory and experiment is found for other antiferromagnetic crystals, which have a cubic or nearly-cubic arrangement of magnetic ions, for example FeF_3 (Moch and Dugautier 1972, Meixner et al. 1973), $FeBO_3$ (Meixner et al. 1973), $KNiF_3$ (Chinn et al. 1971b) and NiO (Perry et al. 1971, Dietz et al. 1971). The Mn^{++} and Fe^{+++} ions have spins of $5/2$, whereas Ni^{++} has $S=1$. It is seen from

(6.82) that the relative shift in peak position caused by magnon interaction should be larger for the smaller spin values, and this is borne out by the experiments on $KNiF_3$ and NiO.

The calculation of magnon-interaction effects is more difficult for MnF_2, where a reasonable theoretical treatment requires inclusion of the ferromagnetic exchange interaction between nearest neighbors on the same sublattice in addition to the stronger antiferromagnetic interaction. The Green-function calculation by Thorpe (1970) includes both exchange couplings, and the results are shown in Figure 6.11. Inclusion of the magnon-interaction effects brings the calculated spectra into much closer agreement with experiment.

Some antiferromagnetic structures are essentially two dimensional in that they have planes of normally spaced magnetic ions with wide separations between planes. Such crystals also have relatively large interaction effects because the number z of nearest neighbors tends to be small, for example $z = 4$ in K_2MnF_4 and K_2NiF_4. Second-order magnetic spectra of two-dimensional antiferromagnets are well accounted for by interacting-magnon theory (Parkinson 1969, Fleury and Guggenheim 1970, Chinn et al. 1971a, Lehmann and Weber 1973).

Second-order magnetic spectra often have an interesting temperature dependence. Results for NiF_2 are shown in Figure 6.17, and broadly similar behavior is found in other antiferromagnets, for example, $RbMnF_3$ (Fleury 1970), $KNiF_3$ and K_2NiF_4 (Fleury and Guggenheim 1970, Chinn et al. 1971b), and $KMnF_3$ (Lockwood and Coombs 1975). The frequency

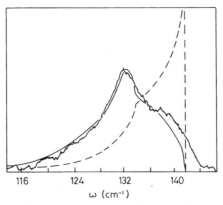

Figure 6.16 Experimental second-order magnetic spectrum of $RbMnF_3$ compared with calculated spectra; dashed line: noninteracting magnon theory; solid line: interacting magnon theory (after Fleury 1968).

of the first-order peak falls with increasing temperature and the peak disappears at the Néel temperature as expected for no applied magnetic field. The second-order peak, however, broadens considerably with increasing temperature but persists as a peak into the paramagnetic phase, with a reduction in energy at T_N (73 K) of only 25% of the zero-temperature value. These results are consistent with the greater dependence of the small wavevector magnons on the long-range order, which disappears at T_N.

A very large amount of theoretical effort has been devoted to the temperature dependence of the second-order spectra (Kawasaki 1970, Davies et al. 1971, Chinn et al. 1971a, Cottam 1972a,b,c, Natoli and Ranninger 1973, Balucani and Tognetti 1973a,b), using various techniques

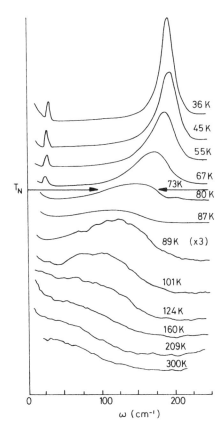

Figure 6.17 Temperature dependence of the first- and second-order magnetic scattering NiF_2. Note a trebling of the scale for $T \geqslant 89$ K (from Fleury 1969).

for calculating the Green function at elevated temperatures. The results are typically in agreement with experiment up to $\frac{2}{3}T_N$, but it is difficult to make a successful theory for higher temperatures, particularly in the region of T_N and above. A better account of experiments extending into the paramagnetic phase is obtained by restricting attention to the moments of the spectrum rather than the complete spectral lineshape (Brya and Richards 1974). It is also possible to construct a theory that is valid in the limit of high temperatures (Richards and Brya 1974). It should be emphasized that the measurement of moments of a light-scattering peak is not straightforward because of the background of light scattered by crystal imperfections.

Finally, other aspects of second-order magnetic scattering worthy of mention are the existence of the exchange-scattering mechanism in more complex magnetic crystals, for example the six-sublattice antiferromagnet $CsMnF_3$ (Chinn 1971) and the six-sublattice ferrimagnet $RbNiF_3$ (Fleury et al. 1969, Chinn et al. 1971b), and the occurrence of resonance scattering effects similar to those in vibrational spectra (Amer et al. 1975, Martin et al. 1977).

6.4 MAGNETIC DEFECT SCATTERING

We are concerned in this section with changes in the spin-wave spectra of pure materials that occur when magnetic ions are replaced by impurity ions, magnetic and nonmagnetic. The concepts and terminology employed are similar to those used in Section 3.3 for the vibrational case. Raman investigations have been largely confined to antiferromagnetic fluorides of transition metal ions, for example MnF_2 and $KMnF_3$, containing magnetic impurities such as Ni^{++} and nonmagnetic impurities such as Zn^{++}. For low concentrations of impurities (~ 1 mole%), the magnetic defects are mostly isolated and their levels can be categorized by the irreducible representations of the paramagnetic site group. If the magnetic interactions between the defect and the host magnetic neighbors are sufficiently strong, as for example in $KMnF_3$:Ni, there may be energy levels higher than the unperturbed magnon frequencies, corresponding to states strongly localized in the vicinity of the defect; these are called localized modes. Excitations in the magnon frequency range with an enhanced amplitude at the defect site are also possible and these are called resonance modes.

As the concentration of magnetic impurity ions is increased, we arrive in the mixed crystal regime. Here, as in the vibrational case, disorder obviates group-theoretical methods, and other theoretical approaches must be adopted. Since magnetic-exchange interactions have a short range, cluster

Magnetic Defect Scattering

models work reasonably well, although more powerful techniques such as the coherent potential approximation (CPA) are available (for a review, see Elliott et al. 1974). Cluster calculations in the magnetic case depend on fewer parameters than in the vibrational case, and more quantitative results can be obtained.

Light scattering by magnetic point defects is treated in Section 6.4.1. Excitations of these defects at low temperatures are reasonably well understood (for a review, see Cowley and Buyers 1971). The more difficult area of mixed crystals is discussed in Section 6.4.2.

6.4.1 Scattering by Point Defects

Consider an antiferromagnet with no applied field in which the crystal field and dipole-dipole forces are negligibly small ($B_A = 0$). If the nearest-neighbor exchange is now taken to be anisotropic, the Hamiltonian from (6.57) is

$$\hat{H} = 2 \sum_{\langle \mu,\nu \rangle} \left\{ I(S_\mu^x S_\nu^x + S_\mu^y S_\nu^y) + J S_\mu^z S_\nu^z \right\}. \tag{6.83}$$

The transverse part I of the exchange interaction is neglected in the Ising approximation, and, as discussed in Section 6.3.2, the energy (6.80) needed to change a spin z component by one unit is equal to the maximum magnon energy.

If a magnetic defect with effective spin S' and exchange interaction J' is now introduced, the spin-wave modes are altered at the defect site. It is assumed that the direction of the impurity spin is the same as that of the host spin it replaces. The defect-induced modes are most simply discussed in the Ising approximation, where the magnetic excitations are localized on particular sites. The excitation energy at the defect site itself is

$$\hbar \omega_D = 2 \sum_\delta J' \langle S^z(\delta) \rangle, \tag{6.84}$$

where the vectors δ defined in (6.48) connect a spin to its z nearest neighbors, and $\langle S^z(\delta) \rangle$ is the expectation value of the spin z component at site δ. For the antiferromagnetic perovskites where $z = 6$ and exchange interactions beyond the nearest neighbors are negligible, we can put

$$\hbar \omega_D = 12 J' S \tag{6.85}$$

at low temperatures where the spins are aligned nearly parallel or antiparallel to the z axis. The excitation energy for the neighbors of a defect is

$$\hbar\omega_N = 10JS + 2J'S'. \tag{6.86}$$

Corresponding expressions for rutile antiferromagnets are readily obtained (Cowley and Buyers 1971).

The magnetic excitations are no longer confined to single sites when the transverse part I of the exchange is included, and it is convenient to classify the modes centered on the defect site by symmetry. The symmetry of the paramagnetic site group in the perovskites is $m3m$ or O_h, and the amplitudes of the seven excitations associated with the defect and its six nearest neighbors transform as

$$2A_{1g} + E_g + T_{1u}. \tag{6.87}$$

One of the A_{1g} modes is located mainly on the defect and the other on the six neighbors. These modes are labeled s_0 and s_1, respectively; labels d and p are used for the E_g and T_{1u} modes. The effect of the transverse exchange I' on the defect-mode energies is discussed by Cowley and Buyers (1971).

It is possible to show in the cluster model that only modes of A_{1g} symmetry give rise to single-excitation Raman scattering. However, scattering by two-mode excitations is generally more intense, as in the second-order scattering by the pure crystal described in Section 6.3.1. The allowed pairs of modes in the perovskite structure are $A_{1g}(D) + A_{1g}(N)$ and $A_{1g}(D) + E_g$. The pairs can be distinguished experimentally, since they correspond, respectively, to the A_{1g} and E_g matrices in Table 1.2.

Some of the earliest studies of magnetic defects were carried out on MnF_2 containing Ni^{++} and Fe^{++} (Oseroff and Pershan 1969, Moch et al. 1969) and also Co^{++} (Moch et al. 1971). Subsequent interest has centered on perovskites such as $KMnF_3$ and $RbMnF_3$, which have the advantages of cubic rather than tetragonal symmetry, a single exchange coupling to nearest magnetic neighbors, and negligible anisotropy. The impurity Ni^{++} has received most attention in these materials, partly because of the almost complete absence of orbital angular momentum in its crystal-field ground state. The free-ion ^3F ground state is split by the octahedral environment to give an orbital singlet as the lowest level, with spin $S' = 1$. The $A_{1g}(D)$ mode of Ni^{++}, corresponding to spin transitions $S'^z = \pm 1 \leftrightarrow S'^z = 0$, has a frequency of ~ 260 cm^{-1} in $KMnF_3$ at helium temperatures. It has been observed in optical fluorescence by Johnson et al. (1966). The top of the spin-wave band in $KMnF_3$ is at ~ 75 cm^{-1} and the impurity mode is therefore highly localized. Its high energy arises because the exchange interaction J' between Ni^{++} and its nearest Mn^{++} neighbors is approximately 3.5 times the pure crystal exchange J. Neutron-scattering measurements (Holden et al. 1971) locate the remaining three defect modes near the top of the spin-wave band.

No first-order Raman scattering has been found in $KMnF_3$:Ni because of lack of intensity. However, the pair mode $A_{1g}(D)+E_g$ was observed at 315 cm^{-1} by Parisot et al. (1971) using Raman scattering. In $RbMnF_3$:Ni Oseroff et al. (1969) have recorded the spectrum shown in Figure 6.18, with the single E_g mode at 73 cm^{-1} and the $A_{1g}(D)+E_g$ mode at 295 cm^{-1}. As in the case of pure materials (see Section 6.3.2), the energy of two-mode excitations is not generally equal to the sum of the energies of the single modes.

6.4.2 Scattering by Mixed Crystals

The most detailed Raman investigations of mixed crystals have also been concerned with second-order scattering (for a brief review, see Hayes and Elliott 1975). In crystals diluted with a nonmagnetic constituent, the two-magnon spectrum remains as a single peak but increases in width and decreases in intensity with increasing dilution. For a crystal of mixed magnetic ions, a single two-magnon peak appears if the constituents are

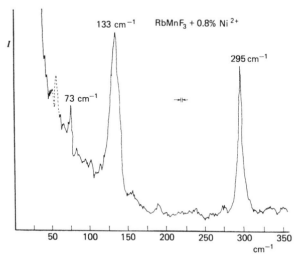

Figure 6.18 Raman spectrum of 0.8% Ni^{++}-doped $RbMnF_3$ at 8 K, including both xz and xy polarizations (from Oseroff et al. 1969).

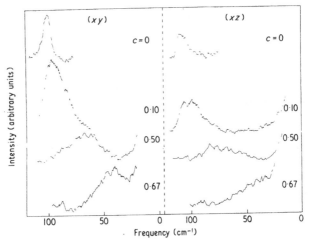

Figure 6.19 Two-magnon Raman spectrum of $Mn_{1-c}Zn_cF_2$ at 5 K for two polarizations and four zinc concentrations (from Buchanan et al. 1972).

magnetically similar, whereas two peaks appear if they are different. Such one- and two-mode behavior is discussed for the phonon case in Section 3.3.2.

The systems $Mn_{1-c}Zn_cF_2$ (Buchanan et al. 1972) and $KNi_{1-c}Mg_cF_3$ (Fleury et al. 1975) are the only magnetic crystals to be studied thus far for a complete range of c. Since Zn and Mg are effectively magnetic vacancies, no defect modes can appear at higher energies than the pure-crystal magnons. However, a magnetic resonance mode has been observed by neutron scattering (Svensson et al. 1969) in $Mn_{1-c}Zn_cF_2$ for $c=0.05$. The transition temperature $T_N(c)$ in such systems falls as c increases, and, where short-range exchange is the dominant magnetic interaction, $T_N(c)$ falls to zero at a critical concentration c_0. This limiting value is called the percolation limit and representative values are $c_0=0.76$ for the rutile structure and $c_0=0.69$ for the perovskite structure.

Figure 6.19 shows the spectrum of $Mn_{1-c}Zn_cF_2$. The two-magnon Raman peak is observable up to $c \approx 0.7$ but the scattering becomes too weak for detection at higher zinc concentrations. The transition temperature falls from 72 K for MnF_2 to 20 K at $c=0.5$, and the low values of $T_N(c)$ at the higher zinc concentrations require a careful interpretation of Raman-scattering experiments. The two-magnon peak frequency falls linearly with increasing c up to $c \approx 0.3$ and thereafter falls more rapidly. The linewidth increases quickly at low c and is already twice that of pure MnF_2 at $c=0.1$. Similar behavior is observed in $KNi_{1-c}Mg_cF_3$.

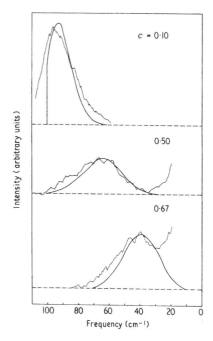

Figure 6.20 Comparison of calculated and measured two-magnon bands of $Mn_{1-c}Zn_cF_2$ for xy polarization and three zinc concentrations (from Buchanan et al. 1972).

The Ising cluster model provides a simple approximation for the mixed-crystal studies of two-magnon excitations (Buchanan et al. 1972). It gives the energy of the two-magnon peak as

$$\hbar\langle\omega\rangle = 4JzS(1-c), \quad (6.88)$$

where percolation effects and the magnon-interaction term in (6.82) are ignored. The Ising cluster model also gives a lineshape for the Raman line, obtained as a histogram by weighting the energy of a nearest-neighbor pair of Mn^{++} ions with z_1 and z_2 magnetic neighbors by the probability of occurrence of z_1 and z_2 for a random distribution of Zn ions. Figure 6.20 compares some smoothed histograms with experiment.

Although a number of two-mode mixed magnetic systems have been studied by neutron scattering (Elliott et al. 1974), Raman studies are limited. Fleury and Guggenheim (1975) measured the spectrum shown in Figure 6.21 for the mixed layered antiferromagnet $Rb_2Mn_{0.5}Ni_{0.5}F_4$. The Raman peaks at 248 and 372 cm^{-1} arise from one Mn- plus one Ni-branch magnon and from the two Ni- branch magnons, respectively. The other peaks are caused by phonons.

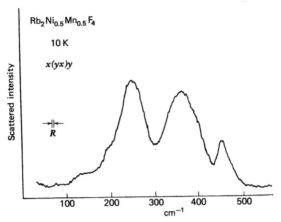

Figure 6.21 Raman spectrum of $Rb_2Ni_{0.5}Mn_{0.5}F_4$ at 10 K in xy polarization (from Fleury and Guggenheim 1975).

The effects of increasing temperature on magnetic scattering are experimentally similar to the effects of disorder produced by dilution with nonmagnetic ions (compare Figures 6.17 and 6.19). Hayes and Elliott (1975) review work on the combined effects of dilution and elevated temperature. In summary, the qualitative behavior of two-magnon scattering with dilution and with increase of temperature conforms to theoretical expectations. However, the detailed forms of the spectra, particularly near the percolation concentration c_0 and the transition temperature T_N are not well understood.

REFERENCES

Abragam A. (1961), *The Principles of Nuclear Magnetism* (Oxford: Clarendon Press).
Amer, N. M., Chiang T., and Shen Y. R. (1975), *Phys. Rev. Lett.* **34**, 1454.
Anda E. (1973), *J. Phys. Chem. Solids* **34**, 1597.
Anda E. (1976), *Solid State Commun.* **19**, 827.
Balucani U. and Tognetti V. (1973a), *Solid State Commun.* **13**, 1811.
Balucani U. (1973b), *Phys. Rev.* **B3**, 4247.
Balucani U. (1976), *Riv. Nuovo Cim.* **6**, 39.
Bass F. G. and Kaganov M. I. (1960), *Sov. Phys.-JETP* **37**, 986.
Brya W. J. and Richards P. M. (1974), *Phys. Rev.* **B9**, 2244.
Buchanan M., Buyers W. J. L., Elliott R. J., Harley R. T., Hayes W., Perry A. M., and Saville I. D. (1972), *J. Phys.* **C5**, 2001.
Chinn S. R. (1971), *Phys. Rev.* **B3**, 121.

References

Chinn S. R., Davies R. W., and Zeiger H. J. (1971a), *Phys. Rev.* **B4**, 4017.
Chinn S. R. and Zeiger H. J. (1971), *AIP Conf. Proc.* **5**, 344.
Chinn S. R., Zeiger H. J., and O'Connor J. R. (1971b), *Phys. Rev.* **B3**, 1709.
Chou H. and Fan H. Y. (1976), *Phys. Rev.* **B13**, 3924.
Cottam M. G. (1972a), *Solid State Commun.* **10**, 99.
Cottam M. G. (1972b), *J. Phys.* **C5**, 1461.
Cottam M. G. (1972c), *Solid State Commun.* **11**, 889.
Cottam M. G. (1975), *J. Phys.* **C3**, 1933.
Cowley R. A. and Buyers W. J. L. (1971), *Rev. Mod. Phys.* **44**, 406.
Cowley R. A., Buyers W. J. L., Martel P., and Stevenson R. W. H. (1973), *J. Phys.* **C6**, 2997.
Cracknell A. P. (1969), *J. Phys.* **C2**, 500.
Davies R. W., Chinn S. R., and Zeiger H. J. (1971), *Phys. Rev.* **B4**, 992.
Dietz R. E., Parisot G. I., and Meixner A. E. (1971), *Phys. Rev.* **B4**, 2302.
Edmonds A. R. (1957), *Angular Momentum in Quantum Mechanics* (Princeton, N.J.: Princeton University Press).
Elliott R. J., Krumhansl J. A., and Leath P. L. (1974), *Rev. Mod. Phys.* **46**, 465.
Elliott R. J. and Loudon R. (1963), *Phys. Lett.* **3**, 189.
Elliott R. J. and Thorpe M. F. (1969), *J Phys.* **C2**, 1630.
Elliott R. J., Thorpe M. F., Imbusch G. F., Loudon R., and Parkinson J. B. (1968), *Phys. Rev. Lett.* **21**, 147.
Fleury P. A. (1968), *Phys. Rev. Lett.* **21**, 151.
Fleury P. A. (1969), *Phys. Rev.* **180**, 591.
Fleury P. A. (1970), *J. Appl. Phys.* **41**, 886.
Fleury P. A. (1971), in M. Balkanski, Ed., *Light Scattering in Solids* (Paris: Flammarion), p. 151.
Fleury P. A. and Guggenheim H. J. (1970), *Phys. Rev. Lett.* **24**, 1346.
Fleury P. A. and Guggenheim H. J. (1975), *Phys. Rev.* **B12**, 987.
Fleury P. A., Hayes W., and Guggenheim H. J. (1975), *J. Phys.* **C8**, 2183.
Fleury P. A. and Loudon R. (1968), *Phys. Rev.* **166**, 514.
Fleury P. A., Porto S. P. S., Cheesman L. E., and Guggenheim H. J. (1966), *Phys. Rev. Lett.* **17**, 84.
Fleury P. A., Porto S. P. S., and Loudon R. (1967), *Phys. Rev. Lett.* **18**, 658.
Fleury P. A., Worlock J. M., and Guggenheim H. J. (1969), *Phys. Rev.* **185**, 738.
Hayes W. and Elliott R. J. (1975), in M. Balkanski, R. C. C. Leite, and S. P. S. Porto, Eds., *Light Scattering in Solids* (Paris: Flammarion), p. 203.
Holden T. M., Cowley R. A., Buyers W. J. L., Svensson E. C., and Stevenson R. W. H. (1971), *J. Phys.* **32**, C1-1184.
Hu H. L. and Morgenthaler F. R. (1971), *Appl. Phys. Lett.* **18**, 307.
Hutchings M. T., Thorpe M. F., Birgeneau R. J., Fleury P. A., and Guggenheim H. J. (1970), *Phys. Rev.* **B2**, 1362.
Ishikawa A. and Moriya T. (1971), *J. Phys. Soc. Jap.* **30**, 117.
Jantz W., Sandercock J. R., and Wettling W. (1976), *J. Phys.* **C9**, 2229.
Johnson L. F., Dietz R. E., and Guggenheim H. J. (1966), *Phys. Rev. Lett.* **17**, 13.
Kawasaki T. (1970), *J. Phys. Soc. Jap.* **29**, 1144.

Khater A. F. (1978a), *J. Phys.* **C11**, 563

Khater A. F. (1978b), *J. Phys.* **C11**, 577

Kinne R. W., O'Sullivan W. J., Ryan J. F., and Scott J. F. (1975), *Phys. Rev.* **B11**, 1960.

Kittel C. (1963), *Quantum Theory of Solids* (New York: Wiley).

Kittel C. (1966), *Introduction to Solid State Physics*, 3rd ed. (New York: Wiley).

Landau L. D. and Lifshitz E. M. (1960), *Electrodynamics of Continuous Media* (Oxford: Pergamon Press).

Le Gall H. (1970), *Grenoble Mag. Conf.* p. 590.

Le Gall H. and Jamet J. P. (1971), *Phys. Stat. Sol.* **46**, 467.

Le Gall H., Jamet J. P., and Desormière B. (1971), in M. Balkanski, Ed., *Light Scattering in Solids* (Paris: Flammarion), p. 170.

Lehmann W. and Weber R. (1973), *Phys. Lett.* **A45**, 33.

Lockwood D. J. and Coombs G. J. (1975), *J. Phys.* **C8**, 4062.

Loudon R. (1968), *Adv. Phys.* **17**, 243.

Loudon R. (1970), *J. Phys.* **C3**, 872.

L'vov V. S. (1968a), *Sov. Phys. JETP* **26**, 113.

L'vov V. S. (1968b) *Sov. Phys. Solid State* **10**, 354.

L'vov V. S. (1968c) *Sov. Phys. Solid State* **9**, 2328.

Macfarlane R. M. (1970), *Phys. Rev. Lett.* **25**, 1454.

Macfarlane R. M. (1974), *Solid State Commun.* **15**, 535.

Macfarlane R. M. and Morawitz H. (1971), in M. Balkanski, Ed., *Light Scattering in Solids* (Paris: Flammarion), p. 133.

Martin T. P., Merlin R., Huffman D. R., and Cardona M. (1977), *Solid State Commun.* **22**, 565.

Meixner A. E., Dietz R. E., and Rousseau D. L. (1973), *Phys. Rev.* **B7**, 3134.

Moch P. and Dugautier C. (1972), *Phys. Lett.* **42A**, 113.

Moch P., Gosso J. P., and Dugautier C. (1971), in M. Balkanski, Ed., *Light Scattering in Solids* (Paris: Flammarion), p. 138.

Moch P., Parisot G., Dietz R. E., and Guggenheim H. J. (1969), in G. B. Wright, Ed., *Light Scattering Spectra of Solids* (New York: Springer-Verlag), p. 231.

Moriya T. (1967), *J. Phys. Soc. Jap.* **23**, 490.

Moriya T. (1968), *J. Appl. Phys.* **39**, 1042.

Natoli C. R. and Ranninger J. (1973), *J. Phys.* **C6**, 345.

Nikotin O., Lindgård P. A., and Dietrich O. W. (1969), *J. Phys.* **C2**, 1168.

Nouet J., Toms D. J., and Scott J. F. (1973), *Phys. Rev.* **B7**, 4874.

Okazaki A., Turberfield K. C., and Stevenson R. W. H. (1964), *Phys. Lett.* **8**, 9.

Oseroff A. and Pershan P. S. (1969), in G. B. Wright, Ed., *Light Scattering Spectra of Solids* (New York: Springer-Verlag), p. 223.

Oseroff A., Pershan P. S., and Kestigian M. (1969), *Phys. Rev.* **188**, 1046.

Parisot G., Dietz R. E., Guggenheim H. J., Moch P., and Dugautier C. (1971), *J. Phys.* **32**, C1-803.

Parkinson J. B. (1969), *J. Phys.* **C2**, 2012.

Perry C. H., Anastassakis E., and Sokoloff J. (1971) *Ind. J. Pure Appl. Phys.* **9**, 930.

Pershan P. S. (1967), *J. Appl. Phys.* **38**, 1482.

References

Rezende S. M. (1973), *J. Phys.* **C6**, L354.
Richards P. M. and Brya W. J. (1974), *Phys. Rev.* **B9**, 3044.
Ryan J. F., Scott J. F., and Nouet J. (1973), *Solid State Commun.* **13**, 793.
Sandercock J. R. (1974), *Solid State Commun.* **15**, 1715.
Sandercock J. R. and Wettling W. (1973), *Solid State Commun.* **13**, 1729.
Shen Y. R. and Bloembergen N. (1966), *Phys. Rev.* **143**, 372.
Shepherd I. W. (1973), *Phys. Lett.* **45A**, 297.
Svensson E. C., Holden T. M., Buyers W. J. L., and Cowley R. A. (1969), *Solid State Commun.* **7**, 1693.
Tanabe Y., Moriya T., and Sugano S. (1965), *Phys. Rev. Lett.* **15**, 1023.
Terakawa S. and Okiji A. (1975), *J. Phys. Soc. Jap.* **39**, 938.
Thorpe M. F. (1970), *J. Appl. Phys.* **41**, 892.
Toms D. J., Ryan J. F., Scott J. F., and Nouet J. (1973), *Phys. Lett.* **44A**, 187.
Wettling W., Cottam M. G., and Sandercock J. R. (1975), *J. Phys.* **C8**, 211.

CHAPTER SEVEN

Raman Scattering by Electrons

7.1 Light Scattering by Rare-Earth Ions — 288

7.2 Light Scattering by Shallow Donors and Acceptors — 293
    7.2.1 Electronic Levels of Impurities in Semiconductors
    7.2.2 Scattering by Donors
    7.2.3 Scattering by Acceptors

7.3 Light Scattering by Conduction Electrons — 301
    7.3.1 Some Properties of Free Carriers
    7.3.2 General Aspects of Light Scattering by a Plasma
    7.3.3 Plasmon Scattering
    7.3.4 Single-Particle Scattering

The spin waves treated in the previous chapter are examples of electronic excitations that produce inelastic light scattering. The excitations involved are linear combinations of transitions to low-lying states of transition-group ions in crystals. More generally, light scattering can be used to study any electronic excitations whose frequencies lie in the accessible range [see (1.24)]. The excitations can be those of electrons bound to particular ions or of electrons that are free to move through crystal bands. In the case of bound charges there are two main kinds of electronic state with suitably placed excitation frequencies, namely states of electrons strongly bound to rare-earth ions and states of electrons and holes weakly bound to donors and acceptors in semiconductors. We treat the former in Section 7.1 and the latter in Section 7.2. These two cases are an interesting contrast. In the former the electronic states are confined to a volume comparable to that of the unit cell, and the effect of the lattice on the electronic levels of the ion can be treated as a perturbation. In the latter the wavefunctions of the impurities extend over a large number of unit cells, their binding energy is a small fraction of the energy gap, and their properties are determined to a considerable extent by the properties of the host material.

Finally, in Section 7.3 we deal with inelastic light scattering caused by single-particle and collective modes of excitation of free carriers, and we discuss effects of applied magnetic fields on this scattering.

7.1 LIGHT SCATTERING BY RARE-EARTH IONS

Both iron-group and rare-earth ions in solids have low-lying electronic excitations that can be studied by Raman scattering. The iron-group ions have received relatively little attention (for a description of the Raman spectrum of Co^{2+} ions in $CdCl_2$, see Christie et al. 1975), and we concentrate here on rare-earth ions. In crystals these ions have groups of excited states that arise from the same $4f^n$ electronic configuration as the ground state. The states most easily studied by light scattering come from the multiplets of varying J associated with the lowest Russell-Saunders term. These are Stark split by the electric field at the crystalline site of the rare earth, usually by a few hundred cm^{-1}. The states are in a difficult spectral region for absorption spectroscopy, and the transitions are in any case forbidden for electric-dipole absorption. The positions of the levels can often be determined by fluorescence measurements, but the light-scattering technique has the advantage of providing information on the symmetry properties of the excited states (Elliott and Loudon 1963). The first successful experiments were carried out on crystals of $PrCl_3$ by Hougen and Singh (1963, 1964).

As in the vibrational case, a great deal of information on the numbers and symmetries of lines in the electronic Raman spectra can be obtained by the use of group theory. The methods are similar to those outlined in Section 3.1.3 for the determination of the crystal vibration symmetries associated with the internal vibrations of constituent molecules. In the electronic case the free-ion $4f^n$ energy levels of the rare earths have degeneracies and symmetries governed by the three-dimensional rotation group appropriate to an ion in free space. The crystalline electric field at the rare-earth site produces splittings of the free-ion energy levels in a manner determined by the compatibility between irreducible representations of the rotation group and the site group, analogous to the kind of compatibility shown in Table 3.2 or the left-hand side of Figure 3.1 for the vibrational case. Details of the effects of site-symmetry properties in crystal-field theory are given by Tinkham (1964) and Cracknell (1968); more extensive discussion and tabulations of rotation and site-group compatibilities are given in the original paper of Bethe (1929) (reprinted in translation by Cracknell 1968).

The wavefunctions for an energy level of a single rare-earth ion in a crystal field are characterized by an irreducible representation of the site group.* Let the ionic ground-state representation be Γ_i and consider excitation to a final state of symmetry Γ_f. The excitation symmetry Γ_X is then given by (1.129) as before. However, electronic transitions are subject to a possible complication that cannot occur in the vibrational case, where an unexcited molecule in its vibrational ground state always has its symmetry Γ_i given by the identity representation Γ_1 or Γ_1^+ of the site group. By contrast, the electronic ground-state representation Γ_i can in principle be any irreducible representation of the site group. Consider, for example, a site of point group $4mm$ or C_{4v} where both ground and excited states have Γ_5 symmetry, so that

$$\Gamma_X = \Gamma_5 \times \Gamma_5^* = \Gamma_1 + \Gamma_2 + \Gamma_3 + \Gamma_4. \tag{7.1}$$

The transition symmetry thus contains simultaneously four contributions, and the polarization dependence of the single-ion cross section obtained from Table 1.2 is

$$\left|(\varepsilon_S^x \varepsilon_I^x + \varepsilon_S^y \varepsilon_I^y)a + \varepsilon_S^z \varepsilon_I^z b\right|^2 + |c|^2 (\varepsilon_S^x \varepsilon_I^y - \varepsilon_S^y \varepsilon_I^x)^2$$
$$+ |d|^2 (\varepsilon_S^x \varepsilon_I^x - \varepsilon_S^y \varepsilon_I^y)^2 + |e|^2 (\varepsilon_S^x \varepsilon_I^y + \varepsilon_S^y \varepsilon_I^x)^2. \tag{7.2}$$

---

*In this chapter, to conform with custom, we generally use the Bethe notation (see Table 1.2 for conversion to the Mulliken notation).

The excitation symmetries considered so far correspond to the so-called single-group irreducible representations. However, in the electronic case only states of ions with even numbers of electrons can be characterized by single-group representations. Ions with odd numbers of electrons have states that are characterized by the so-called double-group irreducible representations (Tinkham 1964, Cracknell 1968). The double-group representations are all at least two dimensional, corresponding physically to the Kramers degeneracy of odd-electron ions. Bethe (1929) includes the case of odd electron numbers in his discussion of the splitting of free-ion energy levels by crystal fields and the double-group representations are included in the tabulations of Koster et al. (1963). Consider again a site of $4mm$ or C_{4v} symmetry now occupied by an ion with an odd number of electrons. For the transition from a Γ_6 ground state to a Γ_7 excited state the transition symmetry is

$$\Gamma_7 \times \Gamma_6^* = \Gamma_3 + \Gamma_4 + \Gamma_5. \tag{7.3}$$

Since the product of two double-group irreducible representations like Γ_6 and Γ_7 is always expressible as a sum of single-group representations, Table 1.2 is adequate for both odd and even numbers of electrons.

The site-group excitation symmetries determine the selection rules and polarization dependence for the scattering by a single ion. The scattering crystal in an experiment contains of course a large number of similar ions. The ions under consideration may be a normal constituent of the crystal so that they occupy all equivalent sites of the lattice, but another case of practical interest is that in which the ions form a dilute concentration of impurities distributed randomly over equivalent sites. The ions on equivalent sites have the same basic energy-level structure in both types of distribution but the methods for combining the single-ion excitations to form excitations of the entire crystal are different for the two cases. Ions on nonequivalent sites normally have different energy-level structures and can be considered separately.

Consider first a dilute distribution where individual ions are sufficiently well separated for the interactions between them to be very small. The different ions then scatter light independently and the total scattering cross section is given by a sum of the individual contributions. If the primitive cell contains just one equivalent site, then the crystal selection rules and scattering symmetries are exactly the same as for a single ion. More generally, where the primitive cell contains several equivalent sites, their contributions to the crystal cross section must be summed with equal weight, taking account of the different orientations of site-group symmetry axes. Kaplyanskii and Negodyiko (1973) tabulate the polarization dependences of the cross sections of cubic crystals for transitions of ions distributed on sites of various symmetries within the primitive cell.

The other case, where the rare-earth ions fully occupy sites of the regular crystal lattice, is treated by methods identical to those used for the vibrational case in Section 3.1.3. Interaction between the different ions is now assumed to be sufficiently strong that the ionic excitations extend through the crystal and have symmetries appropriate to the space group of the lattice. Such electronic excitations of the crystal are called excitons; they have a well-defined wavevector $\mathbf{q}$ and a frequency that varies over the Brillouin zone in the general manner common to other types of crystal excitation. Only the small-wavevector excitons are observed in light-scattering experiments. The symmetry properties of the excitons that result from given single-ion transitions are determined by a factor-group analysis; compatibility and correlation tables analogous to Tables 3.3 and 3.4 connect the site- and space-group irreducible representations and these enable the exciton states associated with single-ion transitions to be determined, as on the right-hand side of Figure 3.1. The scattering properties of the various $\mathbf{q}=0$ excitons are determined by their space-group irreducible representations in the usual way.

Most experiments on electronic scattering by rare-earth ions have been made on crystals where the rare earth is a normal constituent of the lattice. However, the distinction between this and the dilute case is in many experiments of academic interest only. For rare earths in most crystal lattices the size of the interaction between neighboring ions is so small that light-scattering measurements do not resolve the Davydov splittings between the frequencies of those excitons that are the off-spring of a common single-ion transition (for an example of a resolved Davydov splitting, see Section 5.2.4). In this situation, the composite cross section of all the unresolved excitons has the same polarization dependence as the composite cross section of the corresponding single-ion excitations on all the equivalent sites in the primitive cell. The factor-group analysis is then unnecessary and the scattering symmetry is the same as in the dilute case.

The size of the single-ion cross section is determined in a general way by the analysis of Section 1.5.2 and expressions for the differential cross section can be taken from (1.117) or (1.121). The intermediate states $|l>$ come from the $4f^{n-1}5d$ configuration for scattering by rare earths. The sums over intermediate states can be performed with the use of tensor-operator techniques (Axe 1964, Mortensen and Koningstein 1968, Kiel et al. 1969), giving good qualitative agreement with the measured cross sections in some cases. These more accurate calculations show that the rough estimate (1.125) of the size of the cross section is too large by one or more orders of magnitude for scattering by rare-earth ions.

The differential cross sections (1.117) and (1.121) for electronic scattering show neither the symmetry property (2.37) characteristic of vibrational scattering nor the antisymmetry property (2.38) characteristic of scattering by simple spin waves. The matrices of Table 1.2 therefore apply to electronic scattering without further simplification and the scattering can be symmetric, antisymmetric, or a mixture of the two. The presence of antisymmetric contributions in the electronic scattering has been clearly demonstrated in experiments, as for example in the measurements of Kiel et al. (1969) on $CeCl_3$.

A large number of measurements of electronic Raman spectra of rare earths has been made. In addition to the work on $PrCl_3$ and $CeCl_3$ mentioned above, good representative measurements include those of Wadsack et al. (1971) (see Figure 7.1), Argyle et al. (1971), and Boal et al. (1973) on various rare-earth garnets and of Nathan et al. (1975) on the Sm^{2+} ion in SmS, SmSe, and SmTe. The light-scattering spectra often provide more complete information on the electronic energy levels than absorption and fluorescence spectra and almost always resolve some levels not seen at all in these spectra. No Davydov splittings were detected in these measurements, even though the rare-earth ions are normal constituents of the crystals. An example of the work on dilute rare earths is the measurement of scattering by Ce^{2+} in a concentration of only 10^{-5} mole % in crystals of CaF_2 (Kiel and Scott 1970).

For further details of work in this area readers should consult the review by Koningstein and Mortensen (1973).

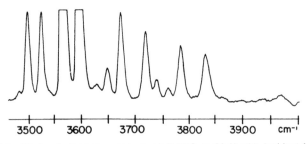

Figure 7.1 Electronic Raman spectrum of DyA1G at 80 K with incident laser light horizontally polarized and the scattered light vertically polarized. The Raman transitions are between the ground $^6H_{15/2}$ manifold and the upper $^6H_{13/2}$ manifold Dy^{3+} (after Wadsack et al. 1971).

7.2 LIGHT SCATTERING BY SHALLOW DONORS AND ACCEPTORS

7.2.1 Electronic Levels of Impurities in Semiconductors

Before we consider details of light scattering by shallow donors and acceptors, we give a brief discussion of the energy levels of impurities in semiconductors (for reviews, see Kohn 1957 and Bassani et al. 1974). In general, the impurity potential in semiconductors introduces electronic states with a fairly large probability density at the impurity site. Depending on the strength and sign of the impurity potential, one obtains localized states close to the upper or lower band edges or deep in the band gap. It is customary to write the impurity potential as

$$U(\mathbf{r}) = U_c(\mathbf{r}) - \frac{e^2}{(4\pi\epsilon_0\kappa_0)r}, \tag{7.4}$$

where $U_c(\mathbf{r})$ is referred to as the central cell contribution and κ_0 is the static relative permittivity. The theoretical approach to impurity levels depends on the relative magnitude of the two terms in (7.4). For isoelectronic substitutional impurities $U_c(\mathbf{r})$ is the only term in (7.4). Some impurities of this type, for example, oxygen in ZnTe and N and Bi in GaP, lead to bound exciton states. These have been studied extensively using recombination luminescence (Dean 1970, Dean et al. 1971) and are not dealt with here. Other impurities for which $U_c(\mathbf{r})$ is relatively large are the interstitial impurities Cu, Fe, and Mn in Ge. These introduce levels deep in the band gap (Bassani et al. 1974), but again they have not been studied by light scattering.

Raman studies have concentrated almost entirely on shallow donor and acceptor states. Here the second term in (7.4) is dominant and the effective mass approximation may be used (Kohn 1957). For very extended states with a slow spatial variation, $U_c(\mathbf{r})$ may be neglected and the Schrödinger equation

$$\left(-\frac{\hbar^2}{2m^*}\nabla^2 - \frac{e^2}{(4\pi\epsilon_0\kappa_0)r} \right) F(\mathbf{r}) = EF(\mathbf{r}) \tag{7.5}$$

applies, where m^* is an effective mass. Equation (7.5) is hydrogen-like and predicts a bound ground state and a series of excited states converging on an ionization continuum (beginning in this case at a band edge). The excited states of (7.5) are even more extended than the ground state so that the effective-mass approximation is even more applicable to them.

In Si and Ge, which have indirect band gaps at room temperature of 1.12 and 0.66 eV, shallow donors are generally group V elements (e.g., P) and shallow acceptors generally belong to group III (e.g., B). The ionization energies of shallow donors and acceptors in Ge (i.e., the energy difference between the ground state and the corresponding band edge) are rather similar, being about 0.011 eV, whereas the corresponding ionization energies in Si show a spread, covering the range 0.03 to 0.16 eV.

A shallow donor state may be regarded as a wave packet composed mainly of Bloch waves from the bottom of the conduction band. In silicon the minimum of the conduction band occurs along a $\langle 100 \rangle$ direction, about three-quarters of the way to the zone boundary (Figure 7.2), and there are six such minima, all equivalent. The wavefunctions for donor states in silicon can be expressed in the form

$$\psi(\mathbf{r}) = \sum_{f=1}^{N} \alpha_f F_f(\mathbf{r}) Q_f(\mathbf{r}), \qquad (7.6)$$

where N is the number of equivalent minima, the α_f are numerical coefficients and $Q_f(\mathbf{r}) = u_f(\mathbf{r}) e^{i\mathbf{k}_f \cdot \mathbf{r}}$ is the Bloch wave [as in (4.76)] at the position $\mathbf{k}_f$ of the fth minimum. In contrast with (7.5), which refers to materials with the conduction band at the center of the Brillouin zone (for example, InSb, GaAs), the hydrogen-like envelope functions $F(\mathbf{r})$ now satisfy

$$\left[-\frac{\hbar^2}{2m_l} \frac{\partial^2}{\partial z^2} - \frac{\hbar^2}{2m_t} \left(\frac{\partial^2}{\partial x^2} + \frac{\partial^2}{\partial y^2} \right) - \frac{e^2}{(4\pi\epsilon_0 \kappa_0) r} \right] F(\mathbf{r}) = E F(\mathbf{r}), \quad (7.7)$$

where, at the bottom of the conduction band, $m_l = 0.98 m_0$ and $m_t = 0.19 m_0$. The lowest solution of (7.7) has a mean Bohr radius of the order of $a^* = a_0 \kappa_0 (m_0/m^*)$, where a_0 is the normal Bohr radius and m^* is an appropriate average of m_l and m_t. Since $m^*/m_0 < 1$ for Si and the relative permittivity κ_0 has the value 12, one finds that $a^*(\mathrm{Si}) \simeq 20 \mathrm{\AA}$. To avoid overlap broadening in such systems, Raman scattering is carried out with

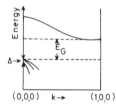

Figure 7.2 Schematic representation of the electronic energy-band structure of silicon along a $\langle 100 \rangle$ direction.

Light Scattering by Shallow Donors and Acceptors

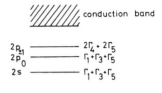

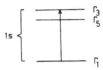

Figure 7.3 Schematic energy level diagrams for donors in semiconductors in sites with point symmetry T_d.

dopant concentrations in the range 10^{-5} to 10^{-6} mole % (for further discussion see Section 7.2.2).

Since (7.7) has tetragonal symmetry, the solutions are different from simple hydrogenic ones. Nevertheless we label them with the standard notation $1s$, $2s$, $2p$, etc. into which they go in the limit $m_l/m_t = 1$. However, an additional axial quantum number is required so that $2p$ states are specified as $|2p, 0\rangle$ and $|2p, \pm 1\rangle$. These substates have quite different energies in Si. The lowest $|1s\rangle$ state has no orbital degeneracy, but there is a sixfold spatial degeneracy corresponding to the six equivalent minima. This degeneracy, however, is raised when corrections to the effective-mass approximation are taken into account. The degeneracy that remains is determined by the symmetry of the impurity site alone (T_d) and the allowed states are Γ_1, Γ_3, and Γ_5 with Γ_1 lowest (Figure 7.3). In Si the overall splitting due to this valley-orbit mixing is of the order of 10 meV. The situation for excited states is similar but the valley-orbit splitting in these more extended states is usually much smaller.

It should perhaps be said that the symmetries shown in Figure 7.3 do not include the electron-spin transformation properties that convert the single-group representations to those of the double group. However, the spin does not change in the Raman transitions and the polarization dependence of the cross section is determined by the representations shown in Figure 7.3.

In Ge the conduction-band minimum occurs at the L point of the Brillouin zone, i.e., on the boundary in a $\langle 111 \rangle$ direction. Since the points $1, 1, 1$ and $-1, -1, -1$ are identical (they differ only by a reciprocal lattice vector), there are only four equivalent minima. The envelope functions $F(\mathbf{r})$ are again given by an equation of the form (7.7) with $z \parallel [111]$ and with $m_l = 1.6\ m_0$ and $m_t = 0.08\ m_0$. There are four equivalent $1s$ solutions for the ground state and valley-orbit splitting results in a singlet ground state Γ_1 and an excited Γ_5 state.

Shallow acceptor states are wave packets made from Bloch waves chosen from near the top of the valence band. The highest point in the valence band in Si and Ge is located at $\mathbf{k}=0$ (Figure 7.2). In the tight-binding limit this point goes into an atomic p function so that the total impurity degeneracy at this point is six. However, this degeneracy is partly raised by the spin-orbit coupling leaving an uppermost quartet $p_{3/2}$ (Γ_8) and a lower doublet $p_{1/2}$ (Γ_7) separated at $\mathbf{k}=0$ by the spin-orbit energy Δ (Figure 7.2). This has a value of 0.048 eV for Si and 0.29 eV for Ge. Because of the valence-band degeneracy the effective-mass theory for acceptor states is more complicated than for donor states (Baldereschi and Lipari 1973, 1974). However, the situation in Ge is somewhat less complex than in Si because Δ is more than an order of magnitude greater than acceptor binding energies, and defect levels associated with the two valence bands can be treated separately.

7.2.2 Scattering by Donors

The first spectroscopic investigation of impurity levels in semiconductors was carried out by Burstein et al. (1953) using infrared absorption techniques and many detailed investigations of this sort have been carried out subsequently (for references, see Bassani et al. 1974). Where parity is a good quantum number, infrared and Raman techniques are complementary. Lattice points in materials like Si and GaP do not have inversion symmetry, so that parity is not a good classification for states localized at substitutional impurity sites. However, for shallow donors and acceptors the wavefunctions are sufficiently extended to make parity classification useful so that some transitions are more readily observed in light scattering than by infrared absorption. The polarization properties of the scattered light also provide additional useful information. In any case, more symmetries of transition are generally observable in light scattering.

The differential Stokes Raman cross section given by (1.117) and (1.121) was evaluated for donors by Colwell and Klein (1972) for a two-band model with an indirect gap at $\mathbf{k}_f$ of amount $\hbar\omega_G$. They conclude that the most intense Raman transition involves an electron in the ground state of the $1s$ manifold undergoing a virtual transition to the valence band, followed by return to another level in the $1s$ manifold. This excitation is referred to as a valley-orbit transition (see Section 7.2.1). The cross section is

$$\frac{d^2\sigma}{d\Omega d\omega_S} = r_e^2 \frac{\omega_S}{\omega_I} \left(\frac{\omega_G^2}{\omega_G^2 - \omega_I^2} \right)^2 \left| \sum_{ijf} \varepsilon_S^i \varepsilon_I^j (\mu^{ij} - \delta^{ij}) \alpha_f^{(u)*} \alpha_f^{(v)} \int F_f^{(u)*} F_f^{(v)} d^3r \right|^2, \quad (7.8)$$

Light Scattering by Shallow Donors and Acceptors

with $\mu^{ij} = (m_0/m^*)^{ij}$ where i and j are the principal axes of the reciprocal effective mass tensor $1/m^*$, $\alpha_f^{(u)}$i are the coefficients of (7.6) for the group representation u and $F^{(u)}$ is an envelope function. The overlap integral in (7.8) is nonzero only if the hydrogenic principal quantum numbers are the same so that transitions from the 1s to the 2s manifold are expected to be weak. To the extent that parity is a good quantum number, the states $1s(\Gamma_1)$ and $1s(\Gamma_5)$ in Si have opposite parity so that Raman transitions between these levels are also expected to be weak. We therefore expect the $1s(\Gamma_1) - 1s(\Gamma_3)$ transition to be predominant in Raman scattering in Si.

The conclusions outlined above are in general agreement with an earlier discussion of Wright and Mooradian (1967). These authors worked on single crystals of Si doped with P using a Nd:YAG laser at 1.065 μm. Figure 7.4 shows the unpolarized Raman spectrum for Si:P. The line at 13.1 meV is the only electronic Raman transition observed and it is assigned to $1s(\Gamma_1) - 1s(\Gamma_3)$. The lines at 64.8 and 37.9 meV are vibrational in origin.

Manchon and Dean (1970) studied the Raman spectrum of donors in the indirect-gap material GaP using 514.5 nm light. The ground state of P-site donors in this material is 1s (Morgan 1968). The conduction-band minima are generally assumed to occur in ⟨100⟩ directions at the boundary of the Brillouin zone, so that there are three equivalent minima (for a discussion of a conduction-band minimum in GaP, see Dean and Herbert 1976). This threefold degeneracy of the 1s donor is lifted by the valley-orbit splitting producing a singlet $1s(\Gamma_1)$ ground state and a doublet $1s(\Gamma_3)$ excited state. Raman transitions between these states have been observed by Manchon and Dean for the P-site donors S, Se, and Te at 53.4, 54.0, and 40.5 meV, respectively. The transition splits symmetrically into two components under uniaxial stress, with no effect of thermalization on the relative intensities, indicating that the excited state is $1s(\Gamma_3)$. For uniaxial stress parallel to [111], no splitting of the Raman line occurs because the conduction-band minima all respond equally.

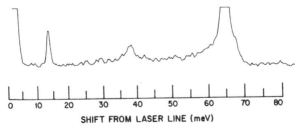

Figure 7.4 Raman spectrum of phosphorus donors in silicon at 4.2 K. The line at 13.1 meV is electronic in origin; the other lines are vibrational (after Wright and Mooradian 1967).

Transmission Raman scattering of As donors in Ge was studied by Doehler et al. (1974) and Doehler (1975) using the 2.098-μm line of the ABC/YAG laser and a cooled PbS detector (see Sections 2.1.3 and 2.3). They found a line at 35 cm^{-1} in a crystal containing 6.3×10^{21} m^{-3} donors that they assigned to the valley-orbit transition $1s(\Gamma_1) - 1s(\Gamma_5)$ (this line was also observed by infrared methods by Reuszer and Fisher 1964).

Investigations of donors have also been carried out in semiconductors with lower symmetry. In CdS, which has the symmetry C_{6v} (or $6mm$), there is only one conduction-band minimum at $\mathbf{k}=0$, and there is no increase in the number of states over the hydrogen-like model. Here Henry and Nassau (1970) observed Raman scattering by $1s-2s$ and $1s-2p_0$ transitions of chlorine donors. However, the 6H polytype of SiC, which has the same hexagonal structure, shows three Raman lines associated with transitions of N donors occupying the three inequivalent carbon sites of the crystal lattice (Colwell and Klein 1972).

We have so far confined our discussion to semiconductors containing donors and acceptors at sufficiently low concentrations to avoid appreciable overlap of their wavefunctions. However, as the concentration of impurity atoms increases, a metal-insulator transition (generally known as the Mott transition) takes place at a certain critical concentration n_c. The $1s(\Gamma_1) - 1s(\Gamma_3)$ valley-orbit transition has been studied as a function of the donor concentration n_d by Jain et al. (1976) in Si:P and by Doehler et al. (1976) in Ge:As. In the former system $n_c \cong 3 \times 10^{24}$m^{-3} and in the latter $n_c \cong 2.8 \times 10^{23}$m^{-3}. In both cases they find that the valley-orbit transition broadens beyond recognition before n_d reaches n_c and that a continuum develops below n_c due to intervalley fluctuations. The origin of the continuum, which completely dominates the Raman spectrum above n_c, is discussed at some length by Jain et al.

7.2.3 Scattering by Acceptors

The first observation of impurity-level scattering in semiconductors was made by Henry et al. (1966) who described electronic scattering by the neutral acceptors Zn and Mg dissolved in Ga sites in GaP. Various transitions observed for GaP:Zn are shown in Figure 7.5, and some are represented on the transition diagram Figure 7.6. We have already pointed out (Section 7.2.1) that according to the effective-mass theory of acceptor states the acceptor ground state is fourfold degenerate and has the double-group representation Γ_8 of the crystal point group T_d; this is also the representation for the top of the valence band.

Transition A is a zero-energy transition of finite width (~ 1 meV) within the strain-broadened ground manifold. By stressing the crystal it is possible to split the Γ_8 state into two Kramers doublets, causing A to move away

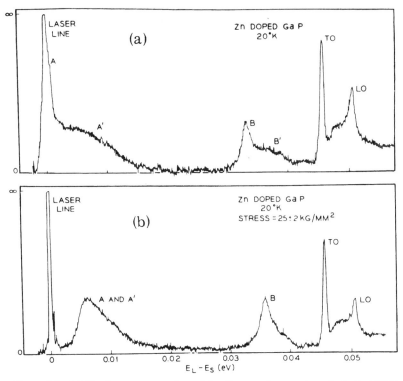

Figure 7.5 (a) The low-energy portion of the Zn acceptor spectrum in GaP at 20 K; (b) the same with uniaxial stress applied in a ⟨111⟩ direction (after Henry et al. 1966).

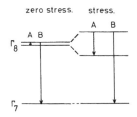

Figure 7.6 Partial energy-level diagram for acceptors in GaP showing two transitions observed in Raman scattering.

from the unshifted laser line (Figure 7.5). An acoustic-phonon wing A' accompanies A, but we shall not concern ourselves with these higher-order transitions (for a discussion, see Chase et al. 1977). Transition B to the first excited state occurs at 33.7 meV for Zn and 34.7 meV for Mg. The transition B of the Zn acceptor splits into two components under uniaxial stress along [211] (Manchon and Dean) and thermalization of intensities is consistent with the ground state A being a Γ_8 quartet. It appears from these investigations that the excited state of the transition B is a Γ_7 Kramers doublet. The energy of transition B for Zn is about one-half of the acceptor binding energy (64 meV).

In view of the large discrepancy between the B line energy and the separation $\Delta = 80$ meV between the $p_{3/2}$ and $p_{1/2}$ valence bands of GaP, it seems unlikely that the B line could arise from transitions between the Γ_8 and Γ_7 acceptor ground states associated with the two bands. This discrepancy, which also holds for the B line in other materials (see below), has been discussed by Chase et al. (1977). These authors investigated the effects of uniaxial stress and high magnetic fields on the Raman spectrum of the Zn acceptor in GaP. They suggested that the B line may be a $\Gamma_8 - \Gamma_7$ transition between levels associated with the $p_{3/2}$ band alone, the Raman intensity arising from admixture of Γ_7 acceptor bound states of the two valence bands by the acceptor potential. This suggestion is so far not confirmed by theory.

Wright and Mooradian (1967) also found a single electronic Raman line in Si at 23.4 meV associated with neutral boron acceptors. This B line was subsequently studied in detail by Cherlow et al. (1973) who placed it at 22.7 meV. They investigated the Zeeman effect and deformation potentials. The results of the Zeeman measurements were not consistent with earlier epr measurements. The stress splitting was consistent with a Γ_8 ground state, but again experiment could not clearly establish the origin of the excited state. Doehler et al. (1974) and Doehler (1975) studied Ge:Ga but did not find any low-lying acceptor transitions of the type B (Figure 7.6) found in Si and GaP.

Wright and Mooradian (1968) found broad acceptor lines in GaAs:Cd at ~ 25 meV and in GaAs:Zn at ~ 22 meV in contrast with the sharp B line of boron in Si (width at half height ~ 2 cm^{-1}) at 22.7 meV. They attribute this difference to the stronger electron-phonon coupling in the polar crystal and hence to a shorter lifetime.

It should be pointed out, in passing, that donors and acceptors may also affect the vibrations of lattice ions in their vicinity, because of their electronic properties, giving rise to localized vibrational modes (see Section 3.3.1). Such modes were first observed by Dean et al. (1970) using Raman-scattering methods. They occurred at an energy of about 1.3 meV lower than the LO phonon energy at $\mathbf{k} = 0$ in crystals of GaP containing S, Se, or

Te donors at ~20K. At this temperature the donors are completely frozen out, and since the bound electron is polarizable it changes the dielectric properties of the medium in its vicinity. The localized mode is caused by this extra polarizability. The model for these modes, which are a fairly general phenomenon (see also Barker 1973), is in contrast with that discussed earlier, which involves a mass change at the impurity site and possibly force-constant changes between the impurity and neighboring lattice ions (Section 3.3.1).

In concluding this section we refer readers for further details of Raman scattering by impurities in semiconductors to a review by Klein (1975).

7.3 LIGHT SCATTERING BY CONDUCTION ELECTRONS

7.3.1 Some Properties of Free Carriers

The scattering of light by the excitations of a free-electron gas is a weak effect. For electrons in solids, however, the cross section is enhanced by band-structure effects and light scattering has been used extensively to study the excitations of free electrons and holes in doped semiconductors. Interest has centered predominantly on conduction-band electrons, and this is reflected in our coverage here. Mobile electrons interacting with each other via Coulomb forces are referred to as a plasma, and, because of the long range of these forces, the excitations fall into two categories: (a) collective excitations, or plasmons, and (b) single-particle excitations. We briefly review some relevant properties of electron plasmas before considering the scattering of light by the two kinds of excitation. A more detailed account of solid-state plasmas is provided by Platzman and Wolff (1973), and their light-scattering properties are also reviewed by Yafet (1973) and Klein (1975).

Consider a sample of semiconductor that contains N_e mobile electrons in a volume V. In the approximation where the electrons are treated as free, except for the modification of their mass to an effective value m^*, the electron energy associated with the electron wavevector $\mathbf{k}$ is $\hbar\omega_\mathbf{k}$ where

$$\omega_\mathbf{k} = \frac{\hbar k^2}{2m^*}. \tag{7.9}$$

The occupancy of this electron state at low temperatures is determined by the Fermi-Dirac distribution

$$n_\mathbf{k} = \left\{ \exp\frac{\hbar(\omega_\mathbf{k} - \omega_F)}{k_B T} + 1 \right\}^{-1}, \tag{7.10}$$

where the Fermi energy $\hbar\omega_F$ is determined at $T=0$ by

$$\omega_F = \frac{\hbar k_F^2}{2m^*} \tag{7.11}$$

and

$$k_F = (3\pi^2 n_e)^{1/3}, \tag{7.12}$$

where $n_e = N_e/V$ is the electron density.

The sharp zero-temperature cut off in the electron distribution at the Fermi energy is smeared out with increasing temperature, and, at temperatures higher than $\hbar\omega_F/k_B$, the Fermi-Dirac distribution goes over to the Maxwell-Boltzmann form

$$n_\mathbf{k} = (4\pi)^{3/2} \frac{n_e}{k_T^3} \exp\frac{-k^2}{k_T^2}, \tag{7.13}$$

where

$$k_T = \frac{(2m^* k_B T)^{1/2}}{\hbar}. \tag{7.14}$$

The concept of a plasma in a solid is valid when the potential energy U of interaction between electrons is much smaller than their kinetic energy K. There is no long-range electronic order in this weak-coupling limit. The Coulomb energy is of order $e^2/4\pi\epsilon_0 \kappa r_0$ where $r_0 \cong n_e^{-1/3}$ is an average electron spacing and κ is an appropriate background relative permittivity. The weak-coupling criterion may therefore be expressed as

$$\frac{U}{K} \cong \frac{e^2 n_e^{1/3}}{4\pi\epsilon_0 \kappa \hbar \omega_F} \ll 1, \tag{7.15}$$

for a degenerate low-temperature plasma, and as

$$\frac{e^2 n_e^{1/3}}{4\pi\epsilon_0 \kappa k_B T} \ll 1 \tag{7.16}$$

for a high-temperature plasma. For semiconductors with $\kappa \approx 10$, (7.16) holds at room temperature for electron concentrations up to about 10^{23} m^{-3}. The condition becomes more stringent with reduction of temperature but semiconductor plasmas usually become degenerate at low temperatures and the alternative criterion (7.15) may be satisfied.

The Coulomb potential energy of a charge-density wave of wavevector **q** and amplitude $e\rho(\mathbf{q})$ is

$$U(\mathbf{q}) = \frac{e^2 V [\rho(\mathbf{q})]^2}{\epsilon_0 \kappa q^2}. \tag{7.17}$$

This expression diverges at zero wavevector as a result of the long range of the Coulomb interaction. Collective effects in the electron plasma are correspondingly more important for small-wavevector excitations. For large-wavevector properties the Coulomb energy is small according to (7.17) and the system tends to behave like an assembly of noninteracting particles.

The most important collective effect for our purposes is the existence at small wavevectors of a longitudinal charge-density oscillation of the plasma. A quantum of this oscillation is called a plasmon, and the dispersion relation is (Pines 1963)

$$\omega_P^2(\mathbf{q}) = \omega_P^2 + \frac{3\hbar^2 k_F^2}{5m^{*2}} q^2, \tag{7.18}$$

where

$$\omega_P^2 = \frac{n_e e^2}{\epsilon_0 \kappa m^*}. \tag{7.19}$$

The plasmon dispersion relation is represented schematically by the dashed line at small wavevectors in Figure 7.7.

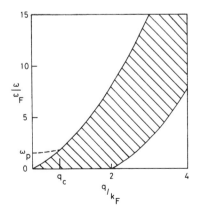

Figure 7.7 Schematic representation of plasmon dispersion [dashed line representing (7.18)] and single-particle dispersion [shaded area representing (7.25)]. q_c is given by (7.26).

The relative permittivity at zero wavevector has a simple frequency dependence in the region of the plasma frequency ω_P

$$\kappa(\omega, 0) = \kappa_\infty \left\{ 1 - \left(\frac{\omega_P}{\omega} \right)^2 \right\}, \qquad (7.20)$$

similar to the general form (4.49) for the special case where the longitudinal frequency is ω_P and the transverse frequency is zero. Thus electromagnetic waves with $\omega > \omega_P$ are transmitted by the plasma but waves with $\omega < \omega_P$ are strongly reflected. The plasma frequency of a metal is typically larger than 10 eV because of the high values of n_e and such plasmas are studied by electron energy-loss experiments or by inelastic X-ray scattering (Eisenberger et al. 1973). However, the possible values of ω_P for semiconductors cover the entire infrared region of the spectrum for varying levels of impurity doping. Such low-energy plasmas are amenable to study by conventional light-scattering techniques using lasers such as Nd:YAG.

Another collective manifestation of the electron interactions is the screening of the normal Coulomb potential of a charge inside the plasma, which becomes

$$\frac{e}{4\pi\epsilon_0 \kappa r} \exp(-q_S r), \qquad (7.21)$$

where the screening wavevector has the Fermi-Thomas form

$$q_S^{FT} = \frac{3^{1/2} m^* \omega_P}{\hbar k_F} \qquad (7.22)$$

for a degenerate plasma at low temperatures and the Debye form

$$q_S^D = \frac{2^{1/2} m^* \omega_P}{\hbar k_T} \qquad (7.23)$$

for a high-temperature plasma in the Maxwell-Boltzmann regime. The screening length $1/q_S$ is a measure of the penetration distance of an external electrostatic field into the plasma, its neutralization at larger distances being a result of polarization of the plasma. The screening effect is equivalent to a wavevector-dependent static ($\omega = 0$) relative permittivity of the form

$$\kappa(0, q) = \kappa_\infty \left\{ 1 + \left(\frac{q_S}{q} \right)^2 \right\}, \qquad (7.24)$$

and the q-dependent term dominates for $q < q_S$.

The screening wavevector separates the domains of collective ($q < q_S$) and single-particle ($q > q_S$) behavior. The single-particle excitation spectrum is similar to that obtained for free electrons in Section 1.5.3, and the excitation frequencies are related to the wavevector $\mathbf{q}$ by (1.126) (with m replaced by m^*). The excitation frequencies occupy a sharply defined band at zero temperature with its upper and lower bounds given by the frequencies

$$\omega_{\min}^{\max} = \frac{\hbar}{m^*}\left\{\tfrac{1}{2}q^2 \pm k_F q\right\}. \tag{7.25}$$

The shaded part of Figure 7.7 shows the region of the ω versus $\mathbf{q}$ plane covered by the single-electron excitations. The plasmon ceases to be a well-defined excitation at the cut-off wavevector

$$q_C \approx \frac{m^*\omega_P}{\hbar k_F}, \tag{7.26}$$

where its dispersion curve enters the band of single-particle excitations (Figure 7.7). For $q > q_C$ the plasmon suffers Landau damping because of the presence of electrons with the same phase velocity as the plasma wave and hence the ability to extract energy from it. Note that the cut-off wavevector has the same order of magnitude as the screening wavevector in accordance with the role of the latter in fixing an upper q limit for collective effects. The sharp confines of the single-particle spectrum in Figure 7.7 are of course smeared out as the temperature increases from zero and the electron distribution changes from Fermi-Dirac to Maxwell-Boltzmann.

7.3.2 General Aspects of Light Scattering by a Plasma

The regions of the electronic excitation spectrum accessible to a light-scattering experiment are determined in the usual way by superposition of the appropriate form of Figure 1.5 on the form of Figure 7.7 applicable to the crystal under investigation. Some part of the single-particle excitation spectrum is always covered. At low temperatures the region extends from zero-frequency shift to a maximum determined by (7.25) for wavevector transfers smaller than $2k_F$, but for larger wavevector transfers the band of observed frequency-shifts detaches from the origin and extends between the minimum and maximum frequencies given by (7.25). At high temperatures, where the electron distribution is given by (7.13), the scattered intensity peaks at zero-frequency shift and its shape gives a measure of the electron velocity distribution (Figure 7.8a). The single-particle scattering is suppressed for sufficiently small wavevectors because of phase interference between the contributions to the scattering from the individual particles and the spectrum is dominated by the collective plasmon excitation (Figure 7.8b).

In deriving an expression for the light-scattering cross section of a plasma it is convenient to treat first the case of an assembly of free electrons. The effects of the crystal electronic energy-band structure on the scattering by a solid-state plasma are inserted subsequently. The cross section for a single free electron is derived in Section 1.5.3. To treat the case of a free-electron plasma, it is necessary to allow for a distribution of a large number of interacting electrons among the available states. The derivation that follows is based on the work of DuBois and Gilinsky (1964), Platzman (1965), and McWhorter (1965).

The cross section is formally given by (1.120), and as in Section 1.5.3 only the A^2 part of the electron-radiation interaction need be retained for scattering by free electrons. If the initial states $|i\rangle$ of the many-electron system are populated according to a thermal distribution n_i and scattering to a variety of final states $|f\rangle$ occurs, then the cross section given by the A^2 part of (1.120) is

$$\frac{d^2\sigma}{d\Omega d\omega_S} = \frac{e^4 \omega_S V^2 (\varepsilon_I \cdot \varepsilon_S)^2}{(4\pi\epsilon_0)^2 c^4 m^2 \omega_I} \sum_{i,f} n_i |\langle f|\rho^*(\mathbf{q})|i\rangle|^2 \delta(\omega - \omega_f + \omega_i), \quad (7.27)$$

where all the sample volume V contributes to the scattering. Here

$$\rho(\mathbf{q}) = \frac{1}{V} \sum_j \exp(-i\mathbf{q} \cdot \mathbf{r}_j) \quad (7.28)$$

is the $\mathbf{q}$ Fourier component of the electron-density distribution, and the delta-function restriction on the excitation frequency converts (1.120) to a

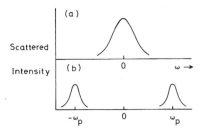

Figure 7.8 Schematic representation of free-electron scattering in semiconductors by (a) single particles and (b) plasmons.

spectral differential cross section. Thus with the use of (1.115), (7.27) becomes

$$\frac{d^2\sigma}{d\Omega\,d\omega_S} = \frac{e^4\omega_S V^2 (\varepsilon_I\cdot\varepsilon_S)^2 \langle\rho(\mathbf{q})\rho^*(\mathbf{q})\rangle_\omega}{(4\pi\epsilon_0)^2 c^4 m^2 \omega_I}. \qquad (7.29)$$

The scattering cross section is therefore proportional to the power spectrum of the fluctuations in electron density, which can be evaluated by the methods of linear response theory. According to one of Maxwell's equations, the magnitude E of the longitudinal electric field associated with a Fourier component of the electron charge density is given by

$$i\epsilon_0 q E = e\rho(\mathbf{q}). \qquad (7.30)$$

The density fluctuations are therefore related to the electric-field fluctuations evaluated with the help of the fluctuation-dissipation theorem in (4.29), and use of the isotropic case of the response function (4.28) leads to

$$\langle\rho(\mathbf{q})\rho^*(\mathbf{q})\rangle_\omega = \left(\frac{\epsilon_0 q}{e}\right)^2 \langle EE^*\rangle_\omega$$

$$= -\frac{\epsilon_0 \hbar q^2}{\pi e^2 V}\{n(\omega)+1\}\,\mathrm{Im}\left\{\frac{1}{\kappa(\omega,\mathbf{q})}\right\}. \qquad (7.31)$$

Using (7.31) the cross section becomes

$$\frac{d^2\sigma}{d\Omega\,d\omega_S} = -\frac{e^2\omega_S \epsilon_0 \hbar V q^2 (\varepsilon_I\cdot\varepsilon_S)^2}{(4\pi\epsilon_0)^2 \pi c^4 m^2 \omega_I}\{n(\omega)+1\}\,\mathrm{Im}\frac{1}{\kappa(\omega,\mathbf{q})}. \qquad (7.32)$$

It is seen that the properties of the many-electron system enter the cross section through the form of the relative permittivity, which can in principle be measured independently or can be calculated by suitable approximations of many-body theory.

The above calculations refer to free electrons. In treating the scattering by electrons in crystals we follow the same route as for free electrons and consider first the scattering by a single electron. The modifications required in the free-electron theory of Section 1.5.3 were derived by Wolff (1966). The main changes result from the existence of several energy bands for a given wavevector $\mathbf{k}$ in a crystal. These lead to a whole new set of interband contributions to Figures 1.9b, and 1.9c for which the band in state l differs from that in states i and f. The interband $\mathbf{A}\cdot\mathbf{p}$ part of the scattering matrix element often exceeds the A^2 contribution for electron scattering in crystals.

The cross section is once more obtained from (1.120). Consider the scattering by an electron in the conduction-band state $c\mathbf{k}$ with a Bloch wavefunction as described in Section 4.2.1. The cross section is

$$\frac{d\sigma}{d\Omega} = \frac{e^4 \omega_S \eta_S}{(4\pi\epsilon_0)^2 c^4 m^2 \omega_I \eta_I} \left| \boldsymbol{\varepsilon}_I \cdot \boldsymbol{\varepsilon}_S + \frac{1}{\hbar m} \sum_l \frac{\langle c\mathbf{k}|\boldsymbol{\varepsilon}_S \cdot \hat{\mathbf{p}}|l\mathbf{k}\rangle \langle l\mathbf{k}|\boldsymbol{\varepsilon}_I \cdot \hat{\mathbf{p}}|c\mathbf{k}\rangle}{\omega_{c\mathbf{k}} + \omega_I - \omega_{l\mathbf{k}}} \right.$$

$$\left. + \frac{\langle c\mathbf{k}|\boldsymbol{\varepsilon}_I \cdot \hat{\mathbf{p}}|l\mathbf{k}\rangle \langle l\mathbf{k}|\boldsymbol{\varepsilon}_S \cdot \hat{\mathbf{p}}|c\mathbf{k}\rangle}{\omega_{c\mathbf{k}} - \omega_{l\mathbf{k}} - \omega_S} \right|^2, \qquad (7.33)$$

where small changes in wavevector from the initial value $\mathbf{k}$ are ignored in the interband matrix elements. Note that the band index l is summed over all valence and conduction bands except for the initial band c; in calculating the contributions that involve transitions of valence electrons, it is necessary to keep careful track of minus signs to arrive at the same form of second-order matrix element as that for conduction-band intermediate states.

There are two limiting cases that illuminate the role of the interband contributions. Suppose first that ω_I is much smaller than any interband transition frequencies, and consider the scattering by an electron of small wavevector $\mathbf{k}$, close to the center of the Brillouin zone. The reciprocal effective mass of a conduction electron [as defined in (7.8)] has tensor components given by (for example, see Chapter 14 of Kittel 1963)

$$\mu^{ij} \equiv \left(\frac{m}{m^*}\right)^{ij} = \delta^{ij} + \frac{2}{\hbar m} \sum_l \frac{\langle c0|\hat{p}^i|l0\rangle \langle l0|\hat{p}^j|c0\rangle}{\omega_{c0} - \omega_{l0}}. \qquad (7.34)$$

The limiting form of the cross section (7.33) is thus

$$\frac{d\sigma}{d\Omega} = \frac{e^4 \omega_S \eta_S}{(4\pi\epsilon_0)^2 c^4 m^2 \omega_I \eta_I} |\boldsymbol{\varepsilon}_I \cdot \boldsymbol{\mu} \cdot \boldsymbol{\varepsilon}_S|^2. \qquad (7.35)$$

This is the solid-state analog of the Thomson cross section (1.14), modified by the crystal refractive indices and the effective electron mass.

The other interesting limit occurs when the incident frequency ω_I approaches the frequency splitting ω_G between the uppermost valence band and the conduction band. A two-band approximation can then be made, similar to that used for phonon resonance scattering in Section 4.2.3. The

Light Scattering by Conduction Electrons 309

matrix element in (7.33) becomes

$$\varepsilon_I \cdot \varepsilon_S + \frac{2\omega_G}{\hbar m} \frac{\langle c0|\varepsilon_I \cdot \hat{\mathbf{p}}|v0\rangle \langle v0|\varepsilon_S \cdot \hat{\mathbf{p}}|c0\rangle}{\omega_G^2 - \omega_I^2}, \qquad (7.36)$$

where the difference between ω_S and ω_I has been ignored. Then, using the two-band approximation to the effective-mass expression (7.34), this matrix element can be recast as

$$\frac{\omega_G^2}{\omega_G^2 - \omega_I^2} \left\{ \varepsilon_I \cdot \boldsymbol{\mu} \cdot \varepsilon_S - \frac{\omega_I^2}{\omega_G^2} \varepsilon_I \cdot \varepsilon_S \right\}, \qquad (7.37)$$

a result first derived by Wolff (1966). In crystals with an isotropic effective mass that is much smaller than the free electron mass, the first term in the matrix element leads to a cross section

$$\frac{d\sigma}{d\Omega} = \frac{e^4 \omega_S \eta_S (\varepsilon_I \cdot \varepsilon_S)^2}{(4\pi\epsilon_0)^2 c^4 m^{*2} \omega_I \eta_I} \left(\frac{\omega_G^2}{\omega_G^2 - \omega_I^2} \right)^2. \qquad (7.38)$$

Comparison with the low-frequency cross section (7.35) shows the large resonance-enhancement effects to be expected when ω_I approaches ω_G. The enhancement is caused entirely by the $\mathbf{A} \cdot \mathbf{p}$ contributions to the cross section. Resonant enhancement has been studied in some detail in GaAs by Pinczuk et al. (1971); they used a back-scattering geometry with incident red light of frequency resonant with the bandgap involving the split-off valence band.

The above cross section refers to scattering by single electrons in crystals. We now wish to generalize the result to apply to a solid-state plasma with a large number of interacting electrons, analogous to the cross section (7.32) for a free-electron plasma. The general result is complicated (Wolff 1969, Blum 1970) and the derivation is lengthy. We consider here only the case of a cubic semiconductor with spherical parabolic energy bands and negligible spin-orbit interaction. Then, in a two-band approximation, the modification of the free-electron plasma cross section (7.32) brought about by solid-state effects is essentially the same as that produced in the single-electron cross section, given by (7.38) for a small effective mass. The result is (Blum 1970)

$$\frac{d^2\sigma}{d\Omega d\omega_S} = -\frac{e^2 \omega_S \eta_S \epsilon_0 \hbar V \kappa^2 q^2}{(4\pi\epsilon_0)^2 \pi c^4 m^{*2} \omega_I \eta_I} \left(\frac{\omega_G^2}{\omega_G^2 - \omega_I^2} \right)^2 (\varepsilon_I \cdot \varepsilon_S)^2 \{n(\omega) + 1\} \, \text{Im} \, \frac{1}{\kappa(\omega, \mathbf{q})}. \qquad (7.39)$$

The shape of the spectrum is mainly determined by the final factor in the cross section. There are two distinguishable contributions to be considered in detail in the following two subsections. One of these, at small q, corresponds to plasmon excitations while the other, at large q, corresponds to single-particle excitations. Electron densities in actual semiconductors cover the range 10^{20} to 10^{26} m^{-3} giving rise to plasma frequencies in the range 4 to 4000 cm^{-1}. The corresponding range of screening wavevectors is of order 10^7 m^{-1} to over 10^9 m^{-1}, depending on the temperature. The wavevectors of Nd:YAG ($\lambda = 1.06$ μm) and CO_2 ($\lambda = 10.6$ μm) lasers in a medium with relative permittivity ~ 10 are, respectively, of orders 2×10^7 m^{-1} and 2×10^6 m^{-1}. It is possible therefore to satisfy the conditions $q < q_S$ and $q > q_S$ by using controlled doping levels. Hence one can study both collective and single-particle excitations using standard Raman techniques.

7.3.3 Plasmon Scattering

In semiconductors and insulators there is a high-frequency plasma oscillation that involves motion of all the electrons in the valence band. This type of plasmon has an energy comparable with that found in metals, and we shall not be concerned with it here. Semiconductors also support a low-energy plasmon involving motion of the free carriers only. For semiconductors like n-GaAs and n-InSb, which have a single zone-center conduction-band minimum, the behavior of the plasma is similar to that of a free-electron plasma and the plasmon frequency is given by (7.19). This behavior is in contrast to materials like n-Si and n-Ge, which have multivalleyed conduction bands and can support a richer variety of collective modes than a single-component plasma (e.g., see Platzman and Wolff 1973).

For a simple band structure it is a good approximation to use the zero-wavevector form (7.20) of the relative permittivity in the collective region. With insertion of a plasma-damping constant γ (for a discussion of damping, see Tell and Martin 1968) we can write

$$\kappa(\omega, 0) = \kappa_\infty \left\{ 1 - \frac{\omega_P^2}{\omega^2 + i\omega\gamma} \right\}, \qquad (7.40)$$

and the factor that occurs in the cross section (7.39) is

$$-\mathrm{Im}\frac{1}{\kappa(\omega,0)} = \frac{\pi\omega_P}{2\kappa_\infty} \frac{\gamma/2\pi}{(\omega_P - \omega)^2 + (\gamma/2)^2} \qquad (7.41)$$

in the Lorentzian approximation of (1.90), valid for small damping. The cross section is thus expected to show a Lorentzian line of width γ centered on the plasma frequency ω_P.

The first direct observation of a low-frequency plasmon in a doped semiconductor was made by Mooradian and Wright (1966) who investigated n-GaAs using a Nd:YAG laser. This radiation has an energy (1.16 eV), considerably below the bandgap of the crystal (1.4 eV) and well above the plasmon frequency ($\lesssim 0.1$ eV), a suitable situation for conventional light scattering. Figure 7.9 shows the scattered spectra for a variety of doping levels. The plasmon is not detectable for $n_e < 10^{22} \mathrm{m}^{-3}$, because it

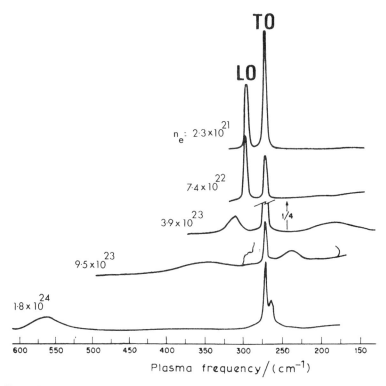

Figure 7.9 Raman scattering of n-type GaAs showing the dependence of plasmon frequency (cm^{-1}) on electron density n_e per cubic meter and effects of coupling of the plasmon to the LO vibrational mode. The position of the TO vibrational mode is also indicated (after Mooradian and Wright, 1966).

merges with the single-particle contribution to the scattering centered on zero frequency (this type of scattering is considered in the following subsection). The zone-center transverse- and longitudinal-optic phonons are also observed at 272 and 296 cm^{-1} (compare Table 4.3).

The experimental results for the peak positions are summarized in Figure 7.10, where they are plotted against the square root of the electron density. According to (7.41) the spectrum should show a single plasma peak at the frequency ω_P, which, by (7.19), is proportional to the square root of the electron density. The departures of the experimental peak positions from the corresponding oblique straight line in Figure 7.10 are a result of coupling between the longitudinal-optic phonon and the plasmon. This interaction occurs because each produces a longitudinal electric field that interacts with the charge density of the other. The resulting coupled excitations have mixed plasmon and longitudinal-phonon character (Varga 1965), and are somewhat analogous to the polaritons discussed in Section 4.3.

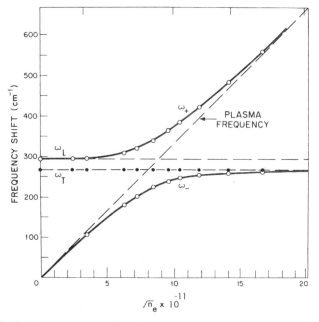

Figure 7.10 Measured frequencies of coupled plasmon and longitudinal-optic phonon modes in n-type GaAs plotted against the square root of electron density per cubic meter. The solid lines are calculated using (7.43) (after Mooradian and McWhorter 1969).

Light Scattering by Conduction Electrons

The frequency dependence of the coupled modes in the region of strong coupling ($\omega_P \sim \omega_L$) can be investigated by considering the contributions of both phonons and plasmons to the relative permittivity of the crystal. For a cubic crystal in the small q limit one finds with the help of (4.49) and (7.40)

$$\kappa(\omega, 0) = \kappa_\infty \left\{ 1 + \frac{\omega_L^2 - \omega_T^2}{\omega_T^2 - \omega^2 - i\omega\Gamma} - \frac{\omega_P^2}{\omega^2 + i\omega\gamma} \right\}. \quad (7.42)$$

Substitution into (7.39) gives the shape of the spectrum for scattering by the coupled modes. Similar to the polariton case of Section 4.3, the scattering peaks for modest damping occur at frequencies obtained in the limit of zero damping ($\Gamma = \gamma = 0$). Thus, if the right-hand side of (7.42) is set equal to zero in this limit, the solutions for ω are given by

$$\omega_\pm^2 = \tfrac{1}{2}(\omega_L^2 + \omega_P^2) \pm \tfrac{1}{2}\left[(\omega_L^2 + \omega_P^2)^2 - 4\omega_P^2 \omega_T^2 \right]^{1/2}, \quad (7.43)$$

and these are compared with experiment in Figure 7.10. It is apparent that for small values of n_e, that is small ω_P, the lower root ω_- has the plasmon-like value ω_P. As n_e increases the lower branch increasingly takes on the character of a longitudinal phonon. In the present case, however, the frequency of the lower branch tends to ω_T, because the free carriers effectively screen out the electrostatic interaction that raises ω_L above ω_T in the insulating solid. On the other hand the higher-frequency root ω_+ starts out as ω_L for low values of n_e and changes to a plasmon-like excitation at high values of n_e. The transverse-optic phonon is not affected by the presence of the electron plasma.

Both the plasmon and phonon parts of the coupled excitations contribute to the scattering. Three scattering mechanisms are involved in general and their contributions are discussed in some detail by Klein (1975). Consider the scattering geometry of Figure 2.22 used in the experiments of Mooradian and Wright (1966). When both incident and scattered beams are polarized parallel to z, the polarization factor in (7.39) is unity and the observed scattering is entirely caused by the plasmon contribution resulting from the electron-density fluctuations.

With the same scattering geometry but for zy or xz polarizations, the polarization factor in (7.39) is zero and the above mechanism makes no contribution. However, these are just the polarizations for which scattering by the longitudinal-phonon part is a maximum (cf. Table 4.2). The theoretical cross section is given by (4.57) with substitution of the relative permittivity from (7.42). The basic scattering mechanisms in this case are the modulation of the electronic susceptibility by the phonon normal coordinate and electric field, that is, the deformation-potential and electro-optic mechanisms.

Finally, for xy polarization in the same scattering geometry, neither part of the excitation makes any contribution to the cross section and no scattering occurs. All the above features of the scattering have been observed experimentally and interpreted theoretically by Mooradian and McWhorter (1967, 1969, see also Platzman and Tzoar 1969). The plasmon and phonon contributions to the scattering cross section are comparable in GaAs for 1.06-μm laser light.

Other crystals in which coupled plasmon-longitudinal phonon excitations have been investigated by light scattering include InSb (Blum and Mooradian 1970), CdS (Scott et al. 1970), GaP (Hon and Faust 1973), and SiC (Klein et al. 1972). These crystals have increasing values of the plasma damping γ, causing the zero-damping peak frequencies (7.43) to become poor approximations. Plasmon scattering has also been observed in InAs (Patel and Slusher 1968) at frequencies below the region of significant coupling to the longitudinal optic phonon. More recent work is directed toward study of plasmon scattering at larger wavevectors, where the dependence of plasmon frequency on q [see (7.18)] and the effects of Landau damping (see Figure 7.7) become important (Buchner and Burstein 1974, Murase et al. 1974, Zemski et al. 1975, Pinczuk et al. 1977).

It seems appropriate to point out here that free carriers have a more general effect on lattice force constants than the coupling between plasmons and longitudinal-optic phonons discussed above. In materials like Si and Ge, the strain produced by a uniaxial stress or a lattice phonon may shift the conduction-band valleys relative to one another and split the top of the valence band at $\mathbf{k}=0$. Carriers in valleys whose energy is raised by strain transfer to lowered valleys, making a smaller contribution to the crystal free energy. Hence part of the work done in deforming the crystal is regained through carrier redistribution, thereby reducing the elastic constant associated with the particular strain. This effect has been studied in Si and Ge containing high concentrations of free carriers (up to $4 \times 10^{26} \mathrm{m}^{-3}$) by Cerdeira and Cardona (1972) and Cerdeira et al. (1973), using Raman scattering.

Interaction between phonons and free carriers also affects the shape of phonon excitations observed in Raman scattering. For example, Cerdeira et al. (1974) found that the shape of the localized vibrational mode (Section 3.3.1) of boron in p-type silicon ($n_h = 4 \times 10^{26} \mathrm{m}^{-3}$), which occurs at $\sim 620 \mathrm{~cm}^{-1}$, is asymmetric and depends on the wavelength of the laser light. These results are explained as interference between the localized vibrational mode and a continuum of Raman-active electronic excitations.

In concluding this section we mention the application of light scattering to the study of electron-hole drops in crystals of Ge and Si. These drops arise from a gas-to-liquid type of phase transition and involve the condensation of excitons (the gas) to a metallic plasma (the liquid). The excitons are created with a high density by shining an intense beam of light, e.g., from an argon-ion laser, on the crystal maintained at liquid-helium temperatures. In materials such as Ge and Si with anisotropic masses and multiple band extrema, the energy of the high-density electron-hole plasma may be less than that of free excitons giving rise to a phase separation (Keldysh 1968). This results in the formation of droplets of the liquid phase (radius ~ 5 μm) in equilibrium with the gas phase. The liquid phase has a relative permittivity slightly different from that of the host material [see (7.20)] and hence scatters light. Worlock et al. (1974) found the droplet radius in Ge by measuring the angular dependence of the elastic scattered intensity of 3.39-μm light from a He-Ne laser (see Section 2.1.3). These and other light-scattering experiments on electron-hole droplets in Si and Ge have been briefly reviewed by Shaklee (1976, this paper contains references to Russian work. See also Doehler et al. 1977).

7.3.4 Single-Particle Scattering

Scattering in the Absence of a Magnetic Field

Single-particle scattering was first observed by Mooradian (1968, 1969) in n-GaAs; detailed investigations have been carried out on this material under conditions where the velocity distribution varies from classical Maxwell-Boltzmann to degenerate Fermi-Dirac. Figure 7.11 shows the light scattered from n-GaAs with $n_e = 10^{22}$m^{-3} at a variety of temperatures. The scattering within ± 100 cm^{-1} of the unshifted line is caused by free carriers. The anti-Stokes component is equal in intensity to the Stokes component at room temperature because of fall off with decreasing frequency of the S-1 response of the phototube used in the experiment.

The experiments of Figure 7.11 used right-angle scattering of Nd:YAG laser light and the excitations sampled thus have wavevectors close to

$$q = 2.8 \times 10^7 \text{m}^{-1}$$

for GaAs. The screening wavevector at room temperature given by (7.23) is

$$q_S^D = 2.2 \times 10^7 \text{m}^{-1}.$$

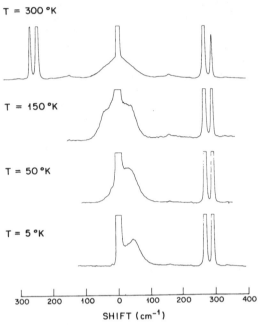

Figure 7.11 Raman spectra of n-type GaAs showing the change from single-particle scattering at room temperature to plasmon scattering at low temperature. TO and LO phonons are also observable between 250 and 300 cm^{-1} (after Mooradian 1969).

Thus the experimental scan is just into the single-particle region and the $T = 300$ K part of the figure is a single-particle spectrum. At low temperatures, however, the screening wavevector gradually increases from the Debye to the Fermi-Thomas limit given by (7.22) as

$$q_S^{FT} = 9.2 \times 10^7 \text{m}^{-1}.$$

The observed modes of the electron plasma thus become mostly collective in nature as the temperature drops and the lower parts of Figure 7.11 show the emergence of a peak at the plasma frequency of about 40 cm^{-1}, clearly evident at 5 K.

An expression for the single-particle cross section can be obtained from (7.39) using a standard result for the imaginary part of the electronic

Light Scattering by Conduction Electrons

relative permittivity (for example, see Chapter 6 of Kittel 1963)

$$\kappa''(\omega,\mathbf{q}) = \frac{\pi e^2}{\epsilon_0 \hbar q^2 V} \sum_{\mathbf{k}} (n_{\mathbf{k}} - n_{\mathbf{k}+\mathbf{q}}) \delta\left\{\omega - \frac{\hbar}{m^*}\left(\tfrac{1}{2}q^2 + \mathbf{k}\cdot\mathbf{q}\right)\right\}. \quad (7.44)$$

It is tedious but fairly straightforward to evaluate this expression in the limits of low and high temperatures where the explicit forms of the thermal distribution are given by (7.10) and (7.13). The result in the high-temperature limit is

$$\kappa''(\omega,\mathbf{q}) = \frac{\pi e^2 n_e \omega}{\epsilon_0 k_B T q^3}\left(\frac{m^*}{2\pi k_B T}\right)^{1/2} \exp\left(-\frac{m^*\omega^2}{2k_B T q^2}\right). \quad (7.45)$$

The low-frequency form of the real part of the relative permittivity is given by (7.24). We assume that q is sufficiently larger than q_S for the real part to be essentially equal to κ_∞. The imaginary part (7.45) is normally small compared to κ_∞.

With these results for the relative permittivity the final factor in (7.39) can be evaluated and the single-particle cross section is

$$\frac{d^2\sigma}{d\Omega\,d\omega_S} = \frac{e^4 \omega_S \eta_S (\boldsymbol{\varepsilon}_I\cdot\boldsymbol{\varepsilon}_S)^2}{(4\pi\epsilon_0)^2 c^4 m^{*2} \omega_I \eta_I}\left(\frac{\omega_G^2}{\omega_G^2 - \omega_I^2}\right)^2$$

$$\times V n_e \left(\frac{m^*}{2\pi k_B T q^2}\right)^{1/2} \exp\left(-\frac{m^*\omega^2}{2k_B T q^2}\right), \quad (7.46)$$

where the thermal factor in (7.39) is approximated for $k_B T$ much larger than $\hbar\omega$. The first two factors in (7.46) reproduce the single-electron cross section (7.38), whereas the remaining factors result from inclusion of the Maxwell-Boltzmann distribution of the n_e electrons at temperature T. Single-particle scattering at low temperatures can be treated by similar methods using the Fermi-Dirac distribution. Yafet (1973) gives a summary of the results, with references.

It is clear from the form of the cross section that light-scattering experiments are capable of investigating electron distributions in semiconductors in a fairly direct way. This may be understood in a general sense by recognizing that an electron moving with velocity **v** scatters light with a

Doppler shift $\Delta\omega = \mathbf{q}\cdot\mathbf{v}$. The scattered spectrum is thus obtained by summing intensities scattered by individual electrons over the velocity distribution.

Figure 7.12 shows a detailed comparison of theory with experiment for GaAs at room temperature with $n_e = 3\times 10^{21}\,\mathrm{m}^{-3}$ (Mooradian 1969). The excitation wavevector q is the same as before but the screening wavevector is now

$$q_S^D = 1.2 \times 10^7 \mathrm{m}^{-1}.$$

The condition for single-particle scattering is thus well satisfied and the observed spectrum closely mirrors the Maxwellian velocity distribution of the electrons.

For smaller wavevectors q the real part of the relative permittivity, given by (7.24) in the static limit, tends to increase and thus produce a fall off in the single-particle cross section. A detailed theory of the electron response for general values of q/q_S due to Salpeter (1960) confirms this expectation and shows that for $q < q_S$ the single-particle scattering should be suppressed and that the plasmon contribution should be dominant.

In contrast to these theoretical expectations Mooradian (1968, 1969) found a very strong single-particle scattering in n-GaAs for $q \ll q_S$. Two sources of the anomalously large strength of the single-particle scattering have been proposed. One of these, suggested by Wolff (1968, 1969), is the nonparabolicity of the conduction band of a semiconductor. A simple case is that of n-InSb, where a better approximation than (7.9) to the electron

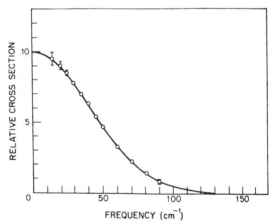

Figure 7.12 Comparison of observed and calculated single-particle scattering in n-type GaAs at room temperature with $n_e = 3\times 10^{21}\,\mathrm{m}^{-3}$ (Mooradian 1969).

energy is (Kane 1957)

$$\omega_k = \frac{\hbar k^2}{2m^*} - \frac{1}{\omega_G}\left(\frac{\hbar k^2}{2m^*}\right)^2. \tag{7.47}$$

When an electron with such an energy is coupled to the electromagnetic field by replacement of the momentum $\mathbf{p} = \hbar\mathbf{k}$ by $\mathbf{p} + e\mathbf{A}$ [compare (1.109)] a variety of additional electron-radiation interactions result. These include terms of the forms $p^2 A^2$ and $(\mathbf{p}\cdot\mathbf{A})^2$ which have no counterparts for free electrons. Since they involve p^2 they give rise to scattering by fluctuations in energy density rather than particle (i.e., charge) density. Such fluctuations are much less affected by electron-electron interactions than are particle-density fluctuations and the resulting light scattering is not suppressed in the limit $q \ll q_S$.

Because of the nearly spherical nature of the conduction band in GaAs the cross section for energy-density scattering has a polarization dependence $(\boldsymbol{\varepsilon}_I \cdot \boldsymbol{\varepsilon}_S)^2$. However, Mooradian (1969) found that scattering also occurs for $\boldsymbol{\varepsilon}_I \perp \boldsymbol{\varepsilon}_S$ in n-GaAs and this contribution can dominate at low temperatures. It is thought to result from a second additional mechanism for single-particle scattering, arising from spin-density fluctuations, which was proposed by Hamilton and McWhorter (1969). These fluctuations are coupled to the radiation field via the spin-orbit coupling in the valence band. The mechanism is analogous to that discussed in Chapter 6 for spin-wave scattering in magnetic crystals and to the bound-electron spin-flip scattering observed in CdS by Thomas and Hopfield (1968). Some further details are given later in this subsection since the mechanism is identical to that which causes spin-flip Raman scattering. Again, as in the case of energy-density scattering, dielectric screening is not important.

Scattering in the presence of a magnetic field

The presence of an applied magnetic field B_0 parallel to the z axis changes the electron energy (7.9) in the parabolic-band approximation to

$$\omega_{l,s^z k^z} = \left(l + \tfrac{1}{2}\right)\omega_C - \omega_O s^z + \left(\frac{\hbar k^{z^2}}{2m^*}\right). \tag{7.48}$$

The three terms represent, respectively, (a) the Landau levels with quantum numbers $l = 0, 1, 2, 3, \ldots$ corresponding to the orbital motion perpendicular to the field, with the cyclotron frequency given by

$$\omega_C = \frac{eB_O}{m^*}, \tag{7.49}$$

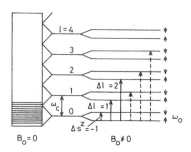

Figure 7.13 Schematic representation of energy levels of an electron in a conduction band with and without an applied magnetic field.

(b) the energy of the electron spin with $s^z = \pm\frac{1}{2}$ in the applied field with the spin transition frequency ω_O defined in (6.30), and (c) the electron kinetic energy parallel to the field. The latter contribution does not significantly change in the transitions to be considered and we can set $k^z = 0$ without loss of generality. A schematic arrangement of the energy levels of a conduction electron both with and without an applied field is shown in Figure 7.13.

Consider the possibility of light-scattering transitions from the magnetic-field ground state, denoted $|0\uparrow\rangle$, to higher levels; the number describing the state is l (7.48) and the arrow represents the spin direction. The first excited state is $|0\downarrow\rangle$, and the transition from the ground state involves a reversal of the electron-spin component but no change in the orbital quantum number l. This is known as a spin-flip transition, and it is the most extensively studied of the magnetic-field effects. The next two transitions indicated in Figure 7.13, to levels $|1\uparrow\rangle$ and $|2\uparrow\rangle$, involve changes of Landau level but no change in the electron-spin component. Finally, the transitions shown dashed in the figure involve changes in both the l and s^z quantum numbers.

The mechanism of light scattering by a spin-flip transition is very similar to that for light scattering by a magnon, considered in Section 6.1.3. For the spin-flip scattering by a conduction electron in a semiconductor like InSb, which has an $L = 0$ conduction band and an $L = 1$ valence band, the transition scheme is similar to that of Figure 6.4. The existence of a spin-orbit interaction is again necessary for nonzero scattering and, with the valence band as the main virtual intermediate state, the cross section is given by (6.51) with ω_P replaced by the energy-gap frequency ω_G and the dipole matrix elements taken between conduction- and valence-band Bloch functions. The matrix element for the spin-flip transition can be represented by a spin Hamiltonian with the same form as (6.54).

It is apparent from the similarity of the theories that spin-flip scattering has the same polarization selection rules as magnon scattering, that is, the incident and scattered beams should be polarized, one parallel to the applied field and one perpendicular to it. Note that the cross section does not vanish in zero applied field, and in this case an additional contribution to the scattering occurs for any polarizations in which ε_S has a component perpendicular to ε_I. The scattering mechanism connects states of the same or opposite spin component depending on the orientation of $\varepsilon_S \times \varepsilon_I$ with respect to s. This is the spin-density mechanism identified by Hamilton and McWhorter (1969) as giving rise to additional zero-field scattering in GaAs (see the earlier part of the subsection).

In the presence of an applied magnetic field, the spin-flip scattering gives rise to inelastic scattering at frequencies $\omega_I \pm \omega_O$. The occurrence of spin-flip scattering was predicted by Yafet (1966) and the first observations were made by Slusher et al. (1967) on InSb (see also Patel and Slusher 1968). Figure 7.14 shows their results for two different applied field strengths. The spin-flip line is the most striking feature of the spectrum, and it corresponds to a differential cross section per conduction electron of order $10^{-27} m^2$, about 200 times larger than the free-electron Thomson cross section (1.14). Large values of ω_O are readily produced (up to 160 cm^{-1} for field strengths up to 10 T) because of large conduction-electron g values, for example, $g = -50$ in InSb (for further discussion, see Wolff 1976).

In passing we mention the recent development of tunable infrared lasers involving spin-flip Raman scattering (for reviews, see Colles and Pidgeon 1975, Scott 1976). Laser action is readily achieved because (a) of the availability of powerful infrared lasers such as CO_2 described in Chapter 2, (b) the scattering cross section is large for spin-flip scattering, and (c) the cross section can be further enhanced by having ω_I close to ω_G. The spin-flip cross section can be up to 10^6 times greater than that of free electrons under suitable experimental conditions.

The other lines in the spectra of Figure 7.14 result from the Landau-level transitions with $\Delta l = 1$ and 2 with no change in the spin components ($\Delta s^z = 0$). The frequency shifts are therefore ω_C and $2\omega_C$ for the energy-level structure of Figure 7.13. Landau-level Raman scattering was first considered as a theoretical possibility by Wolff (1966) and subsequently by Yafet (1966). The calculation follows the same general lines as those for other types of electronic transition and the contributions can again be represented by Figure 1.9. Figure 1.9a makes no contribution, and it turns out (Wolff 1966) that Figures 1.9b and 1.9c make equal and opposite contributions if the electrons are treated as free, except for the effective-mass modification as in (7.9). This cancellation is a consequence of the equal spacing of the Landau levels corresponding to a purely harmonic oscillator, which cannot cause inelastic light scattering (see Section 1.2.2).

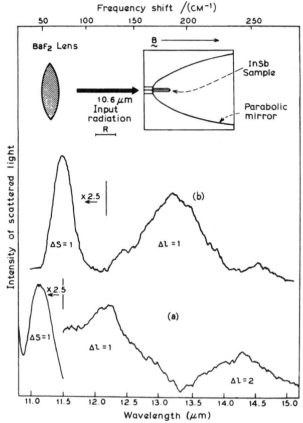

Figure 7.14 Raman scattering of n-type InSb ($n_e = 5 \times 10^{22}$ m^{-3}) showing magnetic spin scattering ($\Delta s^z = 1$) and Landau level scattering ($\Delta l = 1, 2$) in (a) a magnetic field of 2.62 T and (b) 3.67 T (after Slusher et al. 1967).

However, we have already pointed out that crystals like n-InSb have a nonparabolic conduction band whose dependence on wavevector is better represented by (7.47) than by (7.9). This nonparabolicity results in unequal spacing of the Landau levels and makes the transition $\Delta l = 2$, $\Delta s^z = 0$ allowed (Wolff 1966). Subsequent theoretical investigations by Kelley and Wright (1966) and by Yafet (1966) included the effect of the valence-band structure of InSb and found that the combination transition $\Delta l = 2$, $\Delta s^z = \pm 1$ should also be important in light scattering. Finally, the experimental results of Slusher et al. (1967) given in Figure 7.14 show that the $\Delta l = 1$,

$\Delta s^z = 0$ transition produces a significant peak in the cross section, and the theory of this contribution is discussed by Wright et al. (1969). Brueck et al. (1973) have made a very detailed study of spin-flip scattering in InSb and find good agreement between experiment and theory.

Although spin-flip Raman scattering is most intense in InSb, a number of other materials have been investigated. In n-PbTe Patel and Slusher (1969) showed that the g tensor of the electrons in the ellipsoidal conduction band has the values $g_\parallel = 57.5 \pm 2$ and $g_\perp = 15 \pm 1$, parallel and perpendicular to the (111) axis. Spin-flip Raman scattering has also been investigated in CdS by Scott et al. (1970), in CdS and ZnSe by Fleury and Scott (1971) and by Scott et al. (1972) who observed double spin-flip scattering but no Landau-level scattering. In ZnTe, Hollis et al. (1973) observed spin-flip scattering by holes and Hollis et al. (1975) observed simultaneous excitation of a spin-flip and a longitudinal optic phonon. A general feature of many investigations is the observation of changes in the scattered intensity with electron concentration n_e and magnetic field B_O. These result from shifts in the relative positions of the Fermi level and the Landau levels, which lead in turn to changes in the availability of final states for the transitions (Wherrett and Harper 1969, Makarov 1969). However, agreement between theory and experiment in this area is poor, and much more work remains to be done.

Finally, an interesting development in the study of spin-flip cross sections involved measurement of the Faraday rotation of the free carriers in CdS (Romestain et al. 1975). This quantity is theoretically related to the spin-flip cross section as discussed in Section 6.1, and its measurement is less subject to error than direct measurement of the cross section.

REFERENCES

Argyle B. E., Lewis J. L., Wadsack R. L., and Chang R. K. (1971), *Phys. Rev.* **B4**, 3035.
Axe J. D. (1964), *Phys. Rev.* **A136**, 42.
Baldereschi A. and Lipari N. O. (1973), *Phys. Rev.* **B8**, 2697.
Baldereschi A. and Lipari N. O. (1974), *Phys. Rev.* **B9**, 1525.
Barker Jr., A. S. (1973), *Phys. Rev.* **B7**, 2507.
Bassani F., Iadonisi G., and Preziosi B. (1974), *Rep. Prog. Phys.* **37**, 1099.
Bethe H. A. (1929), *Ann. Phys.* **3**, 133.
Blum F. A. (1970), *Phys. Rev.* **B1**, 1125.
Blum F. A. and Mooradian A. (1970), in S. P. Keller, J. C. Hensel, and F. Stern, Eds., *Tenth International Conference on the Physics of Semiconductors*, United States Atomic Energy Commission Division of Technical Information, p. 755.
Boal D., Grunberg P., and Koningstein J. A. (1973), *Phys. Rev.* **B7**, 4757.
Brueck S. R. J., Mooradian A., and Blum F. A. (1973), *Phys. Rev.* **B7**, 5253.

Buchner S. and Burstein E. (1974), *Phys. Rev. Lett.* **33**, 908.
Burstein E., Bell E. E., Davison J. W., and Lax M. (1953), *J. Phys. Chem.* **57**, 849.
Cerdeira F. and Cardona M. (1972), *Phys. Rev.* **B5**, 1440.
Cerdeira F., Fjeldly T. A., and Cardona M. (1973), *Phys. Rev.* **B3**, 4734.
Cerdeira F., Fjeldly T. A., and Cardona M. (1974), *Phys. Rev.* **B9**, 4344.
Chase L. L., Hayes W., and Ryan J. F. (1977), *J. Phys. C, Solid State Phys.* **10**, 2957.
Cherlow, J. M., Aggarwal R. L., and Lax B. (1973), *Phys. Rev.* **B7**, 4547.
Christie J. C., Johnstone I. W., Jones G. D., and Zdansky K. (1975), *Phys. Rev.* **B12**, 4656.
Colles M. J. and Pidgeon C. R. (1975), *Rep. Prog. Phys.* **38**, 329.
Colwell P. J. and Klein M. V. (1972), *Phys. Rev.* **B6**, 498.
Cracknell A. P. (1968), *Applied Group Theory* (Oxford: Pergamon Press).
Dean P. J. (1970), *J. Luminescence* **1**, 398.
Dean P. J., Faulkner R. A., Kimura S., and Ilegems M. (1971), *Phys. Rev.* **B4**, 1926.
Dean P. J. and Herbert D. C. (1976), *J. Lumin.* **14**, 55.
Dean P. J., Manchon Jr., D. D., and Hopfield J. J. (1970), *Phys. Rev. Lett.* **25**, 1027.
Doehler J. (1975), *Phys. Rev.* **B12**, 2917.
Doehler J., Colwell P. J., and Solin S. A. (1974), *Phys. Rev.* **B9**, 636.
Doehler J., Colwell P. J., and Solin S. A. (1976), in M. Balkanski, R. C. C. Leite and S. P. S. Porto, Eds. *Light Scattering in Solids* (Paris: Flammarion) p. 291.
Doehler J., Mattos J. C. V. and Worlock J. M. (1977), *Phys. Rev. Lett.* **38**, 726.
DuBois D. F. and Gilinsky V. (1964), *Phys. Rev.* **133**, A1308.
Eisenberger P., Platzman P. M., and Pandy K. C. (1973), *Phys. Rev. Lett.* **31**, 311.
Elliott R. J. and Loudon R. (1963), *Phys. Lett.* **3**, 189.
Fleury P. A. and Scott J. F. (1971), *Phys. Rev.* **B3**, 1979.
Hamilton D. C. and McWhorter A. L. (1969), in G. B. Wright, Ed. *Light Scattering Spectra of Solids* (New York: Springer-Verlag), p. 309.
Henry C. H., Hopfield J. J., and Luther L. C. (1966), *Phys. Rev. Lett.* **17**, 1178.
Henry C. H. and Nassau K. (1970), *Phys. Rev.* **B2**, 997.
Hollis R. L., Ryan J. F., and Scott J. F. (1973), *Phys. Rev. Lett.* **31**, 1004.
Hollis R. L., Ryan J. F., and Scott J. F. (1975), *Phys. Rev. Lett.* **34**, 209.
Hon D. T. and Faust W. L. (1973), *Appl. Phys.* **1**, 241.
Hougen J. T. and Singh S. (1963), *Phys. Rev. Lett.* **10**, 406.
Hougen J. T. and Singh S. (1964), *Proc. Roy. Soc.* **A277**, 193.
Jain K., Lai S., and Klein M. V. (1976), *Phys. Rev.* **B13**, 5448.
Kane E. O. (1957), *J. Phys. Chem. Solids* **1**, 249.
Kaplyanskii A. A. and Negodyiko V. K. (1973), *Opt Spect.* **35**, 269.
Keldysh L. V. (1968), in S. M. Ryukin and Y. V. Shmartsev, Eds., *Proceedings of the 9th International Conference on the Physics of Semiconductors* (Leningrad: Publishing House Nauka), p. 1303.
Kelley P. L. and Wright G. B. (1966), *Bull. Am. Phys. Soc.* **11**, 812.
Kiel A., Damen T., Porto S. P. S., Singh S., and Varsanyi F. (1969), *Phys. Rev.* **178**, 1518.
Kiel A. and Scott J. F. (1970), *Phys. Rev.* **B2**, 2033.
Kittel C. (1963), *Quantum Theory of Solids* (New York: John Wiley).

References

Klein M. V. (1975), in M. Cardona, Ed., *Light Scattering in Solids* (Heidelberg: Springer-Verlag), p. 147.

Klein M. V., Ganguly B. N., and Colwell P. C. (1972), *Phys. Rev.* **B6**, 2380.

Kohn W. (1957), *Solid State Phys.* **5**, 285.

Koningstein J. H. and Mortensen O. S. (1973), in A. Anderson, Ed., *The Raman Effect*, Vol. 2, (New York: Marcel Dekker), p. 519.

Koster G. F., Dimmock J. O., Wheeler R. G., and Statz H. (1963), *Properties of the Thirty Two Point Groups*, (Cambridge, Mass.: MIT Press).

Makarov V. P. (1969), *Sov. Phys. JETP* **28**, 366.

Manchon D. D. and Dean P. J. (1970), in S. P. Keller, J. C. Hensel, and F. Stern, Eds., *Proceedings of the 10th International Conference on the Physics of Semiconductors* U.S. Atomic Energy Commission, Division of Technical Information, p. 760.

McWhorter A. L. (1965), *Proceedings of the Conference on Quantum Electronics*, (New York: McGraw-Hill), p. 111.

Mooradian A. (1968), *Phys. Rev. Lett.* **20**, 1102.

Mooradian A. (1969), in G. B. Wright, Ed., *Light Scattering Spectra of Solids* (New York: Springer-Verlag), p. 285.

Mooradian A. and McWhorter A. L. (1967), *Phys. Rev. Lett.* **19**, 849.

Mooradian A. and McWhorter A. L. (1969), in G. B. Wright, Ed., *Light Scattering Spectra of Solids* (New York: Springer-Verlag), p. 297.

Mooradian A. and Wright G. B. (1966), *Phys. Rev. Lett.* **16**, 999.

Morgan T. N. (1968), *Phys. Rev. Lett.* **21**, 819.

Mortensen O. S. and Koningstein J. A. (1968), *J. Chem. Phys.* **48**, 3971.

Murase K., Katayama S., Ando Y., and Kawamura H. (1974), *Phys. Rev. Lett.* **33**, 1481.

Nathan M. I., Holtzberg F., Smith J. E., Torrance J. B., and Tsang J. C. (1975), *Phys. Rev. Lett.* **34**, 467.

Patel C. K. N. and Slusher R. E. (1968), *Phys. Rev.* **167**, 413.

Patel C. K. N. and Slusher R. E. (1969), *Phys. Rev.* **117**, 1200.

Pinczuk A., Abstreiter G., Tromner R., and Cardona M. (1977), *Solid State Commun.* **21**, 959.

Pinczuk A., Brillson L., Burstein E., and Anastassakis E., (1971), *Phys. Rev. Lett.* **27**, 317.

Pines D. (1963), *Elementary Excitations in Solids* (New York: Benjamin) p. 148.

Platzman P. M. (1965), *Phys. Rev.* **139**, A379.

Platzman P. M. and Tzoar N. (1969), *Phys. Rev.* **182**, 510.

Platzman P. M. and Wolff P. A. (1973), in H. Ehrenreich, F. Seitz, and D. Turnbull, Eds., *Solid State Phys.* Suppl. 13.

Reuszer J. H. and Fisher P. (1964), *Phys. Rev.* **135**, A1125.

Romestain R., Geschwind S., and Devlin G. E. (1975), *Phys. Rev. Lett.* **35**, 803.

Salpeter E. E. (1960), *Phys. Rev.* **120**, 1528.

Scott J. F. (1976), *Phys. Quantum Elec.* **4**, 325.

Scott J. F., Damen T. C., and Fleury P. A. (1972), *Phys. Rev.* **B6**, 3856.

Scott J. F., Leite R. C. C., Damen T. C., and Shah J. (1970), *Phys. Rev.* **131**, 4330.

Shaklee K. L. (1976), in M. Balkanski, R. C. C. Leite and S. P. S. Porto, Eds., *Light Scattering in Solids* (Paris: Flammarion), p. 160.

Slusher R. E., Patel C. K. N. and Fleury P. A. (1967), *Phys. Rev. Lett.* **18**, 77.

Tell B. and Martin R. J. (1968), *Phys. Rev.* **167**, 381.
Tinkham M. (1964), *Group Theory and Quantum Mechanics* (New York: McGraw-Hill).
Thomas D. G. and Hopfield J. J. (1968), *Phys. Rev.* **175**, 1021.
Varga B. B. (1965), *Phys. Rev.* **A137**, 1896.
Wadsack R. L., Lewis J. L., Argyle B. E., and Chang R. K. (1971), *Phys. Rev.* **B3**, 4342.
Wherrett B. S. and Harper P. G. (1969), *Phys. Rev.* **133**, 692.
Wolff P. A. (1966), *Phys. Rev. Lett.* **16**, 225.
Wolff P. A. (1968), *Phys. Rev.* **171**, 436.
Wolff P. A. (1969), in G. B. Wright, Ed., *Light Scattering Spectra of Solids* (New York: Springer-Verlag), p. 273.
Wolff P. A. (1976), *Phys. Quantum Elect.* **4**, 305.
Worlock J. M., Damen T. C., Shaklee K. L., and Gordon J. P. (1974), *Phys. Rev. Lett.* **33**, 771.
Wright G. B., Kelley P. L., and Groves S. H. (1969), in G. B. Wright, Ed. *Light Scattering Spectra of Solids* (New York: Springer-Verlag), p. 335.
Wright G. B. and Mooradian A. (1967), *Phys. Rev. Lett.* **18**, 608.
Wright G. B. and Mooradian A. (1968), in S. M. Ryukin and Yu. V. Shmartsev, Eds., *Proceedings of the 9th International Conference on the Physics of Semiconductors* (Leningrad: Publishing House Nauka), p. 1067.
Yafet Y. (1966), *Phys. Rev.* **152**, 858.
Yafet Y. (1973), in P. R. Wallace, R. Harris and M. J. Zuckerman, Eds., *New Developments in Semiconductors* (Leiden: Noordhoff International), p. 469.
Zemski V. I., Ivchenko E. L., Mirlin D. N., and Reshina I. I. (1975), *Solid State Commun.* **16**,

CHAPTER EIGHT

Rayleigh and Brillouin Scattering

8.1 Kinematics of Brillouin Scattering and Determination of Elastic Constants — 332
    8.1.1 Centrosymmetric Crystals
    8.1.2 Piezoelectric Crystals

8.2 Brillouin-Scattering Cross Section — 340

8.3 Some Experimental Examples — 345
    8.3.1 Rayleigh Scattering
    8.3.2 Brillouin Scattering

Since the advent of laser spectroscopy a large part of the research effort in Rayleigh and Brillouin scattering (RB scattering) has been devoted to the study of excitations in fluids (for discussions, see Mountain 1966a, Fabelinskii 1968, Benedek 1968, McIntyre and Sengers 1968, Fleury and Boon 1973, Berne and Pecora 1976). Some of the concepts developed in the study of fluids have relevance to solids and will be briefly outlined here. Because the sound waves studied in light scattering are of long wavelength, the microscopic structure of fluids can be ignored to a first approximation. A thermodynamic treatment of fluctuations in a hydrodynamic medium, regarded as a continuous isotropic dielectric, leads to a spectrum of scattered light consisting of two types (Figure 8.1):

1. A quasielastic Rayleigh component centered at ω_I due to nonpropagating entropy fluctuations.
2. A Brillouin doublet symmetrically located about the unshifted line and separated from it by a frequency equal to that of a compressional sound wave propagating through the fluid.

The fluctuations in the susceptibility that give rise to RB scattering are due to variations in thermodynamic quantities, such as density and temperature. The state of a fluid in thermodynamic equilibrium consisting of a single constituent can be described by two variables, for example, the pressure $P(\mathbf{r},t)$ and the entropy $S(\mathbf{r},t)$ so that the effect of fluctuations in these variables on the susceptibility can be expressed as

$$\delta\chi(\mathbf{r},t) = \left(\frac{\partial\chi}{\partial P}\right)_S \delta P(\mathbf{r},t) + \left(\frac{\partial\chi}{\partial S}\right)_P \delta S(\mathbf{r},t). \tag{8.1}$$

Using (1.71) we find that the differential cross section for scattering of light by a fluid is

$$\frac{d^2\sigma}{d\Omega\, d\omega_S} = \frac{\omega_I^4 v V}{16\pi^2 c^4} \langle|\delta\chi|^2\rangle_\omega \cos^2\phi, \tag{8.2}$$

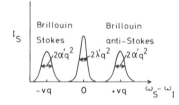

Figure 8.1 Schematic representation of Rayleigh and Brillouin spectra.

where ϕ is the scattering angle. Evaluation of $\langle|\delta\chi|^2\rangle_\omega$ leads to the result (e.g., see Cummins and Gammon 1966)

$$\frac{d\sigma}{d\Omega} = \frac{\omega_I^4 v k_B T}{16\pi^2 c^4}\left[\beta_S\rho^2\left(\frac{\partial\chi}{\partial\rho}\right)_S^2 + \frac{T}{\rho C_P}\left(\frac{\partial\chi}{\partial T}\right)_P^2\right]\cos^2\phi, \qquad (8.3)$$

where β_S is the adiabatic compressibility, ρ is the density, and C_P is the specific heat per unit mass. A formula of the type (8.3) was first obtained by Einstein (1910).

To a good approximation sound waves are pressure (density) fluctuations at constant entropy and the first term in the brackets in (8.3) is associated with the excitation of the Brillouin doublet corresponding to two sound waves with the same frequency travelling in opposite directions (Brillouin 1914, 1922). The second term in (8.3) corresponds to nonpropagating temperature (entropy) fluctuations and gives rise to the Rayleigh component (Landau and Placzek 1934) (these fluctuations propagate in superfluid and solid helium giving rise to second sound; see Section 8.3.1). To a good approximation the ratio of the intensity of the Rayleigh component to the sum of the two Brillouin components is

$$\frac{I_R}{2I_B} = \frac{C_P}{C_V} - 1. \qquad (8.4)$$

This is known as the Landau-Placzek ratio (Landau and Lifshitz 1960); it indicates that the intensity of Rayleigh scattering is proportional to $C_P - C_V$. One weakness of (8.4) arises from neglect of dispersion in thermodynamic properties, since the Brillouin components are measured at relatively high frequencies ($\gtrsim 10^9$ Hz) (Cummins and Gammon 1966).

The calculated line shapes for both Rayleigh and Brillouin components are Lorentzian (Mountain 1966a). For the Rayleigh component we get for the linewidth (Figure 8.1)

$$\Delta\omega_R = 2\lambda' q^2, \qquad (8.5a)$$

where $\lambda' = \lambda/\rho C_P$ is the thermal diffusivity, λ is the thermal conductivity and q is the wavevector transfer. In fluids the linewidth (8.5a) has a value of about 10 MHz for backscattering, decreasing as $\sin^2(\phi/2)$ as one goes to the forward direction, where ϕ is the scattering angle (Figure 8.2). The calculated width of the Brillouin components is (Figure 8.1)

$$\Delta\omega_B = 2\alpha' q^2, \qquad (8.5b)$$

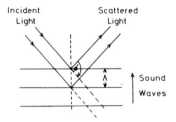

Figure 8.2 Schematic representation of the scattering of light by sound waves.

with $\alpha' = \frac{1}{2}[\zeta' + \lambda'(\gamma - 1)]$ where $\zeta' = (\frac{4}{3}\zeta + \xi)/\rho$ is a generalized kinematic viscosity (ζ is the shear viscosity, and ξ the bulk viscosity coefficient), λ' is the same as in (8.5a) and $\gamma = C_P/C_V$. Observed Brillouin widths for backscattering are $\sim$200 MHz and again they decrease in the forward direction as $\sin^2\phi/2$.

The discussion of RB scattering given so far refers to simple fluids. However, in polyatomic molecules with low vibrational levels a new mode is possible in addition to the thermal and phonon modes that give rise to Rayleigh and Brillouin scattering. This new mode arises from exchange of energy between the internal vibrational modes and translational modes and decays with a lifetime of the order of the relaxation time τ for the energy exchange (Mountain 1966b). It gives rise to density fluctuations and to the appearance of a broad unshifted central mode (see Section 5.2.6) with a width $\Delta\omega \sim 1/\tau$. Such a mode was first observed by Gornall et al. (1966) in the RB spectrum of CCl_4. It is not readily detected, however, since it manifests itself as a broad background under the RB spectrum. Dispersion effects should be kept in mind when comparing ultrasonic and Brillouin-scattering measurements in systems likely to exhibit this type of mode.

Density fluctuations in an isotropic medium can produce only diagonal elements in the polarizability tensor, so that the polarization of the scattered light is in the same direction as the incident light. However, fluids composed of nonspherical molecules, for example, quinoline, show appreciable depolarization of scattered light, indicating the presence of off-diagonal elements in the polarizability tensor. The depolarized spectrum consists primarily of a Lorentzian central component whose width is generally of the order of 1 to 10 cm^{-1}; it is referred to as the Rayleigh wing and is due to reorientation of the molecules in the viscous medium provided by the surrounding fluid. Although liquids do not exhibit a restoring force to the imposition of a static shear, they do exhibit a finite dynamic shear modulus at high frequencies. Indeed, structure has been observed in the depolarized RB spectrum of liquids, such as quinoline, that has been assigned to heavily damped shear waves (e.g., see Stegeman and Stoicheff 1968).

The uncorrelated motion of single particles is one of the simplest dynamical processes that can be studied in a light-scattering experiment. In dilute gases (see below) the particle motions obey a Maxwell-Boltzmann velocity distribution, which produces a Doppler-broadened unshifted line in the scattered light with a Gaussian shape (see also Section 7.3.4). If the scattering particles are embedded in a nonpolarizable fluid with which they have frequent collisions, they scatter light with a Lorentzian profile characteristic of diffusive or Brownian motion. The linewidth now has the value $2Dq^2$, where D is the diffusion coefficient (Lunacek and Benedek 1970). Macromolecules of biological interest have D-values of $\sim 10^{-10}$ m^2 s^{-1} in solution and have linewidths in the kilohertz range ($q \sim 10^7$ m^{-1}). The width of such very sharp lines can be measured with electronic spectrometers, providing information about diffusion of macromolecules in solution (Lunacek and Benedek 1970, Cummins and Pike 1974, Berne and Pecora 1976).

We have already emphasized that the wavelength of fluctuations normally studied in RB scattering ($\sim 10^{-6}$ m) are large compared with molecular spacings so that recourse to hydrodynamics is justified. In dilute gases departures from equilibrium are restored by two distinct relaxation processes (Cohen 1967): (a) a fast process corresponding to the mean free time between molecular collisions and (b) a much slower hydrodynamic process. It was found by Greytak and Benedek (1966) that the wavelengths of the fluctuations that give rise to scattering in gases such as Xe and CO_2 are long enough in the forward direction (for $\phi = 10.6°$, $q = 1.84 \times 10^6$ m^{-1} for 632.8 nm light) to be in the hydrodynamic regime, leading to the usual RB triplet. However, for near-backward scattering the sizes of the fluctuations are no longer small (for $\phi = 169.4°$, $q = 1.98 \times 10^7$ m^{-1}) compared to the mean free path and the spectral components merge with increasing angle into a single line characteristic of single-particle scattering. Perhaps it should be added that even rare gases show depolarized scattering in the central component because of collisions between atoms in the gas (McTague and Birnbaum 1968) (collision-induced scattering has also been investigated in rare-gas liquids by McTague et al., 1969).

Crystals are less optically perfect than fluids and studies of light scattering in the RB region are not so extensive. In particular, static crystal defects give rise to relatively strong scattering at the position of the Rayleigh peak, making the study of some dynamic processes, for example diffusion, quite difficult. Most of the interest in solids has, in fact, centered on measurements of elasticity through Brillouin scattering (we have already dealt with some of the results in Section 5.2.5).

In Section (8.1) we discuss the kinematics of Brillouin scattering and considerations involved in extracting elastic constants from the scattering measurements; piezoelectric crystals receive particular attention. In Section 8.2 we are concerned with selection rules for Brillouin scattering in solids and with the calculation of scattering intensity. As in the Raman case the translational symmetry of the crystal leads naturally to a discussion of scattering in terms of symmetry properties, and since with lasers one can readily explore the polarization properties of the scattered light one is again led to express selection rules in terms of scattering tensors. We shall see that a complete description of the scattering can be given in terms of the Pöckels elastooptic constants p^{mn} (Nye 1957) and elastic constants. Finally in Section 8.3 we discuss some results of experimental investigations in solids.

8.1. KINEMATICS OF BRILLOUIN SCATTERING AND DETERMINATION OF ELASTIC CONSTANTS

8.1.1. Centrosymmetric Crystals

In the previous section we pointed out that a hydrodynamic treatment of fluids predicted the existence of a pair of lines symmetrically spaced about the unshifted line by a frequency equal to that of the sound wave causing the scattering. These Brillouin components may also be properly regarded as a Doppler shift of the incident light caused in the process of Bragg reflection by moving sound waves. If we assume that the density variations caused by the sound waves produce a well-defined diffraction grating for the incident light (Figure 8.2) we can, as in the case of X-ray diffraction by static lattice planes, arrive at the condition for Bragg scattering

$$\lambda_I = \frac{\lambda_0}{\eta_I} = 2\Lambda \sin\frac{\phi}{2}, \tag{8.6}$$

where λ_I is the wavelength of the light in the medium, λ_0 is the wavelength in free space, η_I is the refractive index of the medium, Λ is the wavelength of the sound wave, and ϕ is the scattering angle. But since the diffraction grating is moving the diffracted light undergoes a Doppler shift $\Delta\omega_B$:

$$\Delta\omega_B = \pm 2v\eta_I k_0 \sin\frac{\phi}{2}, \tag{8.7}$$

where v is the phase velocity of the sound waves and k_0 [$=k_I/\eta_I$, see

(1.18)] is the magnitude of the wavevector of the incident light. This relationship was first derived by Brillouin (1922). In deriving (8.7) it should be kept in mind that only the component of the sound velocity parallel to the direction of travel of the light causes a Doppler shift. Combining (8.6) and (8.7) we get

$$\Delta\omega_B = \pm v k_0 \frac{\lambda_0}{\Lambda} = \pm vq,$$
$$= \pm \omega \qquad (8.8)$$

where ω is the angular frequency of the sound wave causing the scattering. In light-scattering experiments the value of ω varies from zero in forward scattering to about 10^{11} Hz in backward scattering. The Bragg condition (8.6) and the Brillouin condition (8.7) are equivalent to the momentum (1.15) and energy (1.2) conservation conditions discussed in Chapter 1. Equation (8.6) is in fact an approximation, equivalent to (1.28), due to neglect of the small difference in energy of the incident and scattered light. This introduces corrections to (8.8) of the order of $\Delta\omega_B/\omega_I \simeq 1/10^5$, too small to be detected experimentally.

The $q=0$ acoustic modes were briefly discussed in Section 3.1.2, where it was shown that they have zero frequencies and that their symmetries are the same as those of the three components of a polar vector. For wavelengths that are not infinite but are still large compared to the primitive-cell dimensions, the acoustic modes in a crystal lattice behave like sound waves in a continuous medium and their properties can be derived by macroscopic elasticity theory.

The crystal can accordingly be represented as a continuous medium of density

$$\rho = \frac{NM}{V}, \qquad (8.9)$$

where M is the primitive-cell mass (3.16). Consider a distortion in which a point (x,y,z) fixed in the medium is shifted to position $(x+u^x, y+u^y, z+u^z)$. It can be shown (Nye 1957, Kittel 1966) that the equations of motion of the medium are

$$\rho \ddot{u}^i = \frac{\partial S^{ix}}{\partial x} + \frac{\partial S^{iy}}{\partial y} + \frac{\partial S^{iz}}{\partial z}, \qquad (8.10)$$

where the S^{ij} are components of the second-rank stress tensor. The stress components are related to components s^{kl} of the second-rank strain tensor in the usual way

$$S^{ij} = C^{ijkl}s^{kl}, \qquad (8.11)$$

where C^{ijkl} is a component of the elastic stiffness tensor and repeated superscripts are assumed to be summed over $x, y,$ and z.

The components of the strain are defined by

$$s^{ij} = \tfrac{1}{2}(u^{ij} + u^{ji}), \qquad (8.12)$$

where the displacement gradients u^{ij} are

$$u^{ij} = \frac{\partial u^i}{\partial x^j} \qquad x^j = x, y, \text{ or } z. \qquad (8.13)$$

The equation of motion (8.10) can thus be written

$$\rho \ddot{u}^i = \frac{\partial S^{ij}}{\partial x^j} = C^{ijkl} \frac{\partial^2 u^k}{\partial x^j \partial x^l}, \qquad (8.14)$$

where we have used the invariance of the elastic stiffness tensor under interchange of its final pair of superscripts. There are three coupled equations of this form for the three components of the displacement.

We seek plane-wave solutions of the equations of motion in which the displacement vector has space and time variations

$$\mathbf{u} \sim \exp i(\mathbf{q} \cdot \mathbf{r} - \omega_\mathbf{q} t), \qquad (8.15)$$

leading to

$$\rho \omega_\mathbf{q}^2 u^i = C^{ijkl} q^j q^l u^k. \qquad (8.16)$$

It is convenient to define the acoustic-mode velocity as

$$v_\mathbf{q} = \frac{\omega_\mathbf{q}}{q}, \qquad (8.17)$$

and (8.16) becomes

$$\rho v_\mathbf{q}^2 u^i = C^{ijkl} \hat{q}^j \hat{q}^l u^k, \qquad (8.18)$$

where the carets denote components of a unit vector parallel to $\mathbf{q}$. The velocities thus depend only on the direction of $\mathbf{q}$ and not on its magnitude. The three solutions of (8.16) and (8.18) are denoted $\omega_{\sigma\mathbf{q}}$ and $v_{\sigma\mathbf{q}}$ ($\sigma = 1, 2, 3$).

Equation (8.11) represents nine equations, each with nine terms on the right-hand side so that there are 81 C^{ijkl} coefficients. These coefficients are generally referred to as elastic constants and form a fourth-rank tensor. Both the stress and strain tensors are symmetric and may be written with a single suffix running from 1 to 6, for example

$$\begin{bmatrix} S^{xx} & S^{xy} & S^{xz} \\ S^{yx} & S^{yy} & S^{yz} \\ S^{zx} & S^{zy} & S^{zz} \end{bmatrix} \rightarrow \begin{bmatrix} S^1 & S^6 & S^5 \\ S^6 & S^2 & S^4 \\ S^5 & S^4 & S^3 \end{bmatrix}. \qquad (8.19)$$

It follows therefore from (8.11) that $C^{ijkl} = C^{jikl} = C^{jilk} = C^{ijlk}$, and this reduces the number of independent components from 81 to 36. Hence (8.11) may be written

$$S^i = C^{ij} s^j \quad (i,j = 1, 2, \ldots, 6). \qquad (8.20)$$

The notation used in (8.20) is called the Voigt notation. The C^{ij} form a 6×6 matrix and have the dimensions of force/area or energy/volume. It should be emphasized that, in spite of their appearance with two suffices, they do not form a second-rank tensor. To transform them to other axes, for example, it is necessary to go back to the tensor notation (8.11).

Finally, it may be shown that the C^{ij} are symmetric, further reducing their number from 36 to 21 (Nye 1957). The number of independent C^{ij} may be reduced even more by crystal symmetry. For example, the C^{ij} matrix for all classes of crystal with cubic symmetry has the form shown in Table 8.1, so that there are only three independent stiffness constants C^{11}, C^{12}, and C^{44}. The form of the C^{ij} matrix for a variety of crystal symmetries has been given by Nye (1957).

The equations of motion for a cubic crystal can be illustrated for a general direction of $\mathbf{q}$ by the x component of (8.18)

$$\rho v_{\mathbf{q}}^2 u^x = C^{11} \hat{q}^x \hat{q}^x u^x + C^{12} \hat{q}^x (\hat{q}^y u^y + \hat{q}^z u^z)$$
$$+ C^{44} (\hat{q}^y \hat{q}^y u^x + \hat{q}^z \hat{q}^z u^x + \hat{q}^x \hat{q}^y u^y + \hat{q}^z \hat{q}^x u^z), \qquad (8.21)$$

with similar equations for the y and z components. The solutions for three propagation directions are shown in Table 8.2. In general, we require solutions of

$$\begin{vmatrix} \lambda^{11} - \rho v_{\mathbf{q}}^2 & \lambda^{12} & \lambda^{13} \\ \lambda^{21} & \lambda^{22} - \rho v_{\mathbf{q}}^2 & \lambda^{23} \\ \lambda^{31} & \lambda^{32} & \lambda^{33} - \rho v_{\mathbf{q}}^2 \end{vmatrix} = 0, \qquad (8.22)$$

Table 8.1 Elastic Constants of a Cubic Crystal

$$\begin{bmatrix} C^{11} & C^{12} & C^{12} & 0 & 0 & 0 \\ C^{12} & C^{11} & C^{12} & 0 & 0 & 0 \\ C^{12} & C^{12} & C^{11} & 0 & 0 & 0 \\ 0 & 0 & 0 & C^{44} & 0 & 0 \\ 0 & 0 & 0 & 0 & C^{44} & 0 \\ 0 & 0 & 0 & 0 & 0 & C^{44} \end{bmatrix}$$

where the λ^{ij} are orientation-dependent linear combinations of elastic constants that have been tabulated, for example, by Cummins and Schoen (1972).

In Brillouin-scattering experiments, one measures sound velocities assuming the refractive index to be known (8.7) and then solves (8.22) for elastic constants. It is of course obvious that solving (8.22) is not straightforward for an arbitrary direction of **q**, especially for low-symmetry crystals. Only for certain simple directions that occur more frequently for high-symmetry crystals is a sound wave purely longitudinal or purely transverse. In general the eigenvectors that specify the modes of vibration are not exactly parallel or perpendicular to the direction of propagation. Such modes are referred to as quasilongitudinal or quasitransverse. From a computational point of view it is desirable to make as many measurements as possible in pure-mode directions.

Only for cubic crystals is it possible to obtain all elastic constants from measurements of pure longitudinal and pure transverse modes. If we consider a sound wave traveling in a $\langle 100 \rangle$ direction of a cubic crystal, then we find from Table 8.2 a pure longitudinal wave with velocity $v=(C_{11}/\rho)^{1/2}$ and two pure transverse modes with velocity $v=(C_{44}/\rho)^{1/2}$. Normally longitudinal modes have the higher velocity. The secular equation is also diagonal for sound propagating in $\langle 111 \rangle$ and $\langle 110 \rangle$ directions, and it is apparent from inspection of Table 8.2 that in cubic crystals it is possible to determine the three elastic constants from measurements in a $\langle 110 \rangle$ direction alone. For further details and for discussions of lower-symmetry crystals, the reader should consult Neighbours and Schacher (1967) and Vacher and Boyer (1972).

8.1.2. Piezoelectric Crystals

The macroscopic elasticity theory described above ignores any effects of the electric fields associated with long-wavelength acoustic vibrations. These effects cannot properly be ignored in piezoelectric crystals where the electrical and elastic properties of the lattice are coupled. The theory must

Table 8.2 Scattering Tensor χ, Eigenvalues $\rho v_{\sigma q}^2$, and Eigenvector $U_{\sigma q}$ of Phonons Traveling in the Direction q in a Cubic Crystal

q	$\rho v_{\sigma q}^2$		χ		$U_{\sigma q}$
[100]	C^{11}		$\begin{pmatrix} p^{11} & 0 & 0 \\ 0 & p^{12} & 0 \\ 0 & 0 & p^{12} \end{pmatrix}$		[100] longitudinal
	C^{44}		$\begin{pmatrix} 0 & p^{44} & 0 \\ p^{44} & 0 & 0 \\ 0 & 0 & 0 \end{pmatrix}$		[010] transverse
	C^{44}		$\begin{pmatrix} 0 & 0 & p^{44} \\ 0 & 0 & 0 \\ p^{44} & 0 & 0 \end{pmatrix}$		[001] transverse
[110]	$\tfrac{1}{2}(C^{11}+C^{12}+2C^{44})$	$\tfrac{1}{2}$	$\begin{pmatrix} p^{11}+p^{12} & 2p^{44} & 0 \\ 2p^{44} & p^{11}+p^{12} & 0 \\ 0 & 0 & 2p^{12} \end{pmatrix}$		[110] longitudinal
	$\tfrac{1}{2}(C^{11}-C^{12})$	$\tfrac{1}{2}$	$\begin{pmatrix} p^{11}-p^{12} & 0 & 0 \\ 0 & p^{12}-p^{11} & 0 \\ 0 & 0 & 0 \end{pmatrix}$		$[1\bar{1}0]$ transverse
	C^{44}	$\tfrac{1}{\sqrt{2}}$	$\begin{pmatrix} 0 & 0 & p^{44} \\ 0 & 0 & p^{44} \\ p^{44} & p^{44} & 0 \end{pmatrix}$		[001] transverse
[111]	$\tfrac{1}{3}(C^{11}+2C^{12}+4C^{44})$	$\tfrac{1}{3}$	$\begin{pmatrix} p^{11}+2p^{12} & 2p^{44} & 2p^{44} \\ 2p^{44} & p^{11}+2p^{12} & 2p^{44} \\ 2p^{44} & 2p^{44} & p^{11}+2p^{12} \end{pmatrix}$		[111] longitudinal
	$\tfrac{1}{3}(C^{11}-C^{12}+C^{44})$	$\tfrac{1}{\sqrt{6}}$	$\begin{pmatrix} p^{11}-p^{12} & 0 & p^{44} \\ 0 & p^{12}-p^{11} & -p^{44} \\ p^{44} & -p^{44} & 0 \end{pmatrix}$		$[1\bar{1}0]$ transverse
	$\tfrac{1}{3}(C^{11}-C^{12}+C^{44})$	$\tfrac{1}{\sqrt{18}}$	$\begin{pmatrix} p^{11}-p^{12} & 2p^{44} & -p^{44} \\ 2p^{44} & p^{11}-p^{12} & -p^{44} \\ -p^{44} & -p^{44} & 2(p^{12}-p^{11}) \end{pmatrix}$		$[11\bar{2}]$ transverse

be extended to take account of the polarization **P** and electric field **E** that accompany the acoustic vibration.

The piezoelectric properties of a material are described by the elements e^{ijk} of its piezoelectric tensor (Nye 1957). The tensor determines the contribution to a component P^i of the polarization resulting from a component s^{jk} of the strain. The strain defined by (8.12) is invariant under interchange of its superscripts and the piezoelectric tensor is accordingly invariant under interchange of its second and third superscripts. The polarization is thus given by the usual expression (4.1) augmented by the piezoelectric contribution:

$$P^g = \epsilon_0 \chi^{gh} E^h + e^{gkl} s^{kl}, \tag{8.23}$$

where repeated superscripts remain summed and the superscript labels are chosen to produce a convenient final result.

The expression (8.11) for the stress is also modified by the piezoelectric properties to become

$$S^{ij} = C^{ijkl} s^{kl} - e^{mij} E^m. \tag{8.24}$$

The basic equation of motion of the elastic medium is still (8.10) but, with substitution of (8.24) and the plane-wave assumption (8.15) for the displacement and field amplitudes, (8.16) is replaced by

$$\rho \omega_q^2 u^i = C^{ijkl} q^j q^l u^k + i e^{mij} q^j E^m. \tag{8.25}$$

It is necessary to eliminate the electric field before this equation can be solved for the vibrational frequencies.

The wavevectors and frequencies of the acoustic modes always satisfy the inequality (1.29) very well and the relation (4.17) between **E** and **P** is therefore valid. Substitution of the components of **P** from (8.23) into (4.17) gives

$$E^m = -q^m q^g \frac{\left\{ \epsilon_0 \chi^{gh} E^h + e^{gkl} s^{kl} \right\}}{\epsilon_0 q^2}. \tag{8.26}$$

Now according to (4.17), **E** and **q** are parallel so that

$$q^m E^h = q^h E^m. \tag{8.27}$$

Thus (8.26) can be solved for E^m and the result is

$$E^m = -\frac{q^m q^n e^{nkl} s^{kl}}{\epsilon_0 q^g \kappa_0^{gh} q^h}, \tag{8.28}$$

where g has been changed to n in the numerator to avoid confusion with regard to the summation convention and

$$\kappa_0^{gh} = \delta^{gh} + \chi^{gh} \tag{8.29}$$

is the low-frequency relative permittivity tensor. This quantity is evaluated at the frequency of the acoustic vibration considered, assumed to be sufficiently high to avoid any distortion of the crystal sample as a whole by the piezoelectric coupling.

The expression (8.28) for the electric field was derived by Chapelle and Taurel (1955, see also Section 17 of Landau and Lifshitz 1960). Substitution into (8.25) leads to

$$\rho\omega_\mathbf{q}^2 u^i = \left\{ C^{ijkl} + \frac{e^{mij}q^m q^n e^{nkl}}{\epsilon_0 q^g \kappa_0^{gh} q^h} \right\} q^j q^l u^k. \tag{8.30}$$

This form of the equation of motion replaces (8.16) in the case of a piezoelectric crystal. The elastic stiffness tensor now appears augmented by an additional term that depends on the relative permittivity, the piezoelectric tensor, and the direction of the wavevector. The importance of the additional term in contributing to the acoustic-mode velocities is illustrated, for example, by the Brillouin scattering measurements of O'Brien et al. (1969) on $LiNbO_3$.

The condition for a crystal to be piezoelectric is that its symmetry group should lack a center of inversion. The 21 noncentrosymmetric crystal classes are listed in Table 4.2; they are the same crystal symmetries for which polar optic vibrations can participate in light scattering. Nye (1957) lists the restrictions on the components of the piezoelectric tensor that are imposed by the 21 symmetry groups. It is found that only 20 symmetries of crystal can in fact be piezoelectric since the symmetry operations in the

Table 8.3 Piezoelectric Contribution to the Elastic Constants of Cubic Crystal Classes 23 and 43 m

$$\begin{vmatrix} 0 & 0 & 0 & 0 & 0 & 0 \\ 0 & 0 & 0 & 0 & 0 & 0 \\ 0 & 0 & 0 & 0 & 0 & 0 \\ 0 & 0 & 0 & q^x q^x & q^x q^y & q^x q^z \\ 0 & 0 & 0 & q^y q^x & q^y q^y & q^y q^z \\ 0 & 0 & 0 & q^z q^x & q^z q^y & q^z q^z \end{vmatrix}$$

Each component is to be multiplied by $\dfrac{(e^{xyz})^2}{\epsilon_0 q^2 \kappa_0}$

cubic group 432 force all the tensor components to vanish, similar to the susceptibility derivatives listed in Table 4.2 (as discussed by Nye 1957, the symmetries of the piezoelectric and electro-optic tensors are closely related).

Table 8.3 shows the additional contribution to the elastic stiffness tensor in (8.30) for the cubic piezoelectric classes 23 and $\bar{4}3m$. This is to be added to the ordinary cubic stiffness tensor shown in Table 8.1. Note that the symmetry of the piezoelectric contribution is generally different from that of the ordinary stiffness tensor.

8.2 BRILLOUIN-SCATTERING CROSS SECTION

The calculation of the acoustic-phonon scattering cross section follows closely that of the optic-phonon cross section given in Chapters 3 and 4. We give here only an outline derivation, emphasizing those features that are special to Brillouin scattering.

The acoustic-mode normal coordinates at zero wavevector have a form given by (3.15) and (3.16). In this limit the three acoustic vibrational frequencies are all zero corresponding to rigid displacements of the entire crystal lattice. Such rigid displacements lead to no changes in the susceptibility and hence do not contribute to the Stokes polarization. To obtain inelastic scattering of light it is necessary to consider the small but nonzero wavevector $\mathbf{q}$ involved in Brillouin scattering. The appropriate normal-mode coordinates $W_{\sigma\mathbf{q}}$ are now wavevector dependent and the atomic displacements vary with position in the lattice. If the crystal is again represented by an elastic continuum, the continuous displacement vector $\mathbf{u}$ introduced in (8.10) is related to the acoustic normal coordinate by

$$\mathbf{u} = \frac{\boldsymbol{\xi}_{\sigma\mathbf{q}} W_{\sigma\mathbf{q}}}{M^{1/2}}, \qquad \sigma = 1, 2, 3, \tag{8.31}$$

where the unit polarization vector $\boldsymbol{\xi}_{\sigma\mathbf{q}}$ is parallel to the atomic displacements for the normal mode $\sigma\mathbf{q}$.

For a nonpiezoelectric crystal the contributions to the Stokes polarization of first order in the wavevector can be expressed in the susceptibility-derivative approximation as

$$P_S^i = \epsilon_0 \frac{\partial \chi^{ij}(\omega_I)}{\partial u^{kl*}} E_I^j u^{kl*}, \tag{8.32}$$

where the displacement gradients defined in (8.13) are given by

$$u^{kl} = iu^k q^l, \qquad (8.33)$$

for the plane-wave spatial variation (8.15).

The susceptibility derivative introduced in (8.32) can be measured independently by observations of the elastooptic effect (Nye, 1957). Analogous to the electrooptic effect described in Section 4.1.3, the elastooptic effect is the change in relative permittivity at frequency ω_I in the presence of a low-frequency strain. It is described by (4.39) with

$$\Delta\kappa^{ij} = -\kappa^i \kappa^j p^{ijkl} u^{kl*} \quad \text{(no summation on } i \text{ and } j), \qquad (8.34)$$

where the p^{ijkl} are elastooptic coefficients and the expression in this form applies only to crystals which have a diagonal relative permittivity in the absence of any applied strain.

With the help of these equations, the cross section (1.70) takes the form

$$\frac{d^2\sigma}{d\Omega \, d\omega_S} = \frac{\omega_I \omega_S^3 \mathfrak{v} V \eta_S |\epsilon_0 \varepsilon_S^i \varepsilon_I^j \kappa^i \kappa^j p^{ijkl} q^l|^2}{(4\pi\epsilon_0)^2 c^4 \eta_I} \langle u^k u^{k*} \rangle_\omega. \qquad (8.35)$$

An expression for the power spectrum of the displacement fluctuations is obtained, with the help of (8.31), by the same procedure used in Sections 1.4.4 and 3.1.4 for the optic modes. Since the acoustic-mode frequencies are usually very small it is permissible to ignore the difference between incident and scattered frequencies and to make a low-frequency expansion of the Bose-Einstein factor. The acoustic-mode spectral differential cross section is then

$$\frac{d^2\sigma}{d\Omega \, d\omega_S} = \frac{k_B T \omega_S^4 \mathfrak{v} \eta_S |\epsilon_0 \varepsilon_S^i \varepsilon_I^j \kappa^i \kappa^j p^{ijkl} \xi_{\sigma q}^k q^l|^2}{(4\pi\epsilon_0)^2 2 c^4 \eta_I \rho v_{\sigma q}^2} g_\sigma(\omega), \qquad (8.36)$$

where the crystal density ρ is defined in (8.9) and $g_\sigma(\omega)$ is the Lorentzian lineshape defined by (1.90). In general one may write the differential cross section as [see (2.31)]

$$\frac{d\sigma}{d\Omega} = \frac{k_B T \omega_S^4 \mathfrak{v} \eta_S \epsilon_0^2}{(4\pi\epsilon_0)^2 2 c^4 \eta_I \rho v_{\sigma q}^2} [\boldsymbol{\varepsilon}_S \cdot \boldsymbol{\chi} \cdot \boldsymbol{\varepsilon}_I]^2. \qquad (8.37)$$

Values of the scattering tensor χ and of the combinations of elastic constants corresponding to the various $\rho v_{\sigma q}^2$ have been tabulated for phonon directions of interest in a variety of crystal classes by Cummins and Schoen (1972). For the cubic classes $\bar{4}3m(T_d)$ and $m3m(O_h)$ the values of $\rho v_{\sigma q}^2$ are the same and are included in Table 8.2.

Consider as an illustration a cubic crystal of $m3m$ or O_h symmetry where the only nonzero elastooptic coefficients are (Nye 1957)

$$p^{11} = p^{22} = p^{33}$$
$$p^{44} = p^{55} = p^{66}$$
$$p^{23} = p^{32} = p^{31} = p^{13} = p^{12} = p^{21} \qquad (8.38)$$

in the abbreviated notation of (8.20). In an experiment where $\mathbf{k}_I$ is parallel to [100] and $\mathbf{k}_S$ to [0$\bar{1}$0], the phonon wavevector $\mathbf{q}$ is parallel to [110]. The differential cross sections for the three corresponding acoustic phonons in the order shown in Table 8.2 are

$$\frac{d\sigma}{d\Omega} = \frac{k_B T \omega_S^4 \mathfrak{v} \kappa^4 \epsilon_0^2 |\epsilon_S^z \epsilon_I^z p^{12} + \epsilon_S^x \epsilon_I^y p^{44}|^2}{(4\pi\epsilon_0)^2 c^4 (C^{11} + C^{12} + 2C^{44})}$$

$$= 0$$

$$= \frac{k_B T \omega_S^4 \mathfrak{v} \kappa^4 \epsilon_0^2 |(\epsilon_S^z \epsilon_I^y + \epsilon_S^x \epsilon_I^z) p^{44}|^2}{(4\pi\epsilon_0)^2 4c^4 C^{44}}, \qquad (8.39)$$

where (8.17) has been used and κ is the isotropic relative permittivity.

The excitation symmetry Γ_X for scattering by acoustic phonons is effectively determined by the displacement gradients (8.13). Since u^i and x^j are both components of polar vectors (PV)

$$\Gamma_X = \Gamma_{PV} \times \Gamma_{PV}. \qquad (8.40)$$

Thus, in accordance with the discussion of selection rules in Section 1.6, light scattering by acoustic phonons is always an allowed process. A zero cross section may occur for a particular experimental geometry, as in (8.39), but the three acoustic phonons for a given direction of $\mathbf{q}$ can all be observed by appropriate choices of scattering geometry.

If the elastooptic coefficients are invariant under interchange of the final pair of superscripts, as is assumed in the use of the abbreviated notation (8.19), then the change (8.34) in the relative permittivity can be expressed in terms of the strain components (8.12). This assumes that there are no contributions from the antisymmetric combinations of displacement gradients,

$$r^{ij} = \tfrac{1}{2}(u^{ij} - u^{ji}). \quad (8.41)$$

These quantities determine the rotational part of the crystal distortion. In a uniform distortion of the crystal, any rotation of the crystal as a whole can be compensated appropriately in a measurement of the elastooptic coefficients. However, in the presence of finite-wavelength acoustic vibrations, such as participate in Brillouin scattering, nonuniform rotations can occur and there may be contributions to the Stokes polarization proportional to the r^{ij}.

The rotational or antisymmetric contributions to Brillouin scattering had been ignored until Nelson and Lax (1970, 1971) pointed out their existence for those acoustic modes in biaxial and uniaxial crystals that cause a shear distortion. The effect is simply a result of the local rotations of the principal axes of the relative permittivity tensor relative to the fixed axes of a scattering experiment. Consider, for example, a crystal whose relative permittivity tensor is diagonal with principal axes x, y, and z. A counterclockwise infinitesimal rotation through angle $\delta\theta^x$ around the x axis corresponds to a single nonzero antisymmetric combination of displacement gradients given by

$$\delta\theta^x = r^{yz}. \quad (8.42)$$

It is not difficult to show that the changes to order $\delta\theta^x$ in the relative permittivity tensor with respect to the original fixed axes are

$$\Delta\kappa^{yz} = \Delta\kappa^{zy} = -(\kappa^{yy} - \kappa^{zz})\delta\theta^x, \quad (8.43)$$

and there are no changes in the other components. Thus if κ^{yy} and κ^{zz} are different, there is an antisymmetric contribution to the photoelastic coefficients defined in accordance with (8.34). The magnitudes of the antisymmetric contributions are determined solely by the amount of anisotropy in the unperturbed relative permittivity.

Nelson and Lax (1970, 1971) give expressions that generalize the above result to crystal classes whose unperturbed relative permittivity tensors are nondiagonal. They also tabulate the forms of the antisymmetric parts of the elastooptic tensor for the various crystal classes (see also Anastassakis and Burstein 1974), thus complementing the symmetric-part tabulation of Nye (1957). Brillouin-scattering measurements have been used to verify the theory by measuring the antisymmetric contributions to the photoelastic tensor in rutile (Nelson and Lazay 1970) and calcite (Nelson et al. 1972). These are both nonpiezoelectric uniaxial crystals, and they show good agreement between experiment and the theoretical size of the effect determined from (8.43).

In piezoelectric crystals, the Stokes polarization (8.32) resulting from the displacement gradient of the acoustic mode is augmented by a contribution from the electric field of the mode. The electric field contribution is given by (4.38) evaluated at the acoustic frequency. This is much smaller than the optic-mode frequencies, and the low-frequency result (4.41) can be used to express the total Stokes polarization as

$$P_S^i = -\epsilon_0 \kappa^i \kappa^j p^{ijkl} E_I^j u^{kl*} - \epsilon_0 \kappa^i \kappa^j r^{ijm} E_I^j E^{m*} \quad \text{(no summation on } i\text{)}, \quad (8.44)$$

where (8.34) has been used. With the acoustic electric field taken from (8.28)

$$P_S^i = -\epsilon_0 \kappa^i \kappa^j \left\{ p^{ijkl} - \frac{r^{ijm} q^m q^n e^{nkl}}{\epsilon_0 q^g \kappa_0^{gh} q^h} \right\} E_I^j u^{kl*}, \quad (8.45)$$

where the strain appearing in (8.28) has been separated into displacement gradients in accordance with (8.12). The second term in (8.45) remains, however, strictly symmetric in k and l, and the antisymmetric contributions discussed above are confined to the elastooptic coefficients. The Brillouin-scattering cross section for a piezoelectric crystal is given by the same expression (8.36) as for a nonpiezoelectric material, but with p^{ijkl} replaced by the quantity in the bracket of (8.45) (Loudon 1969, Nelson and Lax 1971, Lax and Nelson 1971).

All the above calculations of the scattering cross section are based on macroscopic properties of the crystal. It is possible to make a corresponding microscopic calculation similar to the optic phonon case treated in Chapter 4. Thus the classical displacement vector (8.31) can be replaced by an operator quantity, using the operator normal coordinate (4.82). As in the optic phonon case considered in Section 4.2.1, there are two kinds of interaction between an acoustic phonon and the electrons. The short-range interaction has matrix elements

$$\langle \beta, n(\omega)+1 | \hat{H}'_{EL} | \alpha, n(\omega) \rangle = -\left(\frac{\hbar}{2MN\omega_{oq}}\right)^{1/2} \{n(\omega)+1\}^{1/2} \xi^i_{oq} \Xi^{ij}_{o\beta\alpha} q^j$$

(8.46)

similar to (4.84), except that the deformation potentials are now defined

Some Experimental Examples

with respect to the displacement gradients (Whitfield 1961). The long-range interaction is obtained from the operator version of (8.28), which replaces (4.50) and (4.85), and leads by exactly the same route to the acoustic analog of (4.86).

The microscopic calculation of the cross section proceeds as in Section 4.2.2, and results for a nonpiezoelectric crystal are given by Loudon (1963). The expressions are equivalent to the macroscopic Brillouin-scattering cross section but their more explicit dependence on the incident and scattered frequencies can be useful in the interpretation of resonance-scattering experiments (for example, see the work on GaAs of Garrod and Bray 1972, and on CdS of Pine 1972 and Ando and Hamaguchi 1975). Acoustic-phonon resonance scattering is broadly similar to the optic-phonon resonance scattering considered in Section 4.2.3, but the linear wave-vector dependence of the acoustic frequency leads to additional effects in resonance conditions (Brenig et al. 1972, Ulbrich and Weisbuch 1977).

8.3 SOME EXPERIMENTAL EXAMPLES

8.3.1 Rayleigh Scattering

The fluctuations in temperature that give rise to Rayleigh scattering are expected to decay according to a diffusion equation of the form (Chester 1963)

$$\frac{\partial^2(\delta T)}{\partial t^2} + \frac{1}{\tau}\frac{\partial(\delta T)}{\partial t} = \frac{K}{C_P \tau}\nabla^2 \delta T, \qquad (8.47)$$

where K is the thermal conductivity, C_P is the heat capacity per unit volume, τ is the decay time for the heat current, and ∇ is the spatial gradient operator. The equations give a limiting upper velocity $v_{SS} = [K/(C_P\tau)]^{1/2}$ for the propagation of thermal waves, sometimes referred to as second sound. If we have slow time-variation of temperature so that the first term in (8.47) is negligible, we obtain the normal equation for thermal diffusion. The spectral distribution of the Rayleigh light depends on $\omega_{SS} = qv_{SS}$ and τ, giving rise to two distinct regimes (Enns and Haering 1966, see also Griffin 1968):

1. In the diffusion limit corresponding to $\omega_{SS}\tau \ll 1$, we expect a sharp unshifted Lorentzian peak corresponding to the normal Landau-Placzek regime [see (8.4) and (8.5a)].

2. In the second-sound limit corresponding to $\omega_{SS}\tau \gg 1$, temperature fluctuations are propagated as slightly damped temperature waves. The scattered light now consists of two Lorentzian peaks each of width $\sim 1/\tau$ shifted from the frequency of the incident light by the value given by (8.7) with v replaced by v_{SS}. The ratio of the shift of second-sound peaks to the ordinary Brillouin peaks is estimated to be $v_{SS}/v = (\frac{1}{3})^{1/2}$, where v is the velocity of sound (Chester 1963).

Although resolved second-sound peaks have been observed by light scattering in a ^{3}He^{4}He liquid mixture (Pike et al. 1969) the study of Rayleigh scattering due to spontaneous thermal fluctuations in solids has not been rewarding. The reasons for this may be seen from results of detailed studies in NaF. At low temperatures (~ 10 K) in pure dielectric crystals heat pulses can propagate ballistically as transverse and longitudinal excitations with a speed v. At higher temperatures the energy in the ballistic pulses decreases and an increasing amount of energy travels diffusively with no well-defined time of arrival; this happens when the rate of momentum-destroying phonon collisions becomes high. It was found (McNelly et al. 1970, Jackson et al. 1970) that in chemically purified NaF (this material is isotopically pure) an additional heat pulse appears at about 15 K between the transverse and diffusive components, and this was identified as incipient second sound. Subsequently Pohl and Schwarz (1973) explored the possibility of observing second sound in NaF by Rayleigh light scattering. They used the Landau-Placzek ratio (8.4), modified by Wehner and Klein (1972), with (8.35) for the calculated intensity of Brillouin scattering in a cubic crystal, to calculate the relative intensity of Rayleigh and Brillouin scattering. They concluded that in NaF the Rayleigh scattering would be about a factor of 10^7 weaker than the Brillouin scattering at 10 K and about 10^2 times weaker at 200 K. These results rule out the possibility of observing Rayleigh scattering in NaF in the second-sound temperature region. Indeed it is difficult to investigate Rayleigh scattering (caused by spontaneous thermal fluctuations) in solids at any temperature, because of parasitic scattering. This normally gives rise to an unshifted line that is usually much more intense than the Brillouin components (Figure 8.3). However, a central peak 2.3 GHz wide was recently observed in KTaO$_3$ at 300 K by Lyons and Fleury (1976) using Brillouin-scattering methods and was assigned by them to spontaneous thermal fluctuations (see Section 5.2.6).

A particular form of unwanted scattering occurs in amorphous solids when fluctuations occur due to structure variations that are independent of time (low-frequency diffusive modes are relatively unimportant); these give rise to a strong line of almost zero width and are important in determining

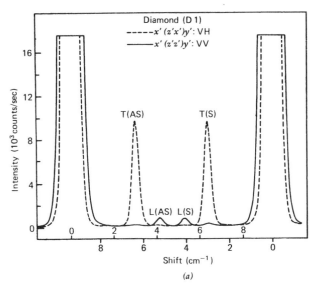

Figure 8.3 (a) Brillouin spectrum of diamond measured at room temperature with incident radiation along $x'\|[110]$ and scattered radiation along $y'\|[\bar{1}10]$; $z'\|[001]$. The letters V, H denote the polarization of the light with respect to the horizontal, (001), scattering plane, the first letter referring to the incident light. T = transverse, L = longitudinal, S = Stokes, AS = anti-Stokes.

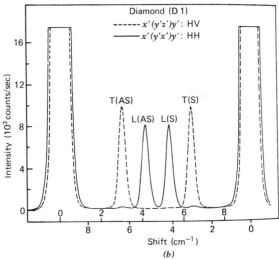

Figure 8.3 (b) Same as (a) except that the incident polarization is horizontal (after Grimsditch and Ramdas 1975).

the transparency of optical fibers for communications purposes (Tauc 1976).

8.3.2 Brillouin Scattering

In the solid state Brillouin scattering has been largely used to determine elastic constants by measuring the velocity of sound (in this context, the application of the technique to the study of second-order structural phase changes has already been discussed in Section 5.2.5). Benedek and Fritsch (1966) made a theoretical study of light scattering by cubic crystals and measured the elastic constants of KCl, RbCl, and KI. The study of Brillouin scattering in diamond has attracted considerable attention in the past because of its good optical quality, its high elastic constants (Table 8.4), and its cubic symmetry. Recently, it has been studied in detail by Grimsditch and Ramdas (1975), and we quote some of their results here to illustrate the method.

We saw in Section 8.2 that the intensity of Brillouin components is determined by changes in refractive index produced by strain due to sound waves and that the change in refractive index is related to the strain through the elastooptic constants p^{mn}. Figures 8.3a and 8.3b show the Brillouin spectrum of diamond excited with 488.0 nm radiation incident along [110] and scattered along [$\bar{1}$10]. With this geometry the scattered phonon propagates along [100] and can be a pure longitudinal phonon or a pure, doubly-degenerate transverse phonon (Section 8.1). The Brillouin intensity is readily calculated for cubic crystals from (8.37) using the scattering tensor χ given in Table 8.2 and the intensities for the above scattering configuration are given in Table 8.5. The letters H, V, T, L, S, and AS in Table 8.5 and Figure 8.3 stand for horizontal, vertical, transverse, longitudinal, Stokes, and anti-Stokes. The disappearance of lines labeled T in HH and VV polarizations and of those labeled L in HV and VH polarizations is consistent with the assignments. The values of the elastic constants of diamond obtained by Grimsditch and Ramdas (1975) are given in Table 8.4 and compared with values given by an ultrasonic

Table 8.4 Elastic Constants of Diamond (10^{11} N/m²)

Method	C^{11}	C^{12}	C^{44}	Ref.
Brillouin scattering 23°C	10.764 ±0.002	1.252 ±0.023	5.774 ±0.014	Grimsditch and Ramdas (1975)
Ultrasonic pulse 25°C	10.79 ±0.05	1.24 ±0.05	5.78 ±0.02	McSkimin and Andreatch (1972)

Table 8.5 Intensities I of Brillouin Components of a Cubic Crystal for Light Incident along [110] and Scattered along [$\bar{1}$10] with Phonon Propagation along [100].

$U_{\sigma q}$	I_V^V	I_V^H	I_H^V	I_H^H
[100] longitudinal	$\dfrac{(p^{12})^2}{C^{11}}$	0	0	$\dfrac{(p^{12}-p^{11})^2}{4C^{11}}$
[010] transverse	0	0	0	0
[001] transverse	0	$\dfrac{(p^{44})^2}{2C^{44}}$	$\dfrac{(p^{44})^2}{2C^{44}}$	0

*Polarizations with respect to the horizontal scattering plane are denoted H (horizontal) and V (vertical). Subscript and superscript on I refer to the incident and scattered polarization respectively.

technique. It is apparent that in this case there is no significant frequency dependence of elastic constants.

Grimsditch and Ramdas (1975) discuss the elastooptic constants of diamond in some detail. Grimsditch and Ramdas (1976) have made an extensive investigation of the elastic and elastooptic constants of rutile. A detailed study of the elastooptic constants of liquid and solid krypton (which forms cubic crystals) has been carried out by Kato and Stoicheff (1975) using absolute intensity measurements of Brillouin spectra. They find $p^{11} = 0.34 \pm 0.04$, $p^{12} = 0.34 \pm 0.05$, $p^{44} = 0.037 \pm 0.005$ for crystalline krypton at 115.5 K. These results are in reasonable agreement with predictions of present theories (Oxtoby and Chandrasekharan 1977). In both investigations mentioned above absolute intensities of scattered light were obtained by comparison with toluene (see Fabelinskii 1968, p. 563).

The use of the backscattering technique in Brillouin scattering has been developed by Sandercock (1975) who used a multipass Fabry-Perot interferometer (Section 2.2.2) to determine the elastic constants of layer compounds such as GaSe. This material has hexagonal symmetry with five independent elastic constants (Nye 1964). Sandercock also showed that Brillouin measurements on materials opaque to the laser light are possible using the backscattering method. When the optical absorption coefficient α [$= 2\eta_I'' \omega_I / c$; see (4.101)] is less than about 10^5 m^{-1} the Brillouin spectrum appears the same as that of a transparent material and has an intrinsic linewidth determined by the phonon lifetime. However, if α is so high that the laser light penetrates only a few wavelengths into the material, the Brillouin lines are broadened by the strong absorption.

The effect of absorption may be understood in an approximate sense by recognizing that scattering occurs within a distance from the surface of

$\sim 1/\alpha$. The uncertainty principle then indicates that q for the phonon can be determined only to an accuracy of $\Delta q \approx \alpha$ and that phonons within the wavevector range Δq contribute to the scattering. Since the Brillouin peaks occur at a phonon frequency $\omega = vq$, they will have a width $\Delta \omega \cong v\Delta q \cong 2v\eta_I'' \omega_I/c$. Hence in situations where $\alpha \gg 10^5$ m^{-1} the Brillouin linewidth is strongly affected by absorption.

Results of backscattering measurements on Si and Ge (Sandercock 1972a) are shown in Figure 8.4. The value of α increases from about 10^6 m^{-1} for Figure 8.4a to about 5×10^7 m^{-1} for Figure 8.4c. The increase in linewidth with increase in α is obvious.

The shape of the Brillouin line in the experiments just discussed depends on the relative magnitudes of η' and η''; it has a symmetric Lorentzian shape only when η'' is appreciably less than η' (Dervisch and Loudon 1976). However, when $\eta'' > \eta'$, as for example with metals, the scattered intensity is expected to become strongly asymmetric (Bennett et al. 1971). As in the case of Raman scattering from metals (Section 3.1.6) the Brillouin-scattering intensity is very weak and experimental studies are not extensive. A careful study of Brillouin scattering by liquid mercury and liquid gallium has been given by Dil and Brody (1976) who observed a strongly asymmetric line shape.

Sandercock (1972b) has observed structure in backscattering of $\lambda = 514.5$ nm light from an unsupported collodion film of thickness $d \simeq \lambda$. The sharp boundaries to the scattering volume lead to broadening of the Brillouin peaks and structure within this broadened envelope arises from the fact that phonons excited with wavevector q^z perpendicular to the film surface can only have discrete values $q^z = p(\pi/d)$ where p is an integer. Thus, instead of the usual Brillouin peak a number of lines is observed (usually three) separated in energy by $\pi v/d$ where v is the velocity of sound; the relative intensities of the lines are determined by the shape of the broadened envelope.

The development of the multipass Fabry-Perot interferometer has also made possible the observation of thermally excited acoustic magnons using Brillouin scattering. Results of such measurements on yttrium iron garnet and $CrBr_3$, obtained using the backscattering method, have already been given in Chapter 6.

The interaction of phonons with free carriers in solids is known as the acoustoelectric effect. This interaction is particularly strong in piezoelectric semiconductors, for example, CdS, and in these materials acoustic waves are readily amplified by carriers with drift velocities exceeding the sound velocity. The frequency distribution of the amplified thermal phonons can be studied by Brillouin scattering methods, and a review of this rather specialized area has been given by Pine (1975).

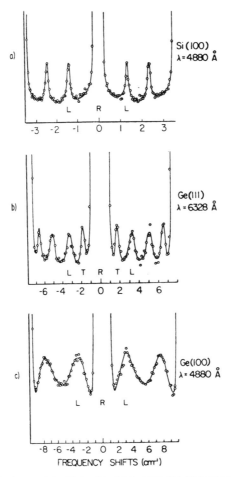

Figure 8.4 Brillouin spectra obtained in backscattering from (a) silicon and (b), (c) germanium. Both absorption and line width increase from (a) to (b) to (c) (after Sandercock 1972a).

REFERENCES

Anastassakis E. and Burstein E. (1974), *J. Phys. C, Solid State Phys.* **7**, 1374.

Ando K. and Hamaguchi C. (1975), *Phys. Rev.* **B11**, 3876.

Benedek G. B. (1968), in M. Chretien, E. P. Gross, and S. Deser, Eds., *Brandeis University Summer Institute of 1966 in Theoretical Physics*, Vol. 1 (Waltham, Mass.: Brandeis University Press), p. 1.

Benedek G. B. and Fritsch K. (1966), *Phys. Rev.* **149**, 647.
Bennett B. I., Maradudin A. A., and Swanson L. R. (1971), in M. Balkanski, Ed., *Light Scattering in Solids* (Paris: Flammarion), p. 443.
Berne B. J. and Pecora R. (1976), *Dynamic Light Scattering* (New York: John Wiley).
Brenig W., Zeyher R., and Birman J. L. (1972), *Phys. Rev.* **B6**, 4617.
Brillouin L. (1914), *Comptes Rendus* **158**, 133.
Brillouin L. (1922), *Ann. Phys. (Paris)* **17**, 88.
Chapelle J. and Taurel L. (1955), *Comptes Rendus* **240**, 743.
Chester M. (1963), *Phys. Rev.* **131**, 2013.
Cohen E. G. D. (1967), in W. E. Brittin, Ed., *Boulder Lectures on Theoretical Physics*, Vol. IXC, (New York: Gordon and Breach), p. 279.
Cummins H. Z. and Gammon R. W. (1966), *J. Chem. Phys.* **44**, 2785.
Cummins H. Z. and Pike E. R., Eds. (1974), *Photon Correlation and Light Beating Spectroscopy* (New York: Plenum Press).
Cummins H. Z. and Schoen P. E. (1972), in F. T. Arecchi and E. O. Schultz-Dubois, Eds., *Laser Handbook*, Vol. 2 (Amsterdam: North Holland), p. 1029.
Dervisch A. and Loudon R. (1976), *J. Phys. C, Solid State Phys.* **C9**, L669.
Dill J. G. and Brody E. M. (1976), *Phys. Rev.* **B14**, 5218.
Einstein A. (1910), *Ann. Phys.* **38**, 1275.
Enns R. H. and Haering R. R. (1966), *Phys. Lett.* **21**, 534.
Fabelinskii I. L. (1968), *Molecular Scattering of Light* (New York: Plenum Press).
Fleury P. A. and Boon J. P. (1973), *Adv. Chem. Phys.* **24**, 1.
Garrod D. K. and Bray R. (1972), *Phys. Rev.* **B6**, 1314.
Gornall W. S., Stegeman G. I., Stoicheff B. P., Stolen R. H., and Volterra V. (1966), *Phys. Rev. Lett.* **17**, 297.
Greytak T. J. and Benedek G. B. (1966), *Phys. Rev. Lett.* **17**, 197.
Griffin A. (1968), *Rev. Mod. Phys.* **40**, 167.
Grimsditch M. H. and Ramdas A. K. (1975), *Phys. Rev.* **B11**, 3139.
Grimsditch M. H. and Ramdas A. K. (1976), *Phys. Rev.* **B14**, 1670.
Jackson H. E., Walker C. T., and McNelly T. F. (1970), *Phys. Rev. Lett.* **25**, 26.
Kato Y. and Stoicheff B. P. (1975), *Phys. Rev.* **B11**, 3984.
Kittel C. (1966), *Introduction to Solid State Physics*, 3rd ed. (New York: Wiley).
Landau L. D. and Lifshitz E. M. (1960), *Electrodynamics of Continuous Media* (New York: Pergamon Press), p. 391.
Landau L. D. and Placzek G. (1934), *Phys. Z. Sowjetunion* **5**, 172.
Lax M. and Nelson D. F. (1971), *Phys. Rev.* **B4**, 3694.
Loudon R. (1963), *Proc. R. Soc.* **A275**, 218.
Loudon R. (1969), *Proceedings of the International School of Physics, E. Fermi* in R. J. Glauber, Ed., Course 42 (New York and London: Academic Press), p. 297
Lunacek J. H. and Benedek G. B. (1970), *Am. J. Phys.* **38**, 575.
Lyons K. B. and Fleury P. A. (1976), *Phys. Rev. Lett.* **37**, 161.
McIntyre D. and Sengers J. V. (1968), in H. N. V. Temperley, J. S. Rowlinson and G. S. Rushbrooke, Eds., *Physics of Simple Fluids* (Amsterdam: North Holland), p. 447.
McNelly T. F., Rogers S. J., Channin D. J., Rollefson R. J., Gouban W. M., Schmidt G. E., Krumhansl J. A., and Pohl R. O. (1970), *Phys. Rev. Lett.* **24**, 100.

References

McSkimin H. J. and Andreatch P. (1972), *J. Appl. Phys.* **43**, 2944.
McTague J. P. and Birnbaum G. (1968), *Phys. Rev. Lett.* **21**, 661.
McTague J. P., Fleury P. A., and du Pre D. B. (1969), *Phys. Rev.* **188**, 303.
Mountain R. D. (1966a), *Rev. Mod. Phys.* **38**, 205.
Mountain R. D. (1966b), *J. Res. Natl. Bur. Standards*, **70A**, 207.
Neighbours J. R. and Schacher G. E. (1967), *J. Appl. Phys.* **38**, 5366.
Nelson D. F. and Lax M. (1970), *Phys. Rev. Lett.* **24**, 379.
Nelson D. F. and Lax M. (1971), *Phys. Rev.* **B3**, 2778.
Nelson D. F. and Lazay P. D. (1970), *Phys. Rev. Lett.* **25**, 1187.
Nelson D. F., Lazay P. D., and Lax M. (1972), *Phys. Rev.* **B6**, 3109.
Nye J. F. (1957), *Physical Properties of Crystals* (Oxford: Clarendon Press).
O'Brien R. J., Rosasco G. J., and Weber A. (1969), *Light Scattering Spectra of Solids* (New York: Springer-Verlag), p. 623.
Oxtoby D. W. and Chandrasekharan V. (1977), *Phys. Rev.* **B16**, 1706.
Pike E. R., Vaughan J. M., and Vinen W. F. (1969), *Phys. Lett.* **30A**, 373.
Pine A. S. (1972), *Phys. Rev.* **B5**, 3003.
Pine A. S. (1975), in M. Cardona, Ed., *Light Scattering in Solids* (Berlin: Springer-Verlag), p. 253.
Pohl D. W. and Schwarz S. E. (1973), *Phys. Rev.* **7**, 2735.
Sandercock J. R. (1972a), *Phys. Rev. Lett.* **28**, 237.
Sandercock J. R. (1972b), *Phys. Rev. Lett.* **29**, 1735.
Sandercock J. R. (1975), *Festkörperprobleme* **15**, 183.
Stegeman G. I. and Stoicheff B. P. (1968), *Phys. Rev. Lett.* **21**, 202.
Tauc J. (1976), in M. Balkanski, R. C. C. Leite and S. P. S. Porto, Eds., *Light Scattering in Solids* (Paris: Flammarion), p. 621.
Ulbrich R. G. and Weisbuch C. (1977), *Phys. Rev. Lett.* **38**, 865.
Vacher R. and Boyer L. (1972), *Phys. Rev.* **B6**, 639.
Wehner R. W. and Klein R. (1972), *Physica.* **62**, 161.
Whitfield G. D. (1961), *Phys. Rev.* **121**, 720.

Index

Acceptors, shallow, 294, 298
  see also electronic scattering
Acoustic anomalies, 206
Acoustic mode normal co-ordinates, 102
Acoustic vibrations, 97
Acoustoelectric effect, 350
Admittance, *see* linear response function
Amorphous solids, 139, 346
Anharmonic lattice potential energy, 102, 111
Anisotropy field, 257
Anomalous low-frequency modes, 141
Antiferromagnet
  defect-induced scattering, 277
  first order experiments, 264
  ground state magnons, 257
  magnon interaction effect, 272
  microscopic cross section, 263
  perovskite structure second-order spectra, 273-274
  rutile structure, 259
  rutile structure second-order spectra, 273-274
  second-order cross section, 270
  second-order scattering, 266
  second-order spectra in two-dimensional structures, 274
  temperature dependence of second-order spectra, 275
Anti-Stokes component, 6
Anti-Stokes polarization, 20
Antisymmetric scattering, 126, 264

Backscattering, 118, 190
  Brillouin, 350
Backscattering geometry, 77
Bethe notation, 47
Birefringent crystal, 88
Bloch equations, 245

Bloch functions, 36, 180
Bond-charge model, 129, 130
Bose-Einstein thermal factor, 7
Brewsters angle, 64
Brillouin doublets, 3
Brillouin scattering
  frequency shift, 15, 332
  magnons, 350
  metals, 350
  piezoelectric crystals, 338
  phonon cross-section,
    by macroscopic approach, 340
    by microscopic approach, 345
  rotational contributions, 343

Central mode, 142, 206, 213, 232
  see also Soft mode
Central peak, 3, 346
Charge-density wave, 205
Classical electron radius, 9, 41
Coherent potential approximation, 277
Collision-induced scattering, 331
Compatibility table, 104
Compton scattering, 10
Correlation table, 106
Coupled-mode spectra, 221
Critical concentration, 280
  see also Percolation limit
Critical exponents
  classical, 205-206
  non-classical, 205-206
Critical fluctuations, 211
Critical opalescence, 3, 214
Critical phenomena, 205
Critical points, 122
Cross section
  atomic scattering, 39
  angular dependence
    for polar modes, 167

for randomly-oriented scatterers, 115
antiferromagnet, 263
  second-order, 270
cubic polar vibrations, 166
destructive interference for random
  distribution of atoms, 11
elastic polarization dependence, 9
ferromagnetic, 244
free electron scattering, 42
general definition, 8
general macroscopic expression, 25, 26, 50
general microscopic expression, 34, 50
longitudinal magnetic, 247
microscopic expression for ferromagnet, 254
polarition scattering, 197
polar modes in powdered crystals, 178
relation between Stokes and anti-Stokes, 7, 31, 186
review of microscopic method, 16-17
second-order vibrational, 127
symmetry properties, 43
transverse magnetic, 247
uniaxial polar vibrations, 174
vibrational, 108
vibrational microscopic expression, 184
Crystalline electric field, 289

Davydov splitting, 106, 228, 262, 292
Debye relaxation, 222
Defect-activated spectrum, 135
Defect-induced scattering, 131
  amorphous solids, 132
  band modes, 132
  gap modes, 132
  localized modes, 132
  magnetic, 276
  mixed crystals, 132
  resonance modes, 132
Deformation potentials, 182
Density of states
  combined or second-order vibrational, 122
  magnetic second-order, 267
Depolarization ratio, 114
Differential cross section
  general definition, 8
  see also cross section, 8
Diffraction grating, 75
Donors, shallow, 294, 296

see also Electronic scattering
Doppler-free scattering, 117

Effective charge, 150
Effective mass approximation, 293
Elastic scattering, classical theory, 9
Elasto-optic constants, 349
Elasto-optic effect, 341
Electric-dipole interaction, 36
Electric-field operators, 35
Electron-hole drops, 315
Electron-hole pair, 111, 180, 188
Electronic scattering, 292
  change from plasmon scattering to single-particle scattering, 316
  Raman spectrum
    of plasmons, 311
    of rare-earth ions, 292
    of shallow acceptors, 299
    of shallow donors, 297
  scattering cross section
    for electrons in solids, 307
    for free electrons, 306
  scattering mechanisms for plasmons, 313
Electron-lattice interaction, 33, 110, 129
  long-range part, 182, 184, 191
  short-range part, 182, 184
Electron-radiation interaction, 50, 110, 181
  A^2 and $A \cdot p$ parts, 35, 38
  electric-dipole approximation, 35
  Hamiltonian, 34, 35
Electro-optic effect, 161
Energy conservation, 6, 121
Energy-density scattering, 319
Equivalent sites, 103, 105
Exchange coupling, 251, 252
Exchange scattering mechanism, 268
Excitation symmetry, 43
  Bethe notation, 47
  degenerate, 49
  Mulliken notation, 47
  two-phonon, 123
  vibrational, 96
Excitons, 188, 291
External vibrations, 103

F center, 135
Fabry-Perot interferometer, 61
  centre spot scanning, 81
  confocal, 83

Index

contrast of, 82
etendue, 81
finesse, 81
multipass scheme, 83
resolving power of, 81
two in series, 82
Factor group, 97, 106
Faraday rotation, 240
  electric contribution, 242
  magnetic contribution, 243
Fast-ion conductors, 223
Fermi-Dirac distribution, 301
Ferrodistortive modes, 208
Ferroelastic materials, 209
Ferromagnet, 240
  cross section, 244
  macroscopic cross section, 247
  magnons, 251
  microscopic cross section, 254
Field-dependent relative permittivity, 157
Fluctuation-dissipation theorem, 27, 29, 50, 108, 155, 247
  quantum-mechanical form, 38
Forbidden scattering, 187
Forward scattering, 78
Free carriers
  effect on elastic constants, 314
  plasma excitations, 301
  single-particle excitations, 301
Free electron scattering
  cross section, 42
Fresnel's equation
  biaxial crystal, 170, 175
  cubic crystal, 162, 192
  large-wavevector approximation, 151
  uniaxial crystal, 169, 170, 195
Fröhlich interaction, 183
  see also Electron lattice interaction, long range part

Gap modes, 134
Generalized force, 27, 30
Green functions, 38, 273

H^- center, 133
Hamiltonian
  anisotropic antiferromagnetic, 277
  antiferromagnetic, 256
  electronic, 180
  electron-lattice, 181-182
  electron-radiation, 34
  electron-vibrational, 33
  ferromagnetic, 251
  radiation, 33
  spin, 255, 270
Hole burning, 63

Improper ferroelectrics, 204
Impurity levels in semiconductors, 293
Incommensurate structures, 105
Index matching, 77
Intensity of scattering
  absolute value, 91
  secondary standard, 92
  temperature dependence, 91
Internal vibrations, 103
Iodine vapor cell, 77
Iron-group ions, 288
Ising model, 272

Jahn-Teller interaction, 111
  cooperative, 202

Kramers degeneracy, 290
Kramers-Heisenberg formula, 16, 39, 41

Landau damping, 305
Landau levels, 319
Landau-level Raman scattering, 321
Landau-Placzek ratio, 329
Laser
  Argon-ion 66
  carbon dioxide-nitrogen, 68
  cavities, 58
  conditions for operation, 54
  dye, 70
  helium neon, 65
  plasma lines, 72
  plasma tube, 64
  neodymium in glass, 72
  neodymium in YAG, 72
  ABC:YAG, 72
  ruby, 72
  tunable infrared, 321
Lattice dynamics, 99, 100-101
  bond-charge model, 127
  polar modes, 148
  rigid-ion model, 127
  shell model, 127
Lead sulfide detector, 88

Line broadening
  Doppler, 54
  homogeneous, 54
  inhomogeneous, 54
  isotopic, 66
Linear response function, 28, 30, 155
  magnetic, 246
  molecular vibrations, 30
Linear susceptibility, 18
  magnetic symmetry properties, 241
  microscopic expression, 181, 185
  symmetry properties for nonmagnetic medium, 108
  tabulation of symmetry properties, 149
Localized vibrational modes, 300
Lorentzian lineshape, 30, 166
Lyddane-Sachs-Teller relationship, 164, 170, 207

Magnetic defect scattering, 276
Magnetic excitons, 256, 261, 265
Magnetic scattering
  antiferromagnetic cross section, 263
  antisymmetric polarization dependence, 249-250
  electric contribution to ferromagnetic cross section, 244
  ferromagnetic temperature dependence, 248
  longitudinal cross section, 247
  magnetic contributions to ferromagnetic cross section, 244
  microscopic cross section for ferromagnet, 254
  microscopic mechanism, 253
  rutile-structure second-order, 270
  second-order, 266
  second-order mechanism, 268
Magnon
  alteration at defect site, 277
  angular dependence of frequency in ferromagnet, 253
  antiferromagnetic, 256
  ferromagnetic, 251
  interaction effects, 271
  Raman scattering, 320
  two-magnon density of states, 267
Mass-weighted displacement coordinates, 100
Mass-weighted force constant, 100

Maxwell-Boltzmann distribution, 302
Mean field magnetic susceptibility, 247
Mercury-vapor cell, 77
Metal-insulator transition, 298
Mixed crystals, 137
  magnetic, 279
Mode structure
  in optical resonators, 59
  single moding, 63-64
Molecular (mean) field magnetic susceptibility, 247
Molecular vibration, 103
  simple harmonic oscillation equation, 29
Momentum conservation, 11, 24, 31, 36, 121, 181
Mulliken notation, 47
Multiple scattering, 187

Nonpolar vibrations, 96
Normal coordinates, 100, 101, 149
Normal modes, 101

One-mode behaviour, 138, 280
Optical fibres, 348
Optical resonator geometries, 62
Optic vibration, 99
Order parameter
  macroscopic, 204
  microscopic, 204
  strain, 231-232
Orientational averaging
  in polycrystalline material, 91
Orthonormality relations, 101

Parasitic scattering, 346
Percolation limit, 280
Phase change
  acoustic anomalies, 228
  cooperative Jahn-Teller, 223
  displacive, 207
  first order, 202
  hydrogen bonded ferroelectrics, 219
  mean field theory, 205
  order-disorder, 219
  perovskites, 214
  quartz, 213
  second-order, 202
Photomultiplier
  quantum efficiency of, 86
Photon counting, 87

Index

Piezoelectric crystals, 212
Plasmons, see Electronic scattering
Point defects, 132
Point group, 43, 97, 105
Polariton, 78, 152
  biaxial crystals, 195
  cubic crystal dispersion relation, 192
  uniaxial crystal dispersion relation, 195
Polarition scattering, 192
  cross section, 197
Polarizability derivatives, see Susceptibility derivatives
Polarization of scattered light
  experimental investigation, 88
  symmetric and antisymmetric degree of, 90, 114
Polarization scrambler, 75
Polar vibrations, 96
  biaxial angular dependence, 175
  cubic cross section, 166
  general cross section, 154
  lifting of degeneracy, 153
  longitudinal, 163, 170
  LST relation, 170
  macroscopic electric field, 151, 152, 164
  powdered crystals, 177
  transverse, 163, 170
  uniaxial angular dependence, 171
  uniaxial cross section, 174
  wurtzite structure crystals, 171
  zinc blende structure crystals, 164
Powdered crystals
  polar-mode cross sections, 177
Power spectrum, 50, 108, 127
  general definition, 18
  macroscopic field, 155
  molecular vibration, 31
  normal coordinate, 155
  polarization fluctuations, 24, 26
  quantum-mechanical form, 28-29, 38
Primitive cell, 97

Quasilongitudinal modes, 336
Quasielastic scattering, 232

Radiation field, Hamiltonian, 33
Raman effect, 3
Raman scattering, 5
  frequency shift, 15
Randomly oriented scatterers, 112, 117

Rare-earth ions, 288
  see also Electronic scattering
Rayleigh-Brillouin scattering cross section
  thermodynamic treatment, 329
Rayleigh formula, 27
Rayleigh's law, 2, 10
Rayleigh scattering, 3
Rayleigh wing, 330
Relative permittivity, 149
  biaxial crystal, 175
  cubic crystal, 163, 165
  general form, 150
  uniaxial crystal, 170
Refractive indices, 13
  magnetic material, 242
Relaxation modes, 211
Resolving power, 74
Resonance modes, 135
Resonance scattering, 10, 33
  appearance of forbidden lines, 190-191
  magnetic, 255
  multiple scattering peaks, 190-191
  refractive index variation, 190
  second-order magnetic, 276
  second-order vibrational, 130
Response function, 38
Rigid-ion model 127
Rule of mutual exclusion, 48

Sample illumination, 79
Scattering cross section, 4, 91
  see also Cross section
Scattering tensor, 184, 186
  resonance form, 187, 189
Second-harmonic generation, 162
Second-order susceptibility, 19, 20, 107
  nonmagnetic symmetry property, 20
  symmetry properties, 43, 47
  table 1.2, 44-46
Second sound, 345
Selection rules, 47
  calcium tungstate, 107
  magnetic defect-induced scattering, 278
  second-order magnetic scattering, 271
  second-order vibrational scattering, 123
Self-energy, 211
Shell model, 127, 129, 135
Signal averaging, 88
Single electrons
  light scattering cross section, 316

velocity distribution in solids, 317
Site group, 104
Sky, color of, 2
Soft mode, 206
  eigenvector, 207
  electronic, 209
  tunneling, 209
Space group, 43, 97, 123
Spectral differential cross section,
    general definition, 6
  see also Cross section
Spectrometer
  contrast of, 75
  double grating, 76
  triple grating, 77
Spin-density fluctuation, 319
Spin-flip Raman scattering, 319, 320
Spin Hamiltonian, 255
Spin-orbit interaction, 253
  antisymmetric scattering, 256
Spin wave, *see* Magnon
Spontaneous emission, 55
Spontaneous scattering, 37
Stimulated scattering, 37
Stokes component, 6
Stokes polarization, 19
  radiated field, 21
  scattered intensity, 24
Stokes scattering
  wavevector versus frequency relation, 13
Susceptibility derivatives
  angular averages, 112
  general definition, 21
  macroscopic field, 153
  magnetic, 241
  magnetic second-order, 250
  measurement by field-induced absorption, 157
  microscopic expressions, 185
  normal coordinate, 107
  relation to electro-optic coefficients, 160-161

relation to second-harmonic generation, 162
  second-order scattering, 127
  tabulation of symmetry properties for polar modes, 158-159
Surface modes, 179

Thomson cross section, 10, 41, 308
Time reversal, 32, 106, 251
Time-reversal symmetry, 43
Two-mode behavior, 138, 280
Two-phonon bound state, 130

Valley-orbit splitting, 295, 297
Van Hove singularities, 122, 124, 267
Vector-potential operator, 35
Vibrational force constants, 100
Vibrational scattering
  cubic polar mode cross section, 165
  diamond-structure crystals, 118, 124
  hexagonal close-packed metals, 118
  in magnetic semiconductors, 111-112
  macroscopic theory, 107
  microscopic cross section, 183
  microscopic mechanisms, 109
  polar mode cross section, 154, 156
  randomly-oriented scatterers, 111-112
  relation between microscopic and macroscopic expressions, 185
  rock salt-structure crystals, 124
  rutile-structure crystals, 118
  second-order, 121, 124
  second-order line in diamond, 129
  second-order macroscopic theory, 127
  second-order microscopic theory, 127
  uniaxial polar mode cross section, 175
  zinc-blende-structure crystals, 124
Vibrational symmetry, 96
  tabulation for common crystals, 98
Vibronic energy levels, 136
Voigt notation, 335

A CATALOG OF SELECTED
DOVER BOOKS
IN ALL FIELDS OF INTEREST

A CATALOG OF SELECTED DOVER BOOKS IN ALL FIELDS OF INTEREST

CONCERNING THE SPIRITUAL IN ART, Wassily Kandinsky. Pioneering work by father of abstract art. Thoughts on color theory, nature of art. Analysis of earlier masters. 12 illustrations. 80pp. of text. 5⅜ x 8½. 23411-8

ANIMALS: 1,419 Copyright-Free Illustrations of Mammals, Birds, Fish, Insects, etc., Jim Harter (ed.). Clear wood engravings present, in extremely lifelike poses, over 1,000 species of animals. One of the most extensive pictorial sourcebooks of its kind. Captions. Index. 284pp. 9 x 12. 23766-4

CELTIC ART: The Methods of Construction, George Bain. Simple geometric techniques for making Celtic interlacements, spirals, Kells-type initials, animals, humans, etc. Over 500 illustrations. 160pp. 9 x 12. (Available in U.S. only.) 22923-8

AN ATLAS OF ANATOMY FOR ARTISTS, Fritz Schider. Most thorough reference work on art anatomy in the world. Hundreds of illustrations, including selections from works by Vesalius, Leonardo, Goya, Ingres, Michelangelo, others. 593 illustrations. 192pp. 7⅛ x 10¼. 20241-0

CELTIC HAND STROKE-BY-STROKE (Irish Half-Uncial from "The Book of Kells"): An Arthur Baker Calligraphy Manual, Arthur Baker. Complete guide to creating each letter of the alphabet in distinctive Celtic manner. Covers hand position, strokes, pens, inks, paper, more. Illustrated. 48pp. 8¼ x 11. 24336-2

EASY ORIGAMI, John Montroll. Charming collection of 32 projects (hat, cup, pelican, piano, swan, many more) specially designed for the novice origami hobbyist. Clearly illustrated easy-to-follow instructions insure that even beginning papercrafters will achieve successful results. 48pp. 8¼ x 11. 27298-2

THE COMPLETE BOOK OF BIRDHOUSE CONSTRUCTION FOR WOODWORKERS, Scott D. Campbell. Detailed instructions, illustrations, tables. Also data on bird habitat and instinct patterns. Bibliography. 3 tables. 63 illustrations in 15 figures. 48pp. 5¼ x 8½. 24407-5

BLOOMINGDALE'S ILLUSTRATED 1886 CATALOG: Fashions, Dry Goods and Housewares, Bloomingdale Brothers. Famed merchants' extremely rare catalog depicting about 1,700 products: clothing, housewares, firearms, dry goods, jewelry, more. Invaluable for dating, identifying vintage items. Also, copyright-free graphics for artists, designers. Co-published with Henry Ford Museum & Greenfield Village. 160pp. 8¼ x 11. 25780-0

HISTORIC COSTUME IN PICTURES, Braun & Schneider. Over 1,450 costumed figures in clearly detailed engravings–from dawn of civilization to end of 19th century. Captions. Many folk costumes. 256pp. 8⅜ x 11¼. 23150-X

CATALOG OF DOVER BOOKS

STICKLEY CRAFTSMAN FURNITURE CATALOGS, Gustav Stickley and L. & J. G. Stickley. Beautiful, functional furniture in two authentic catalogs from 1910. 594 illustrations, including 277 photos, show settles, rockers, armchairs, reclining chairs, bookcases, desks, tables. 183pp. 6½ x 9¼. 23838-5

AMERICAN LOCOMOTIVES IN HISTORIC PHOTOGRAPHS: 1858 to 1949, Ron Ziel (ed.). A rare collection of 126 meticulously detailed official photographs, called "builder portraits," of American locomotives that majestically chronicle the rise of steam locomotive power in America. Introduction. Detailed captions. xi+ 129pp. 9 x 12. 27393-8

AMERICA'S LIGHTHOUSES: An Illustrated History, Francis Ross Holland, Jr. Delightfully written, profusely illustrated fact-filled survey of over 200 American lighthouses since 1716. History, anecdotes, technological advances, more. 240pp. 8 x 10¾. 25576-X

TOWARDS A NEW ARCHITECTURE, Le Corbusier. Pioneering manifesto by founder of "International School." Technical and aesthetic theories, views of industry, economics, relation of form to function, "mass-production split" and much more. Profusely illustrated. 320pp. 6⅛ x 9¼. (Available in U.S. only.) 25023-7

HOW THE OTHER HALF LIVES, Jacob Riis. Famous journalistic record, exposing poverty and degradation of New York slums around 1900, by major social reformer. 100 striking and influential photographs. 233pp. 10 x 7⅞. 22012-5

FRUIT KEY AND TWIG KEY TO TREES AND SHRUBS, William M. Harlow. One of the handiest and most widely used identification aids. Fruit key covers 120 deciduous and evergreen species; twig key 160 deciduous species. Easily used. Over 300 photographs. 126pp. 5⅜ x 8½. 20511-8

COMMON BIRD SONGS, Dr. Donald J. Borror. Songs of 60 most common U.S. birds: robins, sparrows, cardinals, bluejays, finches, more–arranged in order of increasing complexity. Up to 9 variations of songs of each species.
Cassette and manual 99911-4

ORCHIDS AS HOUSE PLANTS, Rebecca Tyson Northen. Grow cattleyas and many other kinds of orchids–in a window, in a case, or under artificial light. 63 illustrations. 148pp. 5⅜ x 8½. 23261-1

MONSTER MAZES, Dave Phillips. Masterful mazes at four levels of difficulty. Avoid deadly perils and evil creatures to find magical treasures. Solutions for all 32 exciting illustrated puzzles. 48pp. 8¼ x 11. 26005-4

MOZART'S DON GIOVANNI (DOVER OPERA LIBRETTO SERIES), Wolfgang Amadeus Mozart. Introduced and translated by Ellen H. Bleiler. Standard Italian libretto, with complete English translation. Convenient and thoroughly portable–an ideal companion for reading along with a recording or the performance itself. Introduction. List of characters. Plot summary. 121pp. 5¼ x 8½. 24944-1

TECHNICAL MANUAL AND DICTIONARY OF CLASSICAL BALLET, Gail Grant. Defines, explains, comments on steps, movements, poses and concepts. 15-page pictorial section. Basic book for student, viewer. 127pp. 5⅜ x 8½. 21843-0

THE CLARINET AND CLARINET PLAYING, David Pino. Lively, comprehensive work features suggestions about technique, musicianship, and musical interpretation, as well as guidelines for teaching, making your own reeds, and preparing for public performance. Includes an intriguing look at clarinet history. "A godsend," *The Clarinet,* Journal of the International Clarinet Society. Appendixes. 7 illus. 320pp. 5⅜ x 8½. 40270-3

HOLLYWOOD GLAMOR PORTRAITS, John Kobal (ed.). 145 photos from 1926-49. Harlow, Gable, Bogart, Bacall; 94 stars in all. Full background on photographers, technical aspects. 160pp. 8⅜ x 11¼. 23352-9

THE ANNOTATED CASEY AT THE BAT: A Collection of Ballads about the Mighty Casey/Third, Revised Edition, Martin Gardner (ed.). Amusing sequels and parodies of one of America's best-loved poems: Casey's Revenge, Why Casey Whiffed, Casey's Sister at the Bat, others. 256pp. 5⅜ x 8½. 28598-7

THE RAVEN AND OTHER FAVORITE POEMS, Edgar Allan Poe. Over 40 of the author's most memorable poems: "The Bells," "Ulalume," "Israfel," "To Helen," "The Conqueror Worm," "Eldorado," "Annabel Lee," many more. Alphabetic lists of titles and first lines. 64pp. 5$\frac{3}{16}$ x 8¼. 26685-0

PERSONAL MEMOIRS OF U. S. GRANT, Ulysses Simpson Grant. Intelligent, deeply moving firsthand account of Civil War campaigns, considered by many the finest military memoirs ever written. Includes letters, historic photographs, maps and more. 528pp. 6⅛ x 9¼. 28587-1

ANCIENT EGYPTIAN MATERIALS AND INDUSTRIES, A. Lucas and J. Harris. Fascinating, comprehensive, thoroughly documented text describes this ancient civilization's vast resources and the processes that incorporated them in daily life, including the use of animal products, building materials, cosmetics, perfumes and incense, fibers, glazed ware, glass and its manufacture, materials used in the mummification process, and much more. 544pp. 6⅛ x 9¼. (Available in U.S. only.) 40446-3

RUSSIAN STORIES/RUSSKIE RASSKAZY: A Dual-Language Book, edited by Gleb Struve. Twelve tales by such masters as Chekhov, Tolstoy, Dostoevsky, Pushkin, others. Excellent word-for-word English translations on facing pages, plus teaching and study aids, Russian/English vocabulary, biographical/critical introductions, more. 416pp. 5⅜ x 8½. 26244-8

PHILADELPHIA THEN AND NOW: 60 Sites Photographed in the Past and Present, Kenneth Finkel and Susan Oyama. Rare photographs of City Hall, Logan Square, Independence Hall, Betsy Ross House, other landmarks juxtaposed with contemporary views. Captures changing face of historic city. Introduction. Captions. 128pp. 8¼ x 11. 25790-8

AIA ARCHITECTURAL GUIDE TO NASSAU AND SUFFOLK COUNTIES, LONG ISLAND, The American Institute of Architects, Long Island Chapter, and the Society for the Preservation of Long Island Antiquities. Comprehensive, well-researched and generously illustrated volume brings to life over three centuries of Long Island's great architectural heritage. More than 240 photographs with authoritative, extensively detailed captions. 176pp. 8¼ x 11. 26946-9

NORTH AMERICAN INDIAN LIFE: Customs and Traditions of 23 Tribes, Elsie Clews Parsons (ed.). 27 fictionalized essays by noted anthropologists examine religion, customs, government, additional facets of life among the Winnebago, Crow, Zuni, Eskimo, other tribes. 480pp. 6⅛ x 9¼. 27377-6

CATALOG OF DOVER BOOKS

FRANK LLOYD WRIGHT'S DANA HOUSE, Donald Hoffmann. Pictorial essay of residential masterpiece with over 160 interior and exterior photos, plans, elevations, sketches and studies. 128pp. 9¼ x 10¾. 29120-0

THE MALE AND FEMALE FIGURE IN MOTION: 60 Classic Photographic Sequences, Eadweard Muybridge. 60 true-action photographs of men and women walking, running, climbing, bending, turning, etc., reproduced from rare 19th-century masterpiece. vi + 121pp. 9 x 12. 24745-7

1001 QUESTIONS ANSWERED ABOUT THE SEASHORE, N. J. Berrill and Jacquelyn Berrill. Queries answered about dolphins, sea snails, sponges, starfish, fishes, shore birds, many others. Covers appearance, breeding, growth, feeding, much more. 305pp. 5¼ x 8¼. 23366-9

ATTRACTING BIRDS TO YOUR YARD, William J. Weber. Easy-to-follow guide offers advice on how to attract the greatest diversity of birds: birdhouses, feeders, water and waterers, much more. 96pp. 5³⁄₁₆ x 8¼. 28927-3

MEDICINAL AND OTHER USES OF NORTH AMERICAN PLANTS: A Historical Survey with Special Reference to the Eastern Indian Tribes, Charlotte Erichsen-Brown. Chronological historical citations document 500 years of usage of plants, trees, shrubs native to eastern Canada, northeastern U.S. Also complete identifying information. 343 illustrations. 544pp. 6½ x 9¼. 25951-X

STORYBOOK MAZES, Dave Phillips. 23 stories and mazes on two-page spreads: Wizard of Oz, Treasure Island, Robin Hood, etc. Solutions. 64pp. 8¼ x 11. 23628-5

AMERICAN NEGRO SONGS: 230 Folk Songs and Spirituals, Religious and Secular, John W. Work. This authoritative study traces the African influences of songs sung and played by black Americans at work, in church, and as entertainment. The author discusses the lyric significance of such songs as "Swing Low, Sweet Chariot," "John Henry," and others and offers the words and music for 230 songs. Bibliography. Index of Song Titles. 272pp. 6½ x 9¼. 40271-1

MOVIE-STAR PORTRAITS OF THE FORTIES, John Kobal (ed.). 163 glamor, studio photos of 106 stars of the 1940s: Rita Hayworth, Ava Gardner, Marlon Brando, Clark Gable, many more. 176pp. 8⅜ x 11¼. 23546-7

BENCHLEY LOST AND FOUND, Robert Benchley. Finest humor from early 30s, about pet peeves, child psychologists, post office and others. Mostly unavailable elsewhere. 73 illustrations by Peter Arno and others. 183pp. 5⅜ x 8½. 22410-4

YEKL and THE IMPORTED BRIDEGROOM AND OTHER STORIES OF YIDDISH NEW YORK, Abraham Cahan. Film Hester Street based on *Yekl* (1896). Novel, other stories among first about Jewish immigrants on N.Y.'s East Side. 240pp. 5⅜ x 8½. 22427-9

SELECTED POEMS, Walt Whitman. Generous sampling from *Leaves of Grass*. Twenty-four poems include "I Hear America Singing," "Song of the Open Road," "I Sing the Body Electric," "When Lilacs Last in the Dooryard Bloom'd," "O Captain! My Captain!"–all reprinted from an authoritative edition. Lists of titles and first lines. 128pp. 5³⁄₁₆ x 8¼. 26878-0

CATALOG OF DOVER BOOKS

THE BEST TALES OF HOFFMANN, E. T. A. Hoffmann. 10 of Hoffmann's most important stories: "Nutcracker and the King of Mice," "The Golden Flowerpot," etc. 458pp. 5⅜ x 8½. 21793-0

FROM FETISH TO GOD IN ANCIENT EGYPT, E. A. Wallis Budge. Rich detailed survey of Egyptian conception of "God" and gods, magic, cult of animals, Osiris, more. Also, superb English translations of hymns and legends. 240 illustrations. 545pp. 5⅜ x 8½. 25803-3

FRENCH STORIES/CONTES FRANÇAIS: A Dual-Language Book, Wallace Fowlie. Ten stories by French masters, Voltaire to Camus: "Micromegas" by Voltaire; "The Atheist's Mass" by Balzac; "Minuet" by de Maupassant; "The Guest" by Camus, six more. Excellent English translations on facing pages. Also French-English vocabulary list, exercises, more. 352pp. 5⅜ x 8½. 26443-2

CHICAGO AT THE TURN OF THE CENTURY IN PHOTOGRAPHS: 122 Historic Views from the Collections of the Chicago Historical Society, Larry A. Viskochil. Rare large-format prints offer detailed views of City Hall, State Street, the Loop, Hull House, Union Station, many other landmarks, circa 1904-1913. Introduction. Captions. Maps. 144pp. 9⅜ x 12¼. 24656-6

OLD BROOKLYN IN EARLY PHOTOGRAPHS, 1865-1929, William Lee Younger. Luna Park, Gravesend race track, construction of Grand Army Plaza, moving of Hotel Brighton, etc. 157 previously unpublished photographs. 165pp. 8⅞ x 11¾. 23587-4

THE MYTHS OF THE NORTH AMERICAN INDIANS, Lewis Spence. Rich anthology of the myths and legends of the Algonquins, Iroquois, Pawnees and Sioux, prefaced by an extensive historical and ethnological commentary. 36 illustrations. 480pp. 5⅜ x 8½. 25967-6

AN ENCYCLOPEDIA OF BATTLES: Accounts of Over 1,560 Battles from 1479 B.C. to the Present, David Eggenberger. Essential details of every major battle in recorded history from the first battle of Megiddo in 1479 B.C. to Grenada in 1984. List of Battle Maps. New Appendix covering the years 1967-1984. Index. 99 illustrations. 544pp. 6½ x 9¼. 24913-1

SAILING ALONE AROUND THE WORLD, Captain Joshua Slocum. First man to sail around the world, alone, in small boat. One of great feats of seamanship told in delightful manner. 67 illustrations. 294pp. 5⅜ x 8½. 20326-3

ANARCHISM AND OTHER ESSAYS, Emma Goldman. Powerful, penetrating, prophetic essays on direct action, role of minorities, prison reform, puritan hypocrisy, violence, etc. 271pp. 5⅜ x 8½. 22484-8

MYTHS OF THE HINDUS AND BUDDHISTS, Ananda K. Coomaraswamy and Sister Nivedita. Great stories of the epics; deeds of Krishna, Shiva, taken from puranas, Vedas, folk tales; etc. 32 illustrations. 400pp. 5⅜ x 8½. 21759-0

THE TRAUMA OF BIRTH, Otto Rank. Rank's controversial thesis that anxiety neurosis is caused by profound psychological trauma which occurs at birth. 256pp. 5⅜ x 8½. 27974-X

A THEOLOGICO-POLITICAL TREATISE, Benedict Spinoza. Also contains unfinished Political Treatise. Great classic on religious liberty, theory of government on common consent. R. Elwes translation. Total of 421pp. 5⅜ x 8½. 20249-6

CATALOG OF DOVER BOOKS

MY BONDAGE AND MY FREEDOM, Frederick Douglass. Born a slave, Douglass became outspoken force in antislavery movement. The best of Douglass' autobiographies. Graphic description of slave life. 464pp. 5⅜ x 8½. 22457-0

FOLLOWING THE EQUATOR: A Journey Around the World, Mark Twain. Fascinating humorous account of 1897 voyage to Hawaii, Australia, India, New Zealand, etc. Ironic, bemused reports on peoples, customs, climate, flora and fauna, politics, much more. 197 illustrations. 720pp. 5⅜ x 8½. 26113-1

THE PEOPLE CALLED SHAKERS, Edward D. Andrews. Definitive study of Shakers: origins, beliefs, practices, dances, social organization, furniture and crafts, etc. 33 illustrations. 351pp. 5⅜ x 8½. 21081-2

THE MYTHS OF GREECE AND ROME, H. A. Guerber. A classic of mythology, generously illustrated, long prized for its simple, graphic, accurate retelling of the principal myths of Greece and Rome, and for its commentary on their origins and significance. With 64 illustrations by Michelangelo, Raphael, Titian, Rubens, Canova, Bernini and others. 480pp. 5⅜ x 8½. 27584-1

PSYCHOLOGY OF MUSIC, Carl E. Seashore. Classic work discusses music as a medium from psychological viewpoint. Clear treatment of physical acoustics, auditory apparatus, sound perception, development of musical skills, nature of musical feeling, host of other topics. 88 figures. 408pp. 5⅜ x 8½. 21851-1

THE PHILOSOPHY OF HISTORY, Georg W. Hegel. Great classic of Western thought develops concept that history is not chance but rational process, the evolution of freedom. 457pp. 5⅜ x 8½. 20112-0

THE BOOK OF TEA, Kakuzo Okakura. Minor classic of the Orient: entertaining, charming explanation, interpretation of traditional Japanese culture in terms of tea ceremony. 94pp. 5⅜ x 8½. 20070-1

LIFE IN ANCIENT EGYPT, Adolf Erman. Fullest, most thorough, detailed older account with much not in more recent books, domestic life, religion, magic, medicine, commerce, much more. Many illustrations reproduce tomb paintings, carvings, hieroglyphs, etc. 597pp. 5⅜ x 8½. 22632-8

SUNDIALS, Their Theory and Construction, Albert Waugh. Far and away the best, most thorough coverage of ideas, mathematics concerned, types, construction, adjusting anywhere. Simple, nontechnical treatment allows even children to build several of these dials. Over 100 illustrations. 230pp. 5⅜ x 8½. 22947-5

THEORETICAL HYDRODYNAMICS, L. M. Milne-Thomson. Classic exposition of the mathematical theory of fluid motion, applicable to both hydrodynamics and aerodynamics. Over 600 exercises. 768pp. 6⅛ x 9¼. 68970-0

SONGS OF EXPERIENCE: Facsimile Reproduction with 26 Plates in Full Color, William Blake. 26 full-color plates from a rare 1826 edition. Includes "The Tyger," "London," "Holy Thursday," and other poems. Printed text of poems. 48pp. 5¼ x 7. 24636-1

OLD-TIME VIGNETTES IN FULL COLOR, Carol Belanger Grafton (ed.). Over 390 charming, often sentimental illustrations, selected from archives of Victorian graphics—pretty women posing, children playing, food, flowers, kittens and puppies, smiling cherubs, birds and butterflies, much more. All copyright-free. 48pp. 9¼ x 12¼. 27269-9

CATALOG OF DOVER BOOKS

PERSPECTIVE FOR ARTISTS, Rex Vicat Cole. Depth, perspective of sky and sea, shadows, much more, not usually covered. 391 diagrams, 81 reproductions of drawings and paintings. 279pp. 5⅜ x 8½. 22487-2

DRAWING THE LIVING FIGURE, Joseph Sheppard. Innovative approach to artistic anatomy focuses on specifics of surface anatomy, rather than muscles and bones. Over 170 drawings of live models in front, back and side views, and in widely varying poses. Accompanying diagrams. 177 illustrations. Introduction. Index. 144pp. 8⅜ x 11¼. 26723-7

GOTHIC AND OLD ENGLISH ALPHABETS: 100 Complete Fonts, Dan X. Solo. Add power, elegance to posters, signs, other graphics with 100 stunning copyright-free alphabets: Blackstone, Dolbey, Germania, 97 more—including many lower-case, numerals, punctuation marks. 104pp. 8⅛ x 11. 24695-7

HOW TO DO BEADWORK, Mary White. Fundamental book on craft from simple projects to five-bead chains and woven works. 106 illustrations. 142pp. 5⅜ x 8. 20697-1

THE BOOK OF WOOD CARVING, Charles Marshall Sayers. Finest book for beginners discusses fundamentals and offers 34 designs. "Absolutely first rate . . . well thought out and well executed."–E. J. Tangerman. 118pp. 7¾ x 10⅝. 23654-4

ILLUSTRATED CATALOG OF CIVIL WAR MILITARY GOODS: Union Army Weapons, Insignia, Uniform Accessories, and Other Equipment, Schuyler, Hartley, and Graham. Rare, profusely illustrated 1846 catalog includes Union Army uniform and dress regulations, arms and ammunition, coats, insignia, flags, swords, rifles, etc. 226 illustrations. 160pp. 9 x 12. 24939-5

WOMEN'S FASHIONS OF THE EARLY 1900s: An Unabridged Republication of "New York Fashions, 1909," National Cloak & Suit Co. Rare catalog of mail-order fashions documents women's and children's clothing styles shortly after the turn of the century. Captions offer full descriptions, prices. Invaluable resource for fashion, costume historians. Approximately 725 illustrations. 128pp. 8⅜ x 11¼. 27276-1

THE 1912 AND 1915 GUSTAV STICKLEY FURNITURE CATALOGS, Gustav Stickley. With over 200 detailed illustrations and descriptions, these two catalogs are essential reading and reference materials and identification guides for Stickley furniture. Captions cite materials, dimensions and prices. 112pp. 6½ x 9¼. 26676-1

EARLY AMERICAN LOCOMOTIVES, John H. White, Jr. Finest locomotive engravings from early 19th century: historical (1804–74), main-line (after 1870), special, foreign, etc. 147 plates. 142pp. 11⅜ x 8¼. 22772-3

THE TALL SHIPS OF TODAY IN PHOTOGRAPHS, Frank O. Braynard. Lavishly illustrated tribute to nearly 100 majestic contemporary sailing vessels: Amerigo Vespucci, Clearwater, Constitution, Eagle, Mayflower, Sea Cloud, Victory, many more. Authoritative captions provide statistics, background on each ship. 190 black-and-white photographs and illustrations. Introduction. 128pp. 8⅞ x 11¾. 27163-3

LITTLE BOOK OF EARLY AMERICAN CRAFTS AND TRADES, Peter Stockham (ed.). 1807 children's book explains crafts and trades: baker, hatter, cooper, potter, and many others. 23 copperplate illustrations. 140pp. 4⅝ x 6. 23336-7

VICTORIAN FASHIONS AND COSTUMES FROM HARPER'S BAZAR, 1867–1898, Stella Blum (ed.). Day costumes, evening wear, sports clothes, shoes, hats, other accessories in over 1,000 detailed engravings. 320pp. 9⅜ x 12¼. 22990-4

GUSTAV STICKLEY, THE CRAFTSMAN, Mary Ann Smith. Superb study surveys broad scope of Stickley's achievement, especially in architecture. Design philosophy, rise and fall of the Craftsman empire, descriptions and floor plans for many Craftsman houses, more. 86 black-and-white halftones. 31 line illustrations. Introduction 208pp. 6½ x 9¼. 27210-9

THE LONG ISLAND RAIL ROAD IN EARLY PHOTOGRAPHS, Ron Ziel. Over 220 rare photos, informative text document origin (1844) and development of rail service on Long Island. Vintage views of early trains, locomotives, stations, passengers, crews, much more. Captions. 8⅞ x 11¾. 26301-0

VOYAGE OF THE LIBERDADE, Joshua Slocum. Great 19th-century mariner's thrilling, first-hand account of the wreck of his ship off South America, the 35-foot boat he built from the wreckage, and its remarkable voyage home. 128pp. 5⅜ x 8½. 40022-0

TEN BOOKS ON ARCHITECTURE, Vitruvius. The most important book ever written on architecture. Early Roman aesthetics, technology, classical orders, site selection, all other aspects. Morgan translation. 331pp. 5⅜ x 8½. 20645-9

THE HUMAN FIGURE IN MOTION, Eadweard Muybridge. More than 4,500 stopped-action photos, in action series, showing undraped men, women, children jumping, lying down, throwing, sitting, wrestling, carrying, etc. 390pp. 7⅞ x 10⅝. 20204-6 Clothbd.

TREES OF THE EASTERN AND CENTRAL UNITED STATES AND CANADA, William M. Harlow. Best one-volume guide to 140 trees. Full descriptions, woodlore, range, etc. Over 600 illustrations. Handy size. 288pp. 4½ x 6⅜. 20395-6

SONGS OF WESTERN BIRDS, Dr. Donald J. Borror. Complete song and call repertoire of 60 western species, including flycatchers, juncoes, cactus wrens, many more–includes fully illustrated booklet. Cassette and manual 99913-0

GROWING AND USING HERBS AND SPICES, Milo Miloradovich. Versatile handbook provides all the information needed for cultivation and use of all the herbs and spices available in North America. 4 illustrations. Index. Glossary. 236pp. 5⅜ x 8½. 25058-X

BIG BOOK OF MAZES AND LABYRINTHS, Walter Shepherd. 50 mazes and labyrinths in all–classical, solid, ripple, and more–in one great volume. Perfect inexpensive puzzler for clever youngsters. Full solutions. 112pp. 8⅛ x 11. 22951-3

CATALOG OF DOVER BOOKS

PIANO TUNING, J. Cree Fischer. Clearest, best book for beginner, amateur. Simple repairs, raising dropped notes, tuning by easy method of flattened fifths. No previous skills needed. 4 illustrations. 201pp. 5⅜ x 8½. 23267-0

HINTS TO SINGERS, Lillian Nordica. Selecting the right teacher, developing confidence, overcoming stage fright, and many other important skills receive thoughtful discussion in this indispensible guide, written by a world-famous diva of four decades' experience. 96pp. 5⅜ x 8½. 40094-8

THE COMPLETE NONSENSE OF EDWARD LEAR, Edward Lear. All nonsense limericks, zany alphabets, Owl and Pussycat, songs, nonsense botany, etc., illustrated by Lear. Total of 320pp. 5⅜ x 8½. (Available in U.S. only.) 20167-8

VICTORIAN PARLOUR POETRY: An Annotated Anthology, Michael R. Turner. 117 gems by Longfellow, Tennyson, Browning, many lesser-known poets. "The Village Blacksmith," "Curfew Must Not Ring Tonight," "Only a Baby Small," dozens more, often difficult to find elsewhere. Index of poets, titles, first lines. xxiii + 325pp. 5⅜ x 8¼. 27044-0

DUBLINERS, James Joyce. Fifteen stories offer vivid, tightly focused observations of the lives of Dublin's poorer classes. At least one, "The Dead," is considered a masterpiece. Reprinted complete and unabridged from standard edition. 160pp. 5³⁄₁₆ x 8¼. 26870-5

GREAT WEIRD TALES: 14 Stories by Lovecraft, Blackwood, Machen and Others, S. T. Joshi (ed.). 14 spellbinding tales, including "The Sin Eater," by Fiona McLeod, "The Eye Above the Mantel," by Frank Belknap Long, as well as renowned works by R. H. Barlow, Lord Dunsany, Arthur Machen, W. C. Morrow and eight other masters of the genre. 256pp. 5⅜ x 8½. (Available in U.S. only.) 40436-6

THE BOOK OF THE SACRED MAGIC OF ABRAMELIN THE MAGE, translated by S. MacGregor Mathers. Medieval manuscript of ceremonial magic. Basic document in Aleister Crowley, Golden Dawn groups. 268pp. 5⅜ x 8½. 23211-5

NEW RUSSIAN-ENGLISH AND ENGLISH-RUSSIAN DICTIONARY, M. A. O'Brien. This is a remarkably handy Russian dictionary, containing a surprising amount of information, including over 70,000 entries. 366pp. 4½ x 6¼. 20208-9

HISTORIC HOMES OF THE AMERICAN PRESIDENTS, Second, Revised Edition, Irvin Haas. A traveler's guide to American Presidential homes, most open to the public, depicting and describing homes occupied by every American President from George Washington to George Bush. With visiting hours, admission charges, travel routes. 175 photographs. Index. 160pp. 8¼ x 11. 26751-2

NEW YORK IN THE FORTIES, Andreas Feininger. 162 brilliant photographs by the well-known photographer, formerly with *Life* magazine. Commuters, shoppers, Times Square at night, much else from city at its peak. Captions by John von Hartz. 181pp. 9¼ x 10¾. 23585-8

INDIAN SIGN LANGUAGE, William Tomkins. Over 525 signs developed by Sioux and other tribes. Written instructions and diagrams. Also 290 pictographs. 111pp. 6⅛ x 9¼. 22029-X

CATALOG OF DOVER BOOKS

ANATOMY: A Complete Guide for Artists, Joseph Sheppard. A master of figure drawing shows artists how to render human anatomy convincingly. Over 460 illustrations. 224pp. 8⅜ x 11¼. 27279-6

MEDIEVAL CALLIGRAPHY: Its History and Technique, Marc Drogin. Spirited history, comprehensive instruction manual covers 13 styles (ca. 4th century through 15th). Excellent photographs; directions for duplicating medieval techniques with modern tools. 224pp. 8⅜ x 11¼. 26142-5

DRIED FLOWERS: How to Prepare Them, Sarah Whitlock and Martha Rankin. Complete instructions on how to use silica gel, meal and borax, perlite aggregate, sand and borax, glycerine and water to create attractive permanent flower arrangements. 12 illustrations. 32pp. 5⅜ x 8½. 21802-3

EASY-TO-MAKE BIRD FEEDERS FOR WOODWORKERS, Scott D. Campbell. Detailed, simple-to-use guide for designing, constructing, caring for and using feeders. Text, illustrations for 12 classic and contemporary designs. 96pp. 5⅜ x 8½.
25847-5

SCOTTISH WONDER TALES FROM MYTH AND LEGEND, Donald A. Mackenzie. 16 lively tales tell of giants rumbling down mountainsides, of a magic wand that turns stone pillars into warriors, of gods and goddesses, evil hags, powerful forces and more. 240pp. 5⅜ x 8½. 29677-6

THE HISTORY OF UNDERCLOTHES, C. Willett Cunnington and Phyllis Cunnington. Fascinating, well-documented survey covering six centuries of English undergarments, enhanced with over 100 illustrations: 12th-century laced-up bodice, footed long drawers (1795), 19th-century bustles, l9th-century corsets for men, Victorian "bust improvers," much more. 272pp. 5⅜ x 8¼. 27124-2

ARTS AND CRAFTS FURNITURE: The Complete Brooks Catalog of 1912, Brooks Manufacturing Co. Photos and detailed descriptions of more than 150 now very collectible furniture designs from the Arts and Crafts movement depict davenports, settees, buffets, desks, tables, chairs, bedsteads, dressers and more, all built of solid, quarter-sawed oak. Invaluable for students and enthusiasts of antiques, Americana and the decorative arts. 80pp. 6½ x 9¼. 27471-3

WILBUR AND ORVILLE: A Biography of the Wright Brothers, Fred Howard. Definitive, crisply written study tells the full story of the brothers' lives and work. A vividly written biography, unparalleled in scope and color, that also captures the spirit of an extraordinary era. 560pp. 6⅛ x 9¼. 40297-5

THE ARTS OF THE SAILOR: Knotting, Splicing and Ropework, Hervey Garrett Smith. Indispensable shipboard reference covers tools, basic knots and useful hitches; handsewing and canvas work, more. Over 100 illustrations. Delightful reading for sea lovers. 256pp. 5⅜ x 8½. 26440-8

FRANK LLOYD WRIGHT'S FALLINGWATER: The House and Its History, Second, Revised Edition, Donald Hoffmann. A total revision–both in text and illustrations–of the standard document on Fallingwater, the boldest, most personal architectural statement of Wright's mature years, updated with valuable new material from the recently opened Frank Lloyd Wright Archives. "Fascinating"–*The New York Times*. 116 illustrations. 128pp. 9¼ x 10¾. 27430-6

PHOTOGRAPHIC SKETCHBOOK OF THE CIVIL WAR, Alexander Gardner. 100 photos taken on field during the Civil War. Famous shots of Manassas Harper's Ferry, Lincoln, Richmond, slave pens, etc. 244pp. 10⅝ x 8¼. 22731-6

FIVE ACRES AND INDEPENDENCE, Maurice G. Kains. Great back-to-the-land classic explains basics of self-sufficient farming. The one book to get. 95 illustrations. 397pp. 5⅜ x 8½. 20974-1

SONGS OF EASTERN BIRDS, Dr. Donald J. Borror. Songs and calls of 60 species most common to eastern U.S.: warblers, woodpeckers, flycatchers, thrushes, larks, many more in high-quality recording. Cassette and manual 99912-2

A MODERN HERBAL, Margaret Grieve. Much the fullest, most exact, most useful compilation of herbal material. Gigantic alphabetical encyclopedia, from aconite to zedoary, gives botanical information, medical properties, folklore, economic uses, much else. Indispensable to serious reader. 161 illustrations. 888pp. 6½ x 9¼. 2-vol. set. (Available in U.S. only.) Vol. I: 22798-7
Vol. II: 22799-5

HIDDEN TREASURE MAZE BOOK, Dave Phillips. Solve 34 challenging mazes accompanied by heroic tales of adventure. Evil dragons, people-eating plants, blood-thirsty giants, many more dangerous adversaries lurk at every twist and turn. 34 mazes, stories, solutions. 48pp. 8¼ x 11. 24566-7

LETTERS OF W. A. MOZART, Wolfgang A. Mozart. Remarkable letters show bawdy wit, humor, imagination, musical insights, contemporary musical world; includes some letters from Leopold Mozart. 276pp. 5⅜ x 8½. 22859-2

BASIC PRINCIPLES OF CLASSICAL BALLET, Agrippina Vaganova. Great Russian theoretician, teacher explains methods for teaching classical ballet. 118 illustrations. 175pp. 5⅜ x 8½. 22036-2

THE JUMPING FROG, Mark Twain. Revenge edition. The original story of The Celebrated Jumping Frog of Calaveras County, a hapless French translation, and Twain's hilarious "retranslation" from the French. 12 illustrations. 66pp. 5⅜ x 8½. 22686-7

BEST REMEMBERED POEMS, Martin Gardner (ed.). The 126 poems in this superb collection of 19th- and 20th-century British and American verse range from Shelley's "To a Skylark" to the impassioned "Renascence" of Edna St. Vincent Millay and to Edward Lear's whimsical "The Owl and the Pussycat." 224pp. 5⅜ x 8½. 27165-X

COMPLETE SONNETS, William Shakespeare. Over 150 exquisite poems deal with love, friendship, the tyranny of time, beauty's evanescence, death and other themes in language of remarkable power, precision and beauty. Glossary of archaic terms. 80pp. 5$\frac{3}{16}$ x 8¼. 26686-9

THE BATTLES THAT CHANGED HISTORY, Fletcher Pratt. Eminent historian profiles 16 crucial conflicts, ancient to modern, that changed the course of civilization. 352pp. 5⅜ x 8½. 41129-X

CATALOG OF DOVER BOOKS

THE WIT AND HUMOR OF OSCAR WILDE, Alvin Redman (ed.). More than 1,000 ripostes, paradoxes, wisecracks: Work is the curse of the drinking classes; I can resist everything except temptation; etc. 258pp. 5⅜ x 8½. 20602-5

SHAKESPEARE LEXICON AND QUOTATION DICTIONARY, Alexander Schmidt. Full definitions, locations, shades of meaning in every word in plays and poems. More than 50,000 exact quotations. 1,485pp. 6½ x 9¼. 2-vol. set.
Vol. 1: 22726-X
Vol. 2: 22727-8

SELECTED POEMS, Emily Dickinson. Over 100 best-known, best-loved poems by one of America's foremost poets, reprinted from authoritative early editions. No comparable edition at this price. Index of first lines. 64pp. 5 3/16 x 8¼. 26466-1

THE INSIDIOUS DR. FU-MANCHU, Sax Rohmer. The first of the popular mystery series introduces a pair of English detectives to their archnemesis, the diabolical Dr. Fu-Manchu. Flavorful atmosphere, fast-paced action, and colorful characters enliven this classic of the genre. 208pp. 5 3/16 x 8¼. 29898-1

THE MALLEUS MALEFICARUM OF KRAMER AND SPRENGER, translated by Montague Summers. Full text of most important witchhunter's "bible," used by both Catholics and Protestants. 278pp. 6⅝ x 10. 22802-9

SPANISH STORIES/CUENTOS ESPAÑOLES: A Dual-Language Book, Angel Flores (ed.). Unique format offers 13 great stories in Spanish by Cervantes, Borges, others. Faithful English translations on facing pages. 352pp. 5⅜ x 8½. 25399-6

GARDEN CITY, LONG ISLAND, IN EARLY PHOTOGRAPHS, 1869–1919, Mildred H. Smith. Handsome treasury of 118 vintage pictures, accompanied by carefully researched captions, document the Garden City Hotel fire (1899), the Vanderbilt Cup Race (1908), the first airmail flight departing from the Nassau Boulevard Aerodrome (1911), and much more. 96pp. 8⅞ x 11¾. 40669-5

OLD QUEENS, N.Y., IN EARLY PHOTOGRAPHS, Vincent F. Seyfried and William Asadorian. Over 160 rare photographs of Maspeth, Jamaica, Jackson Heights, and other areas. Vintage views of DeWitt Clinton mansion, 1939 World's Fair and more. Captions. 192pp. 8⅞ x 11. 26358-4

CAPTURED BY THE INDIANS: 15 Firsthand Accounts, 1750-1870, Frederick Drimmer. Astounding true historical accounts of grisly torture, bloody conflicts, relentless pursuits, miraculous escapes and more, by people who lived to tell the tale. 384pp. 5⅜ x 8½. 24901-8

THE WORLD'S GREAT SPEECHES (Fourth Enlarged Edition), Lewis Copeland, Lawrence W. Lamm, and Stephen J. McKenna. Nearly 300 speeches provide public speakers with a wealth of updated quotes and inspiration–from Pericles' funeral oration and William Jennings Bryan's "Cross of Gold Speech" to Malcolm X's powerful words on the Black Revolution and Earl of Spenser's tribute to his sister, Diana, Princess of Wales. 944pp. 5⅜ x 8⅜. 40903-1

THE BOOK OF THE SWORD, Sir Richard F. Burton. Great Victorian scholar/adventurer's eloquent, erudite history of the "queen of weapons"–from prehistory to early Roman Empire. Evolution and development of early swords, variations (sabre, broadsword, cutlass, scimitar, etc.), much more. 336pp. 6⅛ x 9¼. 25434-8

CATALOG OF DOVER BOOKS

AUTOBIOGRAPHY: The Story of My Experiments with Truth, Mohandas K. Gandhi. Boyhood, legal studies, purification, the growth of the Satyagraha (nonviolent protest) movement. Critical, inspiring work of the man responsible for the freedom of India. 480pp. 5⅜ x 8½. (Available in U.S. only.) 24593-4

CELTIC MYTHS AND LEGENDS, T. W. Rolleston. Masterful retelling of Irish and Welsh stories and tales. Cuchulain, King Arthur, Deirdre, the Grail, many more. First paperback edition. 58 full-page illustrations. 512pp. 5⅜ x 8½. 26507-2

THE PRINCIPLES OF PSYCHOLOGY, William James. Famous long course complete, unabridged. Stream of thought, time perception, memory, experimental methods; great work decades ahead of its time. 94 figures. 1,391pp. 5⅜ x 8½. 2-vol. set.
Vol. I: 20381-6   Vol. II: 20382-4

THE WORLD AS WILL AND REPRESENTATION, Arthur Schopenhauer. Definitive English translation of Schopenhauer's life work, correcting more than 1,000 errors, omissions in earlier translations. Translated by E. F. J. Payne. Total of 1,269pp. 5⅜ x 8½. 2-vol. set.   Vol. 1: 21761-2   Vol. 2: 21762-0

MAGIC AND MYSTERY IN TIBET, Madame Alexandra David-Neel. Experiences among lamas, magicians, sages, sorcerers, Bonpa wizards. A true psychic discovery. 32 illustrations. 321pp. 5⅜ x 8½. (Available in U.S. only.) 22682-4

THE EGYPTIAN BOOK OF THE DEAD, E. A. Wallis Budge. Complete reproduction of Ani's papyrus, finest ever found. Full hieroglyphic text, interlinear transliteration, word-for-word translation, smooth translation. 533pp. 6½ x 9¼. 21866-X

MATHEMATICS FOR THE NONMATHEMATICIAN, Morris Kline. Detailed, college-level treatment of mathematics in cultural and historical context, with numerous exercises. Recommended Reading Lists. Tables. Numerous figures. 641pp. 5⅜ x 8½. 24823-2

PROBABILISTIC METHODS IN THE THEORY OF STRUCTURES, Isaac Elishakoff. Well-written introduction covers the elements of the theory of probability from two or more random variables, the reliability of such multivariable structures, the theory of random function, Monte Carlo methods of treating problems incapable of exact solution, and more. Examples. 502pp. 5⅜ x 8½. 40691-1

THE RIME OF THE ANCIENT MARINER, Gustave Doré, S. T. Coleridge. Doré's finest work; 34 plates capture moods, subtleties of poem. Flawless full-size reproductions printed on facing pages with authoritative text of poem. "Beautiful. Simply beautiful."–*Publisher's Weekly.* 77pp. 9¼ x 12. 22305-1

NORTH AMERICAN INDIAN DESIGNS FOR ARTISTS AND CRAFTSPEOPLE, Eva Wilson. Over 360 authentic copyright-free designs adapted from Navajo blankets, Hopi pottery, Sioux buffalo hides, more. Geometrics, symbolic figures, plant and animal motifs, etc. 128pp. 8⅜ x 11. (Not for sale in the United Kingdom.) 25341-4

SCULPTURE: Principles and Practice, Louis Slobodkin. Step-by-step approach to clay, plaster, metals, stone; classical and modern. 253 drawings, photos. 255pp. 8⅛ x 11. 22960-2

THE INFLUENCE OF SEA POWER UPON HISTORY, 1660–1783, A. T. Mahan. Influential classic of naval history and tactics still used as text in war colleges. First paperback edition. 4 maps. 24 battle plans. 640pp. 5⅜ x 8½. 25509-3

CATALOG OF DOVER BOOKS

THE STORY OF THE TITANIC AS TOLD BY ITS SURVIVORS, Jack Winocour (ed.). What it was really like. Panic, despair, shocking inefficiency, and a little heroism. More thrilling than any fictional account. 26 illustrations. 320pp. 5⅜ x 8½.
20610-6

FAIRY AND FOLK TALES OF THE IRISH PEASANTRY, William Butler Yeats (ed.). Treasury of 64 tales from the twilight world of Celtic myth and legend: "The Soul Cages," "The Kildare Pooka," "King O'Toole and his Goose," many more. Introduction and Notes by W. B. Yeats. 352pp. 5⅜ x 8½.
26941-8

BUDDHIST MAHAYANA TEXTS, E. B. Cowell and others (eds.). Superb, accurate translations of basic documents in Mahayana Buddhism, highly important in history of religions. The Buddha-karita of Asvaghosha, Larger Sukhavativyuha, more. 448pp. 5⅜ x 8½.
25552-2

ONE TWO THREE . . . INFINITY: Facts and Speculations of Science, George Gamow. Great physicist's fascinating, readable overview of contemporary science: number theory, relativity, fourth dimension, entropy, genes, atomic structure, much more. 128 illustrations. Index. 352pp. 5⅜ x 8½.
25664-2

EXPERIMENTATION AND MEASUREMENT, W. J. Youden. Introductory manual explains laws of measurement in simple terms and offers tips for achieving accuracy and minimizing errors. Mathematics of measurement, use of instruments, experimenting with machines. 1994 edition. Foreword. Preface. Introduction. Epilogue. Selected Readings. Glossary. Index. Tables and figures. 128pp. 5⅜ x 8½.   40451-X

DALÍ ON MODERN ART: The Cuckolds of Antiquated Modern Art, Salvador Dalí. Influential painter skewers modern art and its practitioners. Outrageous evaluations of Picasso, Cézanne, Turner, more. 15 renderings of paintings discussed. 44 calligraphic decorations by Dalí. 96pp. 5⅜ x 8½. (Available in U.S. only.)
29220-7

ANTIQUE PLAYING CARDS: A Pictorial History, Henry René D'Allemagne. Over 900 elaborate, decorative images from rare playing cards (14th–20th centuries): Bacchus, death, dancing dogs, hunting scenes, royal coats of arms, players cheating, much more. 96pp. 9¼ x 12¼.
29265-7

MAKING FURNITURE MASTERPIECES: 30 Projects with Measured Drawings, Franklin H. Gottshall. Step-by-step instructions, illustrations for constructing handsome, useful pieces, among them a Sheraton desk, Chippendale chair, Spanish desk, Queen Anne table and a William and Mary dressing mirror. 224pp. 8⅛ x 11¼.
29338-6

THE FOSSIL BOOK: A Record of Prehistoric Life, Patricia V. Rich et al. Profusely illustrated definitive guide covers everything from single-celled organisms and dinosaurs to birds and mammals and the interplay between climate and man. Over 1,500 illustrations. 760pp. 7½ x 10⅛.
29371-8

Paperbound unless otherwise indicated. Available at your book dealer, online at **www.doverpublications.com**, or by writing to Dept. GI, Dover Publications, Inc., 31 East 2nd Street, Mineola, NY 11501. For current price information or for free catalogues (please indicate field of interest), write to Dover Publications or log on to **www.doverpublications.com** and see every Dover book in print. Dover publishes more than 500 books each year on science, elementary and advanced mathematics, biology, music, art, literary history, social sciences, and other areas.